DETOX
DEVELOPMENT

DETOX
DEVELOPMENT
REPURPOSING ENVIRONMENTALLY HARMFUL SUBSIDIES

Richard Damania, Esteban Balseca,
Charlotte de Fontaubert,
Joshua Gill, Kichan Kim,
Jun Rentschler, Jason Russ,
and Esha Zaveri

WORLD BANK GROUP

Contents

FIGURES

TABLES

Acknowledgments

This book was prepared by a World Bank team led by Richard Damania, co-led by Jun Rentschler and Jason Russ, and comprising (in alphabetical order) Esteban Balseca, Charlotte de Fontaubert, Joshua Gill, Kichan Kim, and Esha Zaveri. The book has greatly benefited from the strategic guidance and general direction of Juergen Voegele (Vice President, Sustainable Development Practice, World Bank).

In addition to research completed by the authors, this work draws on background papers, notes, and analyses prepared by the following individuals: Wai Lung Cheung (University of British Columbia), Brian Davidson (University of Melbourne), Claire Donnelley, Ira Dorband (World Bank), Xinming Du (Columbia University), Ebad Ebadi (World Bank), Petra Hellegers (Wageningen University), Hannah Druckenmiller King (Resources for the Future), Christoph Klaiber (World Bank), Wing Yee Lam (University of British Columbia), Nadia Leonova (World Bank), Ruyi Li (World Bank), Kentaro Mayr (University College London), Linh Nguyen (Bates College), Yunsun Park (World Bank), Sebnem Sahin (SERAP LLC), Rashid Sumaila (University of British Columbia), Margaret Triyana (World Bank), Pieter Waalewijn (World Bank), Mahwash Wasiq (World Bank), Eduardo Zegarra (Group for the Analysis of Development—GRADE), and Zoe Zeng (University of British Columbia).

The authors received incisive and helpful advice, inputs, and comments from Hanane Ahmed, Simon Black, Ghada Elabed, Pablo Fajnzylber, Alejandro De la Fuente, Dirk Heine, Valerie Hickey, Jon Jellema, Mimako Kobayashi, Masami Kojima, Knut Korsbrekke, Justice Tei Mensah, John Nash, Astrid Maria Jakobs de Padua, Hrishikesh Prakash Patel, Klas Sander, Kateryna Schroeder, Renaud Seligmann, Avjeet Singh, Stephen Stretton, Bill Young, and Sergiy Zorya.

Invaluable feedback and advice were received from the following peer reviewers at the World Bank: Thomas Flochel (Senior Energy Economist), Madhur Gautam (Lead Agricultural Economist), Martin Heger (Senior Environmental Economist), Daniel Lederman (Lead Economist), Nancy Lozano Gracia (Lead Economist), Aude-Sophie Rodella (Senior Economist), Giovanni Ruta (Lead Environmental Economist), Ernesto Sanchez-Triana (Lead Environmental Specialist), and Yadviga Semikolenova (Lead Energy Specialist).

The authors thank Elizabeth Forsyth, Gwenda Larsen, Lucy Southwood, and Stan Wanat for their excellent editing and proofreading services, as well as the World Bank publishing team consisting of Mark McClure, Jewel McFadden, and Orlando Mota.

Finally, Sreypov Tep provided impeccable administrative support, for which the team is grateful.

This work was made possible by the financial contribution of the Korea Green Growth Trust Fund (see https://www.wbgkggtf.org) of the Sustainable Development Practice, World Bank Group, as well as the PROBLUE Trust Fund (https://www.worldbank.org/en /programs/problue).

About the Authors

Esteban Balseca has worked with the World Bank as a research consultant over the past two years. His interests in economics deal with the environmental and health dimensions of development and the application of econometric methods to assess the impact of policy. As a consultant, he has contributed to the RISE (Resilience, Inclusivity, Sustainability, and Efficiency) framework for identifying country-specific development challenges and to several development reports and studies. Before working with the World Bank, he was an analyst for the Bureau of Labor Statistics. He holds a PhD in economics and has taught economics courses in various universities in Ecuador.

Richard Damania is the chief economist of the Sustainable Development Global Practice at the World Bank. He has held several positions at the World Bank, including as senior economic adviser in the Water Practice and as lead economist in the Africa Region's Sustainable Development Department and in the South Asia and Latin America and the Caribbean Regions. His work has spanned multiple sectors and has helped the World Bank become an acknowledged thought leader on matters relating to the environment, water, and the economy. Before joining the World Bank, he held positions in academia, and he has published extensively, including more than 100 papers in scientific journals.

Charlotte de Fontaubert is global lead for the blue economy and senior fisheries specialist in the World Bank's Global Practice on the Environment, Natural Resources and Blue Economy. Her work focuses on fisheries, the impacts of climate change on marine and coastal ecosystems, and the sustainable development of other oceanic sectors. She has coauthored some of the World Bank's seminal work on fisheries—*The Sunken Billions Revisited* (2017) and *Climate Change and Marine Fisheries in Africa* (2019)—and the blue economy, *Riding the Blue Wave* (2021). Over the past three years, she has led the World Bank's work on fishery subsidies and cowrote the chapter on fishery subsidies in *The Changing Wealth of Nations* (2021). She holds a PhD in marine studies from the University of Delaware with a focus on international fisheries.

Joshua Gill is an agricultural economist with the World Bank's Agriculture and Food Global Practice. His professional interests include agricultural development and public policy and the use of behavioral and experimental methods to understand the decision-making of rural households under external and internal constraints. Before joining the World Bank, he was director of analytics at the Global Innovation Fund. He holds a PhD in agricultural economics from Michigan State University.

Kichan Kim is a junior professional officer in the Office of the Chief Economist of the Sustainable Development Global Practice at the World Bank. His professional interests focus on using geospatial data with statistical analysis to study interactions between the environment and human welfare, and the role of markets in mitigating the links. Before joining the World Bank, he served as a consultant for the International Food Policy Research Institute, working on nutrition-based market quality measures in the context of

African countries. He holds a PhD in agricultural, environmental, and development economics from Ohio State University.

Jun Rentschler is a senior economist in the Office of the Chief Economist of the Sustainable Development Global Practice, working at the intersection of climate change and sustainable resilient development. Before joining the World Bank in 2012, he served as an economic adviser at the German Foreign Ministry. He also spent two years at the European Bank for Reconstruction and Development, working on private sector investment projects in resource efficiency and climate change. Before that, he worked on projects with Grameen Microfinance Bank in Bangladesh and the Partners for Financial Stability Program of the United States Agency for International Development in Poland. He is a visiting fellow at the Payne Institute for Public Policy, following previous affiliations with the Oxford Institute for Energy Studies and the Graduate Institute for Policy Studies in Tokyo. He holds a PhD in economics from University College London, specializing in development, climate, and energy.

Jason Russ is a senior economist in the Office of the Chief Economist of the Sustainable Development Global Practice at the World Bank. His professional interests center on using econometrics and data analytics to diagnose development challenges and quantify the economic and social impacts of environmental externalities. His tenure at the World Bank includes five years in the Water Global Practice, where he helped to develop and coordinate the analytical work program of the Economics Global Solutions Group, including authoring many of its global flagship reports. He has authored numerous publications in academic journals related largely to environmental and development economics. Before joining the World Bank, he was an analyst at PricewaterhouseCoopers. He holds a PhD in economics from George Washington University.

Esha Zaveri is a senior economist with the World Bank's Water Global Practice, with professional interests in water resource management, climate impacts, environmental health, and the use of geospatial data with statistical analysis to study interactions between the environment and social and economic systems. She has published on these topics in leading scientific journals and has coauthored flagship reports of the World Bank on water scarcity (*Uncharted Waters*, 2017), water pollution (*Quality Unknown*, 2019), and migration (*Ebb and Flow*, 2021). Before joining the World Bank, she was a postdoctoral fellow at Stanford University's Center on Food Security and the Environment, where she remains an affiliated scholar. She holds a PhD in environmental economics and demography from Pennsylvania State University.

Main Messages

Detox Development: Repurposing Environmentally Harmful Subsidies examines how subsidy reform can help safeguard the world's foundational natural assets—clean air, land, and oceans. These assets are critical for human health and nutrition and underpin much of the global economy. But subsidies for fossil fuels, agriculture, and fisheries are driving the degradation of these assets and harming people, the planet, and economies. These subsidies exceed US$7 trillion per year—or about 8 percent of global gross domestic product (GDP). This includes both explicit subsidies—which are direct public expenditures totaling about US$1.25 trillion—and implicit subsidies—which measure the societal impacts of externalities and amount to more than US$6 trillion.

Key findings of the report are given below.

Fossil fuel subsidies

- **Fossil fuel usage—incentivized by vast subsidies—is a key driver of the 7 million premature deaths each year due to air pollution.** About 94 percent of the world's population is exposed to unsafe particulate matter ($PM_{2.5}$) concentrations. The health burden of air pollution is particularly high in industrializing middle-income countries. Poor and marginalized groups are often exposed to higher levels of pollution and are less able to afford adequate health care.

- **Countries around the world actively paid about US$577 billion in 2021 to artificially lower the price of polluting fuels such as oil, gas, and coal.** By underpricing fossil fuels, governments not only incentivize overuse, but also perpetuate inefficient polluting technologies and entrench inequality. Of all subsidies to the energy sector, about three-quarters go to fossil fuels.

- **By increasing fossil fuel prices, subsidy reform can reduce the incentives to use polluting fuels—but the effectiveness of this instrument can be limited.** When polluting fuels are expensive, people reduce their consumption. On average, a 10 percent increase in the unit price of energy results in a short-run reduction of consumption of about 2 percent. This means the demand for energy is only sluggishly responsive to prices, especially when cleaner alternatives are unavailable or unaffordable.

- **Fossil fuel subsidy reforms are pro-poor.** In nearly all countries, richer households consume significantly more energy than poorer ones, and thus lose more when subsidies are removed. Even when looked at as a share of income, poor people are not necessarily hit harder by subsidy reform; it depends on the country context.

- **Subsidy reform could reduce air pollution and save up to 360,000 lives by 2035 in 25 high-pollution, high-subsidy countries. But it is more effective when accompanied by complementary policies.** For instance, ensuring the availability and affordability of clean technologies, addressing information and capacity constraints, and addressing behavioral biases are ways to increase the effectiveness of subsidy reform.

Agricultural subsidies

- **Richer countries spend more on agricultural subsidies than poorer countries, even when seen relative to total agricultural production.** The largest subsidizers are China, the European Union, Indonesia, Japan, and the United States. However, low- and middle-income countries spend a larger share of their subsidy budget on coupled subsidies, which are the most distorting and environmentally damaging. Subsidies in high-income countries tend to be uncoupled from production—such as those directed to agricultural research and infrastructure—and thus are less harmful.

- **Agricultural subsidies tend to benefit wealthier farmers—because wealthier farmers use more inputs and produce more outputs—and usually fail to improve productivity or efficiency.** In some countries, this is offset by channeling more subsidies to poorer regions, or by subsidies making up a larger share of poor households' incomes. The report also finds that higher levels of coupled subsidies lead to lower farm-level technical efficiency. Decoupled subsidies, however, which are not linked to production decisions, have no impact on the efficiency of production.

- **Subsidies incentivize excessive fertilizer usage to the extent that it suppresses agricultural productivity, degrades soils and waterways, and damages people's health.** More than half of global agricultural production now occurs in regions where fertilizer is suppressing rather than increasing productivity. This means there is significant room to reduce fertilizer use with positive impacts on crop production. Yet the opposite is achieved by subsidies, as excessive fertilizer application is not absorbed by crops and runs off into waterways. Inefficient subsidy usage is responsible for up to 17 percent of all nitrogen pollution in water in the past 30 years, which has large enough health impacts to reduce labor productivity by up to 3.5 percent.

- **Agricultural subsidies are responsible for the loss of 2.2 million hectares of forest per year, equivalent to 14 percent of global deforestation.** Agricultural subsidies in rich countries are driving significant tropical deforestation around the world. For instance, livestock subsidies in the United States drive deforestation in Brazil by increasing the demand for soybeans as feedstock. In turn, subsidy-driven deforestation causes the spread of vector-transmitted diseases—including 3.8 million additional cases of malaria each year, with an economic impact of up to US$19 billion per year.

Fishery subsidies

- **Subsidies are a key driver of excess fishing capacity, dwindling fish stocks, and lower fishing rents.** The negative impact of subsidies is even greater when fisheries are not managed sustainably and already severely depleted. Repurposing subsidies without incentivizing increased fishing capacity is of paramount importance to safeguarding remaining stocks.

- **Yet, if fisheries remain as open-access regimes, repurposing subsidies may have little impact.** Since much of the overfishing by subsidized fleets occurs in the open seas (a global public good) or in exclusive economic zones in low- and middle-income countries, subsidy reform needs to be coupled with reforms to access regimes.

- **Repurposing all fishery subsidies may cause major harm to small-scale, artisanal fishers.** But well-targeted reforms can lead to triple wins, where ecosystem sustainability improves, fishing fleets of all sizes increase their catches and revenues, and the fishery sector becomes distributionally more progressive.

Principles for repurposing harmful subsidies

Subsidy *reforms* are more than just subsidy *removal* and should consist of a package of measures that mitigate the downside risks of reform—including political opposition and adverse impacts on vulnerable groups—while maximizing their contribution to sustainable development.

- **Building public acceptance and credibility** is key, especially when political opposition threatens to derail reform efforts. Effective communication and transparency are needed to build credibility of assurances to address the adverse consequences of reform.

- **Complementary measures** are necessary when price-based instruments (such as subsidy reductions) are insufficient to solve environmental externalities. For instance, improving public transit can help replace fossil fuels, and laws can protect endangered natural capital.

- **Social protection and compensation** are an imperative in all contexts where subsidy removal may threaten the livelihoods of vulnerable groups and increase poverty.

- **Carefully sequenced reforms** can reduce the disruption from large price shocks due to the one-off removal of subsidies and enable households and firms to adjust gradually.

- **Sound strategies for reinvesting reform revenues** can ensure that subsidy reforms help to deliver on development priorities, such as infrastructure, health, and education—while lending credibility to the public good objectives of subsidy reform.

The world's sustainable development goals are directly undermined by the roughly US$1.25 trillion in explicit subsidies paid every year to fossil fuel, agriculture, and fishery sectors. This report documents the hidden consequences of subsidies. It shows that subsidy reform can remove distorted incentives that obstruct sustainability goals, but it also can unlock significant domestic financing to facilitate and accelerate sustainable development efforts that would have greater, wider, and more equitable benefits.

Executive Summary

Government subsidies today make up an enormous share of public budgets worldwide, perhaps larger than at any point in human history. In many countries, the magnitude of explicit subsidies in the natural resource sectors exceeds that of investments in important public goods such as health and education. This report identifies and quantifies known as well as new channels through which poorly designed subsidies in natural resource sectors, though often well intentioned, deepen inequality, diminish productivity, and drive the destruction of ecosystems. Especially in an era of fiscal constraints and degrading natural capital, reform and repurposing of perverse and harmful subsidies offer an opportunity to promote greater sustainability, inclusion, and shared prosperity.

Subsidies to natural resource sectors date back at least as far as the late eighteenth century. Lamentations about fishery subsidies can be found in *The Wealth of Nations*, the 1776 treatise by Adam Smith, the founder of modern economics. At the time, Holland and Scotland each subsidized its own herring industry in an attempt to outcompete the other. The subsidies were intended to support poor fishermen and help consumers by lowering the price of herring. But as Smith observed, the subsidies had the opposite effect. Wealthier fishermen owned bigger boats and therefore captured more of the subsidy. And the generous subsidy encouraged inefficiencies that offset any downward pressure on prices. Put simply, the subsidy had the unintended effect of inducing a collapse of the Scottish herring industry. "Well intentioned, but counterproductive" are unfortunate characteristics of subsidies that have persisted in modern natural resource subsidy programs.

This report examines the impacts of subsidies on the world's stock of foundational natural capital—clean air, land, and oceans. These natural assets are critical for human health and nutrition and underpin much of the economy. Poor air quality is responsible for approximately one in five deaths globally. And as the new analyses in this report show, some of these deaths can be attributed to fossil fuel subsidies. Agriculture is the largest user of land worldwide, feeding the world and employing 1 billion people, including 78 percent of the world's poor. But agriculture is subsidized in ways that promote inefficiency, inequity, and unsustainability. And oceans, which support the world's fisheries and supply about 3 billion people with almost 20 percent of their intake of animal protein, are in a collective state of crisis: more than 34 percent of fisheries are overfished, and this situation is exacerbated by open-access regimes and capacity-increasing subsidies.

Given the scarcity of public funds and the challenges related to sustainability, reexamining and repurposing environmentally harmful subsidies are especially relevant. In 2020, total global debt reached 263 percent of gross domestic product (GDP), its highest level in half a century. In emerging markets and developing economies, rising debt is particularly concerning. In these economies, government debt rose by 9 percentage points to 63 percent of GDP in 2020, the fastest one-year increase in 30 years (World Bank 2022).

As debt levels rise, countries must devise smarter, more efficient ways to use their scarce public resources. In this context, the report asks three overarching questions:

1. What is the magnitude of total subsidies in the natural resource space?

2. What are the impacts of these subsidies on equity, efficiency, and the environment, and what are the gains from reforming or eliminating them entirely?

3. How can governments reform, repurpose, or eliminate subsidies in ways that are sustainable and politically feasible?

Although the literature on subsidies is large, significant knowledge gaps are embedded in each of these questions. In addressing these gaps, the report contributes new evidence in several related areas: the effects of commodity price changes on tropical forest loss, the responses of agricultural yields to fertilizer use across countries and regions, the distributional incidence of air pollution across countries, and some of the hidden consequences of coal power.

Explicit subsidies that impact air pollution and the agriculture and fishery sectors

Governments spend a large percentage of their budget on subsidies that exacerbate air pollution and affect the agriculture and fisheries sectors. The magnitude of subsidies for fossil fuels, agriculture, and fisheries is vast and likely exceeds US$7 trillion per year in explicit and implicit subsidies—or approximately 8 percent of global GDP. Explicit subsidies are direct fiscal expenditures from governments or taxpayers to producers or consumers; they cost about US$1.2 trillion per year—more than the GDP of Mexico—in these three sectors. Implicit subsidies are measured as unpriced externalities and account for the rest of the burden of subsidies on society and the economy. The distribution of subsidies across sectors and countries is highly skewed and uneven. As the report shows, high-income and upper-middle-income countries are responsible for a disproportionate share of global explicit subsidies. Nevertheless, the budgetary impacts on low- and lower-middle-income countries from explicit subsidies are nontrivial.

For fossil fuels alone, explicit subsidies—that is, direct fiscal support—totaled US$577 billion in 2021. This amount represents about three-quarters of all subsidies in the energy sector and has the triple effect of increasing the consumption of fossil fuels, reducing the incentives for investing in energy-efficient technologies, and making it more difficult for cleaner and renewable forms of energy to compete. For context, under the Paris Agreement on Climate Change, governments committed to raising US$100 billion annually in climate financing—less than one-fifth of what they spend to prop up fossil fuels.

In the agriculture sector, explicit subsidies in countries with available data total US$635 billion per year, or 18 percent of agricultural value added in these countries. The true global number likely exceeds US$1 trillion. More than 60 percent of these subsidies are coupled with production, implying that farmers receive support for buying specific inputs or growing specific crops. This form of subsidy distorts farmers' decisions, often reducing productivity and causing harmful environmental spillovers that encourage deforestation, pollute waterways, and deplete water supplies—often beyond national borders. This report quantifies these effects.

By some estimates, explicit subsidies in the fisheries sector total US$35.4 billion per year, of which US$22.2 billion are considered to enhance capacity and contribute to overfishing.

These subsidies include spending on fuel subsidies, fishing access agreements, boat construction and renewal, fisheries development projects, fishing port development, tax exemptions, and marketing and storage infrastructure. Fishery subsidies are not distributed evenly around the world. Indeed, five entities—China, the European Union, Japan, the Republic of Korea, and the United States—contribute 58 percent of the total estimated subsidy. High-income or upper-middle-income countries spend the most on subsidies, often to support fishing fleets that traverse and deplete fish stocks across the global oceans and often in the exclusive economic zones (EEZs) of low- and middle-income countries.

Implicit subsidies that impact air pollution and the agriculture and fishery sectors

If explicit subsidies are excessive, implicit subsidies are exorbitant (table ES.1). Although explicit subsidies exceed US$1 trillion, they are dwarfed by implicit subsidies to producers and consumers. Implicit subsidies are the price difference between the "undistorted" (socially optimal) price and the actual price that emerges after the subsidy is paid. Such gaps may arise when the subsidy encourages environmentally damaging behavior and often reflects inadequate regulation and policies that promote external damage. The burning of fossil fuels, for instance, emits harmful chemicals into the air, including fine particulate matter ($PM_{2.5}$) and sulfur dioxide, which have enormous impacts on health, as well as greenhouse gases such as carbon dioxide, which contribute to global climate change. These externalities impose costs on others, which are often treated as an implicit subsidy accruing to the polluter.

Implicit subsidies represent some of the most challenging environmental problems of our time. Implicit subsidies for fossil fuels amount to an estimated US$5.4 trillion per year, or more than 6 percent of global GDP, with the local impacts of air pollution and global climate change constituting more than 75 percent of the total. Agriculture emits about 6.8 gigatons of carbon dioxide equivalent (CO_2-eq) per year, or the equivalent of US$272 billion to US$544 billion worth of external damage. According to some estimates, the environmental damage from agriculture exceeds US$3.1 trillion per year, split almost

TABLE ES.1 Estimates of annual explicit and implicit subsidies, by sector

Sector	Explicit subsidy estimates	Implicit subsidy estimates
Fossil fuels	• US$577 billion: estimated fossil fuel subsidies for 191 countries (Parry, Black, and Vernon 2021)	• US$5.4 trillion: estimated impacts from local air pollution, greenhouse gas emissions, road congestion, and forgone tax revenues (Parry, Black, and Vernon 2021)
Agriculture	• US$635 billion: estimated agricultural subsidies for 84 countries (based on data from Gautam et al. 2022)	• US$548 billion to US$1.1 trillion: estimated impacts from greenhouse gas emissions (chapter 1 of this report) • US$5.3 trillion (Pharo et al. 2019), which includes: – US$1.5 trillion from greenhouse gas emissions – US$1.7 trillion from natural capital loss – US$2.1 trillion from pollution, pesticides, and antimicrobial resistance
Fisheries	• US$35.4 billion: estimated fishery subsidies for 152 countries (Sumaila, Ebrahim, et al. 2019; Sumaila, Skeritt, et al. 2019)	• US$83 billion: estimated economic benefits forgone due to open access (World Bank 2017)
Total	• US$1.25 trillion	• US$6 trillion to US$10.8 trillion

Source: World Bank.

equally between damages from greenhouse gases and costs due to the destruction or degradation of other natural capital such as land and water (Pharo et al. 2019). For fisheries, the largest implicit subsidy is the lack of effective regulations to reduce overcapacity and prevent overfishing. This implicit subsidy results in forgone economic benefits of an estimated US$83 billion per year, or nearly 20 percent of the size of the total sector.

Effect of subsidies on air quality and health

While air may be abundant, clean air is remarkably scarce and made scarcer by subsidies. This report demonstrates that about 94 percent of humanity—7.28 billion people—are directly exposed to unsafe average concentrations of fine particulate matter, one of the most pervasive air pollutants. Much of the low- and middle-income world is exposed to damaging levels of $PM_{2.5}$ of more than 15 micrograms per cubic meter ($\mu g/m^3$), a level that the World Health Organization deems to be unsafe (map ES.1).[1] This report also finds that 716 million people living in extreme poverty are directly exposed to unsafe $PM_{2.5}$ concentrations and are consistently located in countries with low quality of and poor access to health care. While estimates vary considerably, the Global Burden of Disease study estimates that air pollution causes about 7 million deaths each year (IHME 2020). Air pollution is not limited to $PM_{2.5}$; it consists of a toxic medley of pollutants, including ozone, nitrogen oxides, and sulfur dioxides, emitted from a wide range of sectors, including transport, power generation, industry, and residential heating, which are powered predominantly by fossil fuel combustion.

MAP ES.1 Percentage of population exposed to $PM_{2.5}$ over 15 $\mu g/m^3$

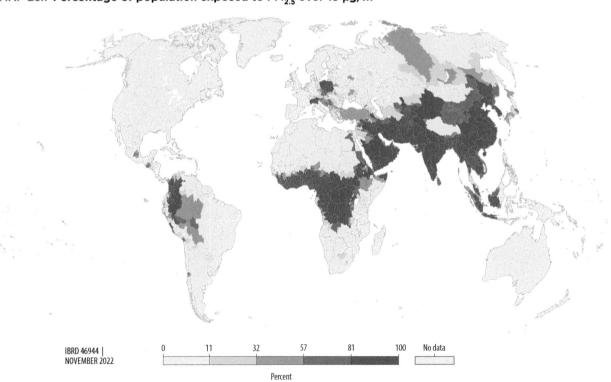

IBRD 46944 | NOVEMBER 2022

0   11   32   57   81   100   No data

Percent

Source: Rentschler and Leonova 2022.
Note: $PM_{2.5}$ = fine particulate matter; $\mu g/m^3$ = micrograms per cubic meter.

New evidence on toxic air pollution from the world's coal power plants illustrates the magnitude of unequal exposure (map ES.2). Richer countries and richer areas in all countries tend to build more coal power plants. Yet about 40 percent of the world's coal plants operate at a loss, with governments spending about US$13.6 billion to lower the price of coal artificially. A new analysis presented in this report, based on 3,800 coal-fired power plants in 71 countries, finds that areas located downwind of coal plants tend to experience higher levels of pollution and be poorer than upwind areas. Thus in countries rich or poor, lower-income groups are affected disproportionately by air pollution. This finding could reflect the fact that poorer people locate in neighborhoods where higher pollution lowers the price of land. It also could be a consequence of the health impacts of air pollution, which are known to lower labor productivity, cognitive performance, and incomes.

Although subsidies are harmful, simply removing them may not be sufficient to tackle pollution. The report estimates that a US$0.10 per liter increase in the average annual retail price of common transport fuels (for example, diesel) may be associated with a decrease of 2.2 $\mu g/m^3$ in the average annual concentration of $PM_{2.5}$ in capital cities.[2] While a notable improvement, this reduction makes barely a dent in cities such as New Delhi that have annual average $PM_{2.5}$ concentrations upward of 150 $\mu g/m^3$. Removing explicit fossil fuel subsidies could reduce $PM_{2.5}$ concentrations enough to prevent about 360,000 deaths between now and 2035—a large number, but only a fraction of overall deaths attributed to air pollution.

The evidence points to the limits of price-based measures to curb pollution, as energy consumption is often price inelastic. A literature of more than 400 empirical studies

MAP ES.2 **Global distribution of coal-fired power plants**

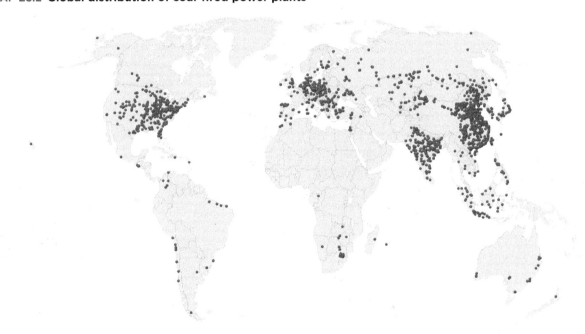

IBRD 46904 |
NOVEMBER 2022

Source: Global Coal Plant Tracker (https://globalenergymonitor.org/projects/global-coal-plant-tracker/).
Note: The map shows the spatial distribution of coal-fired power plants as of January 2013. Each dot denotes an operating unit.

providing 2,000 estimates shows that energy consumption tends to be price inelastic, implying that the response of energy demand to price changes is sluggish. For instance, a meta-analysis conducted for this report suggests that, on average, a 10 percent increase in the unit price of energy results in a short-run reduction in consumption of 2.8 percent. This finding has important implications: removing explicit fossil fuel subsidies—albeit a necessary first step—is insufficient for solving the air pollution challenge. Moreover, since political reality imposes a limit on how far energy prices can be raised, complementary policies are needed to ensure the availability and affordability of clean alternatives, address information and capacity constraints, and influence behavior.

Reforms of fossil fuel subsidies are typically pro-poor. By owning more cars and by heating and lighting bigger houses, the rich consume more energy and benefit disproportionately from energy subsidy schemes. Hence the richest income group always loses more from the removal of subsidies than the poorest—on average, 13 times more in 19 countries examined in this report. And while conventional wisdom holds that subsidies constitute a larger share of the income of the poor, who therefore lose more from subsidy reform, the data offer mixed evidence. In simulations of subsidy reform conducted for this report, the richest income group lost on average 10 percent more, as a share of their income, than the poorest group in most countries.

Subsidies for agriculture

Agricultural subsidies rarely achieve their stated purposes and often wreak havoc on forests, water supplies, and public health. Although agricultural subsidies are often intended to increase the efficiency of production, they usually have the opposite effect, making farming less efficient. A global analysis finds that, when countries increase their coupled subsidies, the technical efficiency of farming declines, even if output increases. These global results are backed up by several sources of evidence. Two meta-analyses and case studies confirm that, while subsidies may raise total agricultural production or even yields, they do so at the expense of efficiency, leading to wasted inputs and greater environmental destruction.

New research in this report also finds that subsidies tend to be poorly targeted to poor farmers and can exacerbate inequalities. Subsidies tend to accrue to wealthier farmers in disproportionately large amounts, even when programs are designed to be targeted to reach the poor. For instance, in Malawi and Tanzania, input subsidy programs designed to reach the poor pay US$5 to the top income quintile for every US$1 paid to the bottom income quintile. Nevertheless, the subsidies make up a substantially larger percentage of the bottom quintile's income, so eliminating these subsidies without compensation would be very harmful.

Agricultural subsidies can also widen the gender and equity gaps in agriculture, disproportionately affecting women and marginalized groups. The role of women in rural agriculture is growing. Yet despite comprising more than 48 percent of the agricultural labor force in low- and middle-income countries, women and some marginalized groups continue to have less access than men to input and output markets as well as to landownership. When a subsidy that is meant to increase agricultural yields or address poverty does not account for such differences, it can magnify inequalities.

Although all coupled subsidies can induce inefficiencies, market price support is found to be less distortive than other types of coupled subsidies. Market price support alters the price that farmers receive on their products. While this may cause farmers to alter the choice of crops grown and reduce the technical efficiency of agriculture, it does so to a lesser degree than other types of coupled subsidies. Likewise, evidence also shows that market price support leads to lower damages to water quality than other forms of coupled subsidies. There are various reasons for this finding, including that the benefits of market price support for farmers are often less certain than direct payments, and the subsidy is less likely to influence and distort methods of production because it is linked to outputs rather than to inputs.

In some geographies, the use of subsidized fertilizers is so excessive that it actually harms yields. New research finds that in subregions of South Asia and East Asia, use of nitrogen fertilizer is well beyond what is considered efficient, exacerbated by subsidies. Figure ES.1 shows the global relationship between fertilizer use and agricultural yields, measured by net primary productivity. It demonstrates that, at low and moderate levels of use, fertilizer has the intended beneficial impact on yields. However, at very high

FIGURE ES.1 Change in global agricultural productivity due to the use of nitrogen fertilizer, by quantile of use and region

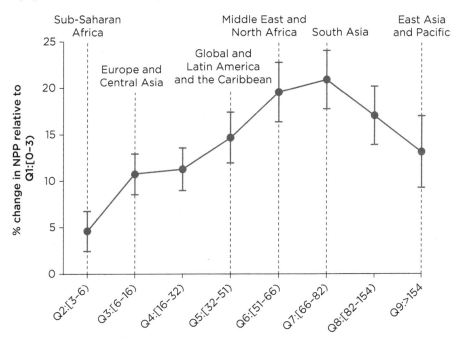

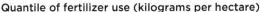

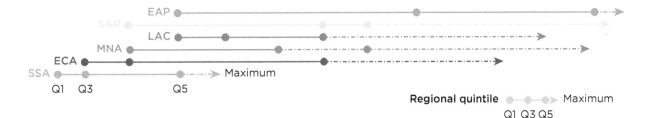

Source: World Bank estimates.
Note: EAP = East Asia and Pacific; ECA = Europe and Central Asia; LAC = Latin America and the Caribbean; MNA = Middle East and North Africa; SAR = South Asia; SSA = Sub-Saharan Africa. NPP = net primary productivity.

applications, the benefits level off and even begin to decline. Strikingly, about 50 percent of the global calories produced occur in areas where nitrogen fertilizers are overused, implying that there is room to reduce the use of fertilizer in these areas without having an adverse impact on yields. Given the recent rise in fertilizer prices, some countries and regions have space to bring fertilizer use closer to optimal levels with a limited or potentially positive impact on food supplies.

Subsidies drive both the deterioration of water quality by inducing the overuse of nitrogen fertilizers and the increase of water scarcity by incentivizing the overextraction of water. Globally, crops absorb only about 45 percent of nitrogen that is applied to fields. Part of the excess fertilizer runs off into waterways, with adverse effects on the environment and human health. A new analysis in this report estimates that input subsidies have been responsible for 17 percent of all nitrogen pollution in recent years. In the areas of the world where input subsidies are highest, subsidy-induced increases in water pollution have health impacts that decrease labor productivity by up to between 2.7 percent and 3.5 percent. Coupled subsidies also promote the abstraction of groundwater supplies for irrigation. New evidence finds that, at the mean level of subsidy exposure, agricultural areas around the world risk losing up to 13.2 cubic kilometers of water per year, roughly equivalent to the total amount of water lost in California between 2011 and 2014 at the height of the drought.

Agricultural subsidies are responsible for the loss of 2.2 million hectares of forest per year, equivalent to 14 percent of annual deforestation and 0.5 percent of global CO_2-eq emissions. Deforestation is sensitive to the price of commodities cultivated near the forest frontier. By increasing the profitability of cultivating such crops, subsidies induce farmers to expand cropland into forest frontiers. This expansion is particularly problematic in the major tropical forests of the world. Today, much of the Amazon Forest lies within a perilous 5 kilometers of the agricultural frontier, where encroaching crops are cultivated (map ES.3).

In addition, there are unseen cross-border spillovers—agricultural subsidies in rich countries drive tropical deforestation in vulnerable biomes around the world. The impact of subsidies is not confined to national borders, and it spills over national boundaries in ways that have not been recognized. For instance, the report shows that livestock subsidies in the United States drive deforestation in Brazil by increasing demand for soybeans as feedstock, a relationship that is likely not isolated to these two countries.

Tropical deforestation not only is linked to environmental losses, but also is implicated in the spread of zoonotic and vector-borne diseases, especially malaria. The report documents a global link between deforestation and the spread of malaria, finding that global agricultural subsidies can be linked to an additional 1.3 million to 3.8 million cases of malaria worldwide, with a total economic cost between US$3 billion and US$19 billion globally. These cases are likely to occur most often in areas with dense forest, where inhabitants are more likely to be poorer, and in Afro-indigenous populations in the Amazon region.

Deforestation affects the poor, indigenous groups, and women especially hard. In the Amazon jungle, for example, indigenous people have long struggled to preserve their land and way of life. Their plight continues to be of concern, as land disputes arising from the expansion into the Amazon basin have led to record numbers of territorial invasions and reports of violence. At the same time, food, medicine, and energy goods provided by forests are dwindling, affecting especially women, who traditionally gather these products, and their families.

MAP ES.3 Distance to the agricultural frontier in South America

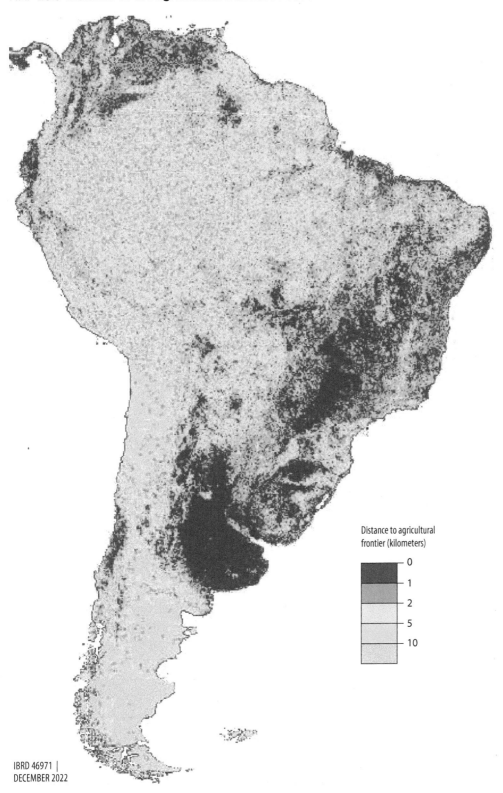

Distance to agricultural
frontier (kilometers)

- 0
- 1
- 2
- 5
- 10

IBRD 46971 |
DECEMBER 2022

Source: Druckenmiller 2022.
Note: Forest cover loss is measured by distance to the agricultural frontier, which is classified by 30-meter pixels.
Data on the extent of current crop production were obtained from the United States Geological Survey's Global
Croplands database (https://www.usgs.gov/apps/croplands/app/map?lat=0&lng=0&zoom=2).

Subsidies for fisheries

Subsidies contribute to the global decline in fisheries, but simply removing them will not be sufficient to stem the decline. Fisheries and the oceans that contain them are critical drivers of long-term environmental stability. However, more than 30 percent of global fish stocks are overfished, driven by inadequate control of access to fish stocks and harmful subsidies. This report focuses on three ecosystems—the Mauritanian EEZ, South China Sea, and the East China Sea—where large harmful subsidies are given to fishing vessels. The analysis finds that repurposing subsidies in ways that do not incentivize increased fishing capacity is critical for reducing fishing effort overall, increasing biomass, and ultimately increasing the rents captured by fishers. However, repurposing subsidies is not a panacea. When fisheries remain as quasi-open-access regimes with inadequate management of harvesting, repurposing subsidies may have little impact. Indeed, the two policy changes of repurposing subsidies and controlling access must be targeted jointly in order to have a meaningful and positive effect.

Reforms of fishery subsidies need to ensure that they do not leave the poorest behind. Results from the Mauritanian EEZ show that, while the aggregate effect of removing all harmful subsidies is an increase in total rents, artisanal fishers—who are often small-scale, poorer fishers—may lose out significantly. However, if subsidies are removed only for the larger pelagic and demersal fleets—many of which fly the flag of richer countries—and are kept in place for artisanal fleets, then fleets of all sizes benefit. Thus smart reforms can bring triple wins, with ecosystem health and sustainability improving, fishing fleets of all sizes increasing their catches and revenues, and the fishery sector becoming distributionally more progressive.

Reforming and repurposing harmful subsidies

If subsidies are so harmful, why are they so persistent? More than 200 years ago Frédéric Bastiat, economist and thinker, warned, "[That] which is seen may be as important as that which is unseen." On the one hand, much of the damage done by subsidies is unseen and emerges cumulatively and with lags, making attribution of damage difficult and weakening public pressure for reform. In addition, given the pervasiveness of subsidies, economies and people adjust to their presence, which builds inertia against change due to behavioral biases that favor the status quo. The benefits of the subsidy (explicit and implicit) also tend to accrue to special interest groups with a strong interest in perpetuating these policies and often commanding outsized influence over policy. On the other hand, damages from the policy are spread across entire nations, regions, and even generations, which makes forming coalitions for change difficult. Together, these characteristics are formidable forces against reforms, even though reforms may benefit society at large.

The reversal of subsidy reforms across the world points to the risks of poorly designed strategies that neglect distributional consequences, the magnitude of resistance, and the need for building a strong coalition in favor of change. The lessons learned from past reform efforts converge toward five guiding principles for designing and implementing successful subsidy reforms:

- *Build public acceptance and overcome credibility gaps*. Although communication has been recommended widely and consistently, communication is routinely neglected in efforts to reform subsidies. Economic efficiency does not imply political feasibility, and even when an existing subsidy system is found to be harmful and inefficient, reforming it and replacing it with an alternative policy framework is difficult. Failing to communicate why the reform is happening and how these programs will be repurposed can lead to public backlash and cause the reform to fail. A challenging unresolved problem is that of credibility and time inconsistency: even if compensation today makes all subsidy beneficiaries better off, committing future governments to sustaining those compensatory policies is difficult. Hence these beneficiaries may regard promises of compensation as little more than "cheap talk." Overcoming the credibility gap is crucial in such circumstances.

• *Implement complementary measures to improve effectiveness and lower the costs of reform.* Oftentimes, subsidy reform on its own will not be sufficient to achieve the intended goals, and complementary measures may be needed. For instance, removing fossil fuel subsidies may not lead to a significant decline in their use if alternatives are not in place. Raising gasoline prices while simultaneously investing in public transportation will be a lot more effective than simply doing the former. Likewise, repurposing fishery subsidies may not be sufficient to address problems of open access and may call for improvements in fishery management practices.

• *Mitigate short-term price shocks through social protection and compensation.* Compensating vulnerable households and firms is crucial for ensuring social stability and generating public support for reform. An important way to establish credibility is to compensate vulnerable households first, before reforming the subsidy, in order to build trust and credibility and assuage fears. Cash transfers offer a flexible and progressive alternative to subsidies. They can increase aggregate welfare and protect livelihoods and are therefore often considered central elements of social protection and revenue redistribution mechanisms.

• *Smooth the transition with carefully phased, step-wise reductions in harmful subsidies.* A gradual approach is typically less disruptive than a rapid one. Timing is crucial for determining not only *when* but also *how* to reform. Although rapid reforms—like shock therapy—may have appeal in terms of immediacy and visibility, there are strong merits to gradualism. Sudden price changes can be disruptive, especially if they are large. More gradual adjustments allow for adaptive changes and improvements—for instance, to safety net programs—and provide an opportunity for people and the economy to adjust to changes in relative prices. Gradual adjustments are perhaps most important for changes with large-scale impacts that cascade through the economy, such as impacts on the price of fossil fuels or food. Effective reform also depends on the careful timing of complementary measures, such as communication and compensation.

• *Redistribute revenue through long-term reinvestments with equitable or progressive benefits.* Subsidies are often substantial relative to GDP, so policy makers must be transparent in their plans for reallocating the revenues from a subsidy reform in a way that is consistent with long-term development strategies. Depending on a country's specific needs, revenues from reform could be used to invest in infrastructure—such as low-carbon electrification, public transit, digitization, or irrigation—or improved health care coverage, public education services, or institutional and tax reform. Even if reinvestment strategies are adjusted later on, formulating them early can lend credibility to the public good objectives of subsidy reform.

With competing needs and stretched budgets, repurposing inefficient and unsustainable spending is among the more cost-effective and economically attractive ways to achieve the global goals of sustainability and inclusivity. Indeed, in an era when public coffers are empty, debts are reaching unsustainable levels, inequalities are rising, and environmental degradation is slowing growth and shortening lives, reevaluating these spending programs and repurposing subsidies that are not working as intended must be a priority. Although doing so will entail demanding policy reforms, the costs of inaction will be far higher.

Notes

1. The World Health Organization recommends an average annual concentration of $PM_{2.5}$ of 5 μg/m^3 as the safe threshold.
2. For comparison, at low levels of concentration, a reduction in $PM_{2.5}$ levels by 5 μg/m^3 corresponds to a reduction in all-cause mortality of about 4 percent. In a sample of 131 countries, the mean price of gasoline is US\$0.99 (US\$0.83 for diesel).

References

Druckenmiller, H. 2022. "The Effect of Agricultural Commodity Prices and Producer Supports on Global Deforestation." Background paper prepared for this report, World Bank, Washington, DC.

Gautam, M., D. Laborde, A. Mamun, W. Martin, V. Piñeiro, and R. Vos. 2022. *Repurposing Agricultural Policies and Support: Options to Transform Agriculture and Food Systems to Better Serve the Health of People, Economies, and the Planet.* Washington, DC: World Bank and International Food Policy Research Institute (IFPRI).

IHME (Institute for Health Metrics and Evaluation). 2020. *Global Burden of Disease Study 2019 Results.* Seattle, WA: IHME. https://vizhub.healthdata.org/gbd-results/.

Parry, I., S. Black, and N. Vernon. 2021. "Still Not Getting Energy Prices Right: A Global and Country Update of Fossil Fuel Subsidies." Working Paper 2021/236, International Monetary Fund, Washington, DC.

Pharo, P., J. Oppenheim, C. R. Laderchi, and S. Benson. 2019. *Growing Better: Ten Critical Transitions to Transform Food and Land Use.* FOLU Report. London: Food and Land Use Coalition (FOLU).

Rentschler, J., and N. Leonova. 2022. "Air Pollution and Poverty: $PM_{2.5}$ Exposure in 211 Countries and Territories." Policy Research Working Paper 10005, World Bank, Washington, DC.

Sumaila, U. R., N. Ebrahim, A. Schuhbauer, D. Skerritt, Y. Li, H. S. Kim, T. G. Mallory, V. W. L. Lam, and D. Pauly. 2019. "Updated Estimates and Analysis of Global Fisheries Subsidies." *Marine Policy* 109 (November): 103695.

Sumaila, U. R., D. Skerritt, A. Schuhbauer, N. Ebrahim, Y. Li, H. S. Kim, T. G. Mallory, V. W. L. Lam, and D. Pauly. 2019. "A Global Dataset on Subsidies to the Fisheries Sector." *Data in Brief* 27 (December): 104706.

World Bank. 2017. *The Sunken Billions Revisited: Progress and Challenges in Global Marine Fisheries.* Washington, DC: World Bank.

World Bank. 2022. *Global Economic Prospects, January 2022.* Washington, DC: World Bank. https://openknowledge.worldbank.org/handle/10986/36519.

Abbreviations

AUC	area under the curve
CEDS	Community Emissions Data System
CGE	computable general equilibrium
CO_2-eq	carbon dioxide equivalent
CPAT	Carbon Policy Assessment Tool
CPI	consumer price index
DALYs	disability-adjusted life years
DEA	data envelopment analysis
DID	difference-in-differences
ECS	East China Sea
EEZ	exclusive economic zone
ENVISAGE	Environmental Impact and Sustainability Applied General Equilibrium
ERP	effective rate of protection
ESRAF	Energy Subsidy Reform Assessment Framework
EU	European Union
EwE	Ecopath with Ecosim
FAO	Food and Agriculture Organization
FERU	Fisheries Economics Research Unit
GDP	gross domestic product
GEMS	Global Environment Monitoring System
GESS	Growth Enhancement Support Scheme
GRACE	Gravity Recovery and Climate Experiment
GSSE	general services support estimate
GWS	groundwater signal
HIV/AIDS	human immunodeficiency virus / acquired immune deficiency syndrome
HYSPLIT	Hybrid Single-Particle Lagrangian Integrated Trajectory

IEA	International Energy Agency
IMF	International Monetary Fund
IUU	illegal, unreported, and unregulated
LPG	liquefied petroleum gas
MAFAP	Monitoring and Analysing Food and Agricultural Policies program
MERS	Middle East Respiratory Syndrome
MPS	market price support
MSY	maximum sustainable yield
NAIVS	National Agricultural Input Voucher Scheme
NDVI	normalized difference vegetation index
NO_2	nitrogen dioxide
NO_x	nitrogen oxide
NPK	nitrogen, phosphorus, and potassium
NPP	net primary productivity
NRP	nominal rate of protection
NSCS	northern South China Sea
OECD	Organisation for Economic Co-operation and Development
$PM_{2.5}$	fine particulate matter
PS	producer support
PSCT	producer single commodity transfer
PSE	producer support estimate
PSM	propensity score matching
R&D	research and development
SAR	special administrative region
SARS	Severe Acute Respiratory Syndrome
SDGs	Sustainable Development Goals
SO_2	sulfur dioxide
SPAM	Spatial Production Allocation Model
TCT	taxpayer-to-consumer transfers
TFP	total factor productivity
TSE	total support estimate

TWS	terrestrial water storage
μg/m³	micrograms per cubic meter
WHO	World Health Organization
WTO	World Trade Organization
ZNFU	Zambia National Farmers Union

CHAPTER 1

Introduction

Global Natural Resource Subsidies

"The Earth is the only thing we all have in common."
—*Wendell Berry*

CHAPTER AT A GLANCE

This report examines the impact of subsidies on the world's stock of foundational natural capital—clean air, land, and oceans.

- These forms of natural capital are critical for human health and nutrition and underpin much of the economy. This chapter describes the definition of subsidies in each of these sectors and summarizes subsequent chapters examining their impacts.

- Subsidies are important tools that governments can use to encourage desirable outcomes, support economically, environmentally, or politically important industries, or achieve particular goals related to economic efficiency or equity.

- But subsidies can also be distortive by reducing economic efficiency, exacerbating negative externalities, and causing significant damage to the environment, human health, and economic productivity.

This chapter describes the many definitions of subsidies and presents data on the magnitude of subsidies in three sectors affecting critical natural resources: fossil fuels, agriculture, and fisheries.

- Of all subsidies to the energy sector, about three-quarters go to fossil fuels. For fossil fuels alone, explicit subsidies—that is, policies or direct fiscal expenditures that lower the price of consumption or production of fossil fuels—totalled US$577 billion in 2021. By comparison, under the Paris Agreement on Climate Change, governments committed to raise only US$100 billion annually in climate financing—just a fifth of what they spend to prop up fossil fuels.

- Agricultural subsidies exceed an estimated US$635 billion per year, approximately 0.9 percent of gross domestic product (GDP) and 18 percent of agricultural value added for the 84 countries with available data. More than 60 percent of these subsidies is in the form of coupled support, which distorts producers' decisions and leads to harmful environmental and economic impacts. In 38 countries with data on irrigation support, spending on irrigation totals approximately 1.8 percent of GDP. It is unclear whether this level of spending is warranted even within a narrow benefit-cost framework.

- Global fishery subsidies are estimated at about US$35 billion per year. Out of this amount, US$22 billion are identified as harmful subsidies, such as fuel subsidies, that can lead to overcapacity and overfishing. For almost all regions of the world, harmful subsidies are higher than beneficial subsidies, except for North America, where a greater share of subsidies supports monitoring and management of fishing activities to ensure sustainable use.

Overview

Properly managing natural resource sectors like agriculture, fisheries, water, energy, and air is critical for ensuring economic growth that is robust, sustainable, and inclusive. But management of these sectors is often challenging given their open-access and common-pool nature and the many unpriced externalities that are by-products of their use or production. As recently highlighted in the Dasgupta Review, despite the fact that cultural, social, economic, and environmental health are closely intertwined, all of the natural assets on which humanity and economies depend are in decline (Dasgupta 2021): ambient air pollution is responsible for an estimated 4.5 million premature deaths each year, while another 2.3 million deaths are caused by indoor air pollution (IHME 2020); polluted water is implicated in stunting and cognitive deficiencies; about 75 percent of global lands are substantially degraded (Montanarella, Scholes, and Brainich 2018), reducing food production and other critical services like flood protection and biodiversity; and 66 million hectares of forests are lost each year due to shifting agriculture (Curtis et al. 2018). According to the United Nations Food and Agriculture Organization, 34 percent of global fish stocks are overfished, which has huge food security, economic, and social consequences (Srinivasan et al. 2010; World Bank 2017).

To what extent do poorly designed subsidies cause and exacerbate these problems? This report attempts to cast some quantitative light on this issue. To do so, it focuses on three foundational natural resources that are critical for human health and nutrition and underpin much of the economy:

1. *Air quality*, which is affected by a range of pollutants from a variety of sources, chief among which is the combustion of fossil fuels. Approximately one in five deaths globally is due to unsafe air, and, as the new analyses in this report show, a significant number of these deaths can be attributed to the burning of underpriced fossil fuels.

2. *Agriculture*, which is responsible for feeding the world and employing 1 billion people, including 78 percent of the world's poor. Agriculture is also responsible for 21 percent of the tree cover lost globally (Curtis et al. 2018), a growing crisis of water quality (Damania et al. 2019), and 26 percent of global carbon dioxide equivalent (CO_2-eq) emissions (Poore and Nemecek 2018). Nevertheless, nearly every country on Earth subsidizes agriculture in some way.

3. *Fisheries*, which supply about 3 billion people with almost 20 percent of their protein intake from animals (Mathiesen 2015). Fisheries are in a collective state of crisis. More than 30 percent of fisheries are overfished, which is generating approximately US$83 billion in lost economic rents. Subsidies in this sector exacerbate overfishing and lead to further depletion of this valuable resource (Sumaila et al. 2010, 2021).

What are subsidies and why do they matter?

There is no universally accepted definition of a subsidy. Particular organizations such as the Organisation for Economic Co-operation and Development (OECD) or the World Trade Organization (WTO) use definitions that align with their specific policy objectives. For instance, the WTO's definition is narrow and contains three basic elements: "(i) a financial contribution (ii) by a government or any public body within the territory of a Member (iii) which confers a benefit."[1] All three of these elements must be satisfied in order for a subsidy to exist in the legal parlance of the WTO. The narrow definition used by the WTO may reflect the need for verifiable and easily quantifiable evidence of support that can withstand legal scrutiny.

Economics, though, admits to a much wider range of definitions that could include nonfinancial policy support (for example, granting free access to a resource or favoring one firm or sector over another) as well as financial support (Sakai, Yagi, and Sumaila 2019). For instance, an uncorrected externality like air or water pollution would, in theory, be treated as a subsidy if the costs fall on some other party. When externalities are included in the definition, it is important to measure the extent of the subsidy correctly and to be precise about the definition being used.

Indeed, different types of expenditures and policies (and lack thereof) may, at different times, be considered subsidies, public good provision, environmental policies, or social safety nets. Rather than focusing on any single definition of what a subsidy is, this report takes a practical view, acknowledging that different definitions are appropriate in different contexts and at different times.

1. A narrow definition of a subsidy would include only direct fiscal outlays from the government to producers or consumers that are intended to affect the production or consumption of goods and services. This definition corresponds to the WTO and other more traditional definitions. It could be broadened to include policies that do not include direct fiscal outlays from government, but result in transfers from producers to consumers or vice versa. These policies would include trade barriers, price ceilings, and price floors.

2. Expenditures on the provision of public goods can also be considered a subsidy if they are intended to benefit producers in a particular industry. For instance, research and development (R&D) of agricultural technologies, construction and maintenance of infrastructure like irrigation systems and ports, and even expenditures on weather observation and early warning systems may be considered subsidies if their uses are provided at below-market price and they benefit private producers.

3. An even broader definition would include external costs that a person or firm generates and some other entity pays for—that is, an externality. This definition would include costs that are monetary (such as expenditures made to mitigate damages) as well as non-monetary (such as health damages from air or water pollution). Box 1.1 elaborates on these definitions.

In principle, there are good reasons for all three definitions, and the appropriate measure would depend on the intended use. Nevertheless, data and estimates of the three definitions vary widely. Data on direct, *explicit subsidies* like those in the first definition are generally more available, as they tend to be detailed in government budgets, and obtaining reliable estimates is usually simply a matter of data collection. Nevertheless, challenges remain in aggregating subsidies from different ministries, levels of government (federal versus state and local), and across sectors that use different definitions. Data on the provision of public goods can also be relatively easier to obtain, however, complicating decisions on what should be considered a subsidy and what should be considered a welfare or a safety net program. Obtaining estimates on external damages, or *implicit subsidies*, can be the most challenging. Accordingly, much of the new analyses presented in this report focuses on these damage estimates.

Subsidies are important tools that governments can use to encourage certain behavior, support economically or politically important industries, or achieve particular goals

Formal definitions of implicit and explicit subsidies

Definitions of subsidies vary considerably. In general, there is more understanding of how explicit subsidies ought to be measured and defined—that is, as the financial value of support provided by the government to a sector. The support could be monetary (for example, a cash transfer or tax exemption) or in-kind (for example, free fertilizers or fuel). In principle at least, the explicit subsidy measures the financial value of such transfers. Measuring implicit subsidies raises a more complex set of issues. An implicit subsidy measures the externality resulting from an explicit subsidy or policy exemption. Since externalities are included in the definition of an implicit subsidy, it is important to distinguish the portion of uncorrected externalities caused by the subsidy from the total external cost of the activity. As an example, the use of a pesticide may have environmental impacts, even without subsidies. If a subsidy induces greater use of the pesticide, the additional impact may also count as an implicit subsidy.

To illustrate more precisely, consider a firm that generates negative externalities (say, pollution) that induce damages costing D per unit of output Q.[a] In the absence of a subsidy, the firm produces output Qn, and the damage to society is DQn.

Now suppose that there is a subsidy of S per unit of output produced. With a subsidy of S per unit of production, let production levels be Qs. Then the external damage is DQs ($Qs > Qn$).

Define the optimal (Pigouvian) damage and corresponding level of output as DQ^*. In this stylized example, there are three measures of "subsidy," each corresponding to a different definition and none of which is theoretically wrong:

1. The (narrow monetary) definition of an explicit subsidy would simply be SQs, which corresponds to the World Trade Organization and other more traditional definitions that are concerned mainly with fiscally tied definitions. In principle, such a definition is incomplete if a subsidy is concerned with all forms of policy support (tacit and explicit).
2. A broader definition of a subsidy would recognize that greater damage has occurred than would have occurred without the subsidy. In this case, the subsidy is measured as $SQs + D(Qs - Qn)$, where SQs is the explicit subsidy and $D(Qs - Qn)$ measures the implicit subsidy.
3. A more complicated definition would take deviations from the optimal level of pollution (EDQ^*) into account and thus be $SQs + D(Qs - Q^*)$. This definition would be the most accurate definition of a subsidy, but it requires estimating the optimal level of damage (DQ^*), which can be complicated. For this reason, the second definition is used more widely in practice.

Finally, ignoring implicit subsidies (that is, externalities) for computational convenience brings problems of consistency: Would all in-kind (nonpecuniary) contributions be excluded? If not, what is the difference between one in-kind contribution (such as free pesticides) and another (the health externality from the pesticide)? One cannot rely on the fact that one kind of contribution affects profits and the other affects health (since a profit function is a subset of a societal welfare function). At the other extreme, it would also be inappropriate to measure the subsidy as $SQs + DQs$, since this approach wrongly assumes that without the subsidy there would be no external costs.

a. The pollution and damage functions are all linear for simplicity, and abatement is not considered.

related to economic efficiency or equity. But they can also be distortive by reducing economic efficiency and exacerbating negative externalities. The litany of problems created by subsidies is widely recognized. Subsidies can reduce total factor productivity by shifting resources to less productive sectors. And, when imprudently applied to natural resource sectors, they can have harmful impacts on the environment. For instance, natural resource subsidies can lead to overcapitalization, which results in more land being devoted to agricultural use or more fishing boats attempting to harvest a shrinking supply of fish (Milazzo 1998). They can also send the wrong economic signals, indicating, for example, that scarce natural resources—like water in a desert—are abundant when, in fact, they are not. The consequence is overuse and inefficient use, which can result in a resource deficit that acts as a drag on economic progress and growth.

Perverse subsidies, especially for the use of natural resources, are also likely one of the most significant sources of inefficient spending (Arguedas and van Soest 2009; Yih et al. 2018). In 2020, total global debt surged to 263 percent of GDP, its highest level in half a century (World Bank 2022). In emerging markets and developing economies, government debt alone increased by 9 percent of GDP in 2020 as countries responded to the COVID-19 crisis by stimulating the economy while dealing with reduced revenues.

In normal circumstances, such levels of debt might be sustainable or even desirable if they were servicing prudent investments. But circumstances are not normal, as the world is dealing with multiple crises, including COVID-19, disrupted supply chains, rising inflation in many countries, and food and energy shocks stemming from the conflict in Ukraine. Credit markets are tightening, tax revenues are declining, and government spending is on the rise. With fiscal space shrinking quickly in many countries, economic stability will depend, at least in part, on better and more effective public spending. Assessing the magnitude and impact of subsidies on renewable natural resources is a key part of bringing greater efficiency and equity in public spending and addressing the unsustainable use of natural resources. Given the regressive nature of many subsidy schemes (Schuhbauer et al. 2020), subsidy reforms can also help to address concerns of rising inequality and poverty as well as enhance environmental sustainability.

The magnitude of subsidies in natural resource sectors

Although subsidies are such a large and important part of government budgets, reliable quantification has proven elusive due to the complexity, interconnections, and scale of support. This complexity arises partially because subsidies come in different forms, are provided by different levels of government (subnational as well as central governments), and can go by several different definitions—price ceilings or floors, direct support to producers or households, support through the subsidizing of inputs, tax expenditures, or unpriced or unaddressed negative externalities. This section presents new analyses combined with reviews of the literature to describe the magnitude of subsidies in the selected sectors.

Fossil fuel subsidies

Economists advocate price-based policy instruments as a central tool for addressing the adverse societal costs associated with fossil fuels, such as air pollution. In principle, this approach calls on governments to reflect the environmental and health costs of polluting activities in their prices—in particular, by taxing the fuels and activities that drive air pollution. However, rather than taxing polluting activities, many governments around the world provide explicit subsidies to lower the cost of using fossil fuels, thus entrenching polluting technologies and practices.

In an effort to promote industrialization and energy affordability—but also to cater to influential political interest groups—governments around the world are actively lowering the cost of polluting forms of energy through "explicit" subsidization schemes. These schemes have grown into expensive support programs for the consumers and producers of oil, gas, and coal products. Globally, explicit fossil fuel subsidies are estimated to have been around US$577 billion in 2021 (Parry, Black, and Vernon 2021). Thus they are almost three times more than global subsidies paid to the renewable energy sector (IRENA 2020); they are also almost six times more than the amount that countries have committed to raise in annual climate financing (US$100 billion) under the Paris Agreement on Climate Change. Online appendix A provides country-level figures for fossil fuel subsidies.[2]

While US$577 billion is a vast amount to spend on propping up polluting fuels in one year, it is likely to be an underestimate. In particular, subsidies paid to polluting industries and the producers of fossil fuels are often far more difficult to define, observe, and quantify. Such producer subsidies can refer to various kinds of preferential treatment of fossil fuel exploration, extraction, or processing firms or other energy-intensive companies, industries, or products (chapter 3). Such producer subsidies could be explicit, such as grants, low-interest loans, or direct payments; they may be in-kind, such as credit subsidies, government guarantees to protect investment, or subsidies through public procurement.[3] The in-kind component of these subsidies is especially difficult to identify and measure, explaining the scarcity of studies. A study of G-20 countries estimates that producer subsidies amounted to US$444 billion in 2014. The largest share of these producer subsidies came in the form of fossil fuel investments by state-owned enterprises, amounting to US$286 billion (Bast et al. 2015).

Even when fossil fuels are not subsidized explicitly, their prices do not fully reflect the vast societal and environmental damages they cause. The polluting activities that drive these externalities are reinforced and incentivized by the underpricing of fossil fuels. The International Monetary Fund (IMF) calls this failure to price externalities "implicit subsidies" to fossil fuels (Parry, Black, and Vernon 2021). The IMF estimates the cost of these implicit fossil fuel subsidies at US$5.4 trillion in 2020, with local air pollution and global climate change impacts constituting more than 75 percent of the total. At US$2.5 trillion a year, local air pollution was the single largest unpriced environmental externality from fossil fuels in 2020—far more than the size of explicit subsidies. An important implication is that removing explicit subsidies alone is unlikely to bring fuel prices to their socially optimal level.

Agricultural subsidies

As with fossil fuel subsidies, agricultural support can be considered in terms of explicit and implicit subsidies. However, unlike air pollution, much of the effort to quantify global subsidies is restricted to identifying explicit support. Even when it comes to explicit support, however, quantifying global magnitudes can be extremely difficult, largely because countries and organizations measure support in different ways. In addition, several definitions exist for agricultural support, which are discussed in more depth in chapter 6. The most comprehensive measure of agricultural support is the total support estimate (TSE). This estimate includes support for both outputs (final crops produced) and inputs (seeds and fertilizer, for example), the sum of which is called the producer support estimate (PSE), as well as any taxpayer-to-consumer transfers (TCTs) and general services support estimates (GSSEs). GSSEs mainly track the provision of public goods like R&D, infrastructure financing and maintenance, and educational programs.

Globally, between 2016 and 2018, annual TSE for 84 countries with available data amounted to US$635 billion per year (see online appendix A for country-level values). This amount equals approximately 0.9 percent of GDP and nearly one-fifth of agricultural value added for these countries. The share of this TSE that was transferred to individual producers—that is, PSE—was about 71 percent, with the remaining share split between GSSE (18 percent) and TCT (11 percent). Thus the bulk of support goes to producers. Around 61 percent of this support is in the form of coupled support, such as market price support or payments for input use, which distort producers' decisions. As discussed in chapters 8 and 9, this type of support is responsible for much of the harmful environmental impacts. Recent trends, however, suggest that some countries have increased their funding for decoupled support, like direct payments and agricultural investments. This information comes from a database assembled by Gautam et al. (2022), which is discussed more in chapter 6. The 84 countries included accounted for 67 percent of the global value of agricultural production in 2016 (FAO 2022). Thus total explicit subsidies are likely to be significantly higher, perhaps approaching US$1 trillion, based on a simple extrapolation.

Estimating the magnitude of implicit agricultural subsidies is much more difficult. Total greenhouse gases from agriculture are estimated to be approximately 13.7 gigatons of CO_2-eq or approximately 26 percent of total annual greenhouse gas emissions (Poore and Nemecek 2018). Given that nearly all of these emissions are untaxed and unregulated, these emissions can be considered an implicit agricultural subsidy. At a shadow price of between US$40 and US$80 per ton of CO_2-eq, this subsidy is the equivalent of US$548 billion to US$1.1 trillion worth of external damages that are not internalized by producers or consumers of agricultural products. Other studies have estimated this value to be much higher, at US$1.5 trillion per year (Pharo et al. 2019).

In addition to greenhouse gas emissions, agriculture is responsible for damages to other types of natural capital, like forests and other natural habitats as well as freshwater stocks, which can be considered an implicit subsidy. In many regions, nitrogen fertilizer is applied in such large quantities that much of it is not absorbed by crops and ends up in water runoff, collecting in water supplies. In countries like India, where nitrogen fertilizer is heavily subsidized, a mere 32 percent of nitrogen is absorbed by plants. Even in regions like Europe and North America, where subsidy rates are much lower, only about 52 percent and 68 percent of nitrogen, respectively, is absorbed by plants (Zhang et al. 2015). This excess nitrogen in waterways has enormous environmental and health impacts, which are discussed in chapter 8. While it is difficult to put a single dollar figure on their impact, environmental, health, and productivity damages can be very significant, as chapter 8 discusses.

Fishery subsidies

Rich countries are subsidizing the destruction of fisheries in all corners of the world's oceans. Explicit subsidies in the fishery sector total an estimated US$35.4 billion per year, of which US$22.2 billion are considered to enhance capacity and contribute to overfishing. This spending includes subsidies for fuel, fishing access agreements, boat construction and renewal, fishery development projects, fishing port development, tax exemptions, and marketing and storage infrastructure. Fishery subsidies are not distributed evenly around the world. Five entities—China, the European Union, Japan, the Republic of Korea, and the United States—contribute 58 percent of the total estimated subsidy. Indeed, the vast majority of subsidies in the fishery sector are spent by high-income or upper-middle-income countries, often to support fishing fleets that traverse the global oceans.

The largest implicit subsidies for fisheries are surely the lack of regulations, which enable open access and lead to overfishing. Overfishing results in a loss of economic benefits estimated at US\$83 billion per year, representing an implicit subsidy that is nearly 20 percent of the size of the total sector.

The remainder of this report

The remainder of this report explores the efficiency, equity, and environmental effects of these subsidies on air, land, and oceans. The report has four parts. Part 1 has four chapters (chapters 2–5) examining the impact of subsidies on air quality. Part 2, also with four chapters (chapters 6–9), studies the economic and environmental impact of agricultural subsidies. Part 3 contains a single chapter (chapter 10), which is devoted to subsidies in the fishery sector. Each of these sections presents new research on the impacts of subsidies on economic production, distributional outcomes, and environmental damages. Finally, part 4 presents a policy framework for reforming and repurposing subsidies (chapter 11) and a concluding chapter (chapter 12).

An overview provided as a separate document presents a chapter-by-chapter summary. Technical appendixes are also provided online at https://openknowledge .worldbank.org/handle/10986/39423.

Notes

1. For the WTO definition of a subsidy, see https://www.wto.org/english/tratop_e/scm_e/subs_e.htm.
2. Online appendix A can be found at https://openknowledge.worldbank.org/handle/10986/39423.
3. Such in-kind fossil fuel subsidies have also been labeled "implicit subsidies," but this definition differs from the International Monetary Fund definition of "implicit subsidies," which reserves this term for the environmental externalities associated with fossil fuel use.

References

Arguedas, C., and D. P. van Soest. 2009. "On Reducing the Windfall Profits in Environmental Subsidy Programs." *Journal of Environmental Economics and Management* 58 (2): 192–205.

Bast, E., A. Doukas, S. Pickard, L. van der Burg, and S. Whitley. 2015. *Empty Promises: G20 Subsidies to Oil, Gas, and Coal Production*. Washington, DC: Oil Change International; London: Overseas Development Institute. https://odi.org/en/publications/empty-promises-g20-subsidies -to-oil-gas-and-coal-production/.

Curtis, P. G., C. M. Slay, N. L. Harris, A. Tyukavina, and M. C. Hansen. 2018. "Classifying Drivers of Global Forest Loss." *Science* 361 (6407): 1108–11.

Damania, R., S. Desbureaux, A. S. Rodella, and J. Russ. 2019. *Quality Unknown: The Invisible Water Crisis*. Washington, DC: World Bank.

Dasgupta, P. 2021. *The Economics of Biodiversity: The Dasgupta Review*. London: HM Treasury.

FAO (Food and Agriculture Organization). 2022. FAOSTAT Statistical Database. Rome: FAO.

Gautam, M., D. Laborde, A. Mamun, W. Martin, V. Piñeiro, and R. Vos. 2022. *Repurposing Agricultural Policies and Support: Options to Transform Agriculture and Food Systems to Better Serve the Health of People, Economies, and the Planet*. Washington, DC: World Bank and the International Food Policy Research Institute (IFPRI).

IHME (Institute for Health Metrics and Evaluation). 2020. *Global Burden of Disease Study 2019 Results*. Seattle, WA: IHME. https://vizhub.healthdata.org/gbd-results/.

IRENA (International Renewable Energy Agency). 2020. "Renewable Capacity Statistics 2020." IRENA, Abu Dhabi.

Mathiesen, Á. M. 2015. "The State of World Fisheries and Aquaculture 2012." FAO, Rome.

Milazzo, M. 1998. *Subsidies in World Fisheries: A Reexamination*. Vol. 23. Washington, DC: World Bank.

Montanarella, L., R. Scholes, and A. Brainich, eds. 2018. *The IPBES Assessment Report on Land Degradation and Restoration*. Bonn: Secretariat of the Intergovernmental Science-Policy Platform on Biodiversity and Ecosystem Services.

Parry, I., S. Black, and N. Vernon. 2021. "Still Not Getting Energy Prices Right: A Global and Country Update of Fossil Fuel Subsidies." Working Paper 2021/236, International Monetary Fund, Washington, DC.

Pharo, P., J. Oppenheim, C. R. Laderchi, and S. Benson. 2019. *Growing Better: Ten Critical Transitions to Transform Food and Land Use*. FOLU Report. London: Food and Land Use Coalition (FOLU).

Poore, J., and T. Nemecek. 2018. "Reducing Food's Environmental Impacts through Producers and Consumers." *Science* 360 (6392): 98792.

Sakai, Y., N. Yagi, and U. R. Sumaila. 2019. "Fishery Subsidies: The Interaction between Science and Policy." *Fisheries Science* 85 (3): 439–47.

Schuhbauer, A., D. J. Skerritt, N. Ebrahim, F. Le Manach, and U. R. Sumaila. 2020. "The Global Fisheries Subsidies Divide between Small- and Large-Scale Fisheries." *Frontiers in Marine Science* (September 29): 792.

Srinivasan, U. T., W. W. Cheung, R. Watson, and U. R. Sumaila. 2010. "Food Security Implications of Global Marine Catch Losses due to Overfishing." *Journal of Bioeconomics* 12 (3): 183–200.

Sumaila, U. R., A. S. Khan, A. J. Dyck, R. Watson, G. Munro, P. Tydemers, and D. Pauly. 2010. "A Bottom-Up Re-estimation of Global Fisheries Subsidies." *Journal of Bioeconomics* 12 (2010): 201–25.

Sumaila, U. R., D. J. Skerritt, A. Schuhbauer, S. Villasante, A. M. Cisneros-Montemayor, H. Sinan, D. Burnside, et al. 2021. "WTO Must Ban Harmful Fisheries Subsidies." *Science* 374 (6567): 544.

World Bank. 2017. *The Sunken Billions Revisited: Progress and Challenges in Global Marine Fisheries*. Washington, DC: World Bank.

World Bank. 2022. *Global Economic Prospects, January 2022*. Washington, DC: World Bank. doi:10.1596/978-1-4648-1758-8.

Yih, H., H. Huang, F. Li, and C. Tsai. 2018. "Environmental Tax Reform, R&D Subsidies, and CO_2 Emissions: View Double Dividend Hypothesis." *International Journal of Energy Economics and Policy* 8 (5): 288.

Zhang, X., E. A. Davidson, D. L. Mauzerall, T. D. Searchinger, P. Dumas, and Y. Shen. 2015. "Managing Nitrogen for Sustainable Development." *Nature* 528 (7580): 51.

PART I

AIR

CHAPTER 2

Toxic Air
Overview

"Saving our planet, lifting people out of poverty,
advancing economic growth ... these are one and the same fight."
—*Ban Ki-moon*

CHAPTER AT A GLANCE

- *Clean air is key for sustaining life, yet 94 percent of the world population is exposed to toxic levels of air pollution.* As this part of the report shows, air pollution is one of the most far-reaching environmental crises facing the world and—like climate change—is linked directly to the underpricing and overuse of polluting fossil fuels.

- *Air pollution is a toxic medley of many different pollutants from many different sources,* including particulate matter ($PM_{2.5}$), nitrogen oxides (NO_x), sulfur dioxide (SO_2), and black carbon, among others. Many of these pollutants are generated directly through the combustion of fossil fuels that are cheaply available. Others are generated through polluting industrial processes, residential uses (such as heating), and transport systems. Dust storms and forest fires add to these anthropogenic sources of pollution, reducing air quality both outdoors and inside people's homes.

- *Accounting for about 7 million premature deaths each year, air pollution is one of the leading causes of death worldwide,* especially affecting poorer people who are both more exposed and more vulnerable to it. The burden of pollution is particularly high in rapidly industrializing middle-income countries, but low-income countries have a window of opportunity to follow a cleaner, more efficient development trajectory.

- *Reforming explicit fossil fuel subsidies is a necessary but insufficient step to tackling the air pollution challenge.* Policy makers must fully reflect the health and societal costs of air pollution in the price of fossil fuels and implement complementary policies that enable the transition to clean and efficient technologies.

Introduction

Few things are as vital to sustaining human life as air. The average adult breathes more than 10,000 liters of air per day, supplying air deep into the body's respiratory and pulmonary system and delivering the oxygen necessary for core cognitive and physical functions.

While air may appear to be abundant, *safe air* is remarkably scarce. Human activities have resulted in such widespread degradation of air quality that, today, the vast majority of the world's population inhales unsafe levels of air pollution. The air in most inhabited areas exhibits unnaturally high concentrations of a range of toxic pollutants, including particulate matter, ozone, NO_2, carbon monoxide, and SO_2 (box 2.1). Many of these

Air pollution: A toxic medley of many different pollutants from many different sources

Ambient air pollution is one of the major causes of diseases such as lung cancer, stroke, heart disease, and chronic and acute respiratory diseases. Based on Taheripour et al. (2022), this box provides a brief overview of the most important air pollutants and their health impacts.

- *Particulate matter (PM$_{2.5}$ and PM$_{10}$)* refers to both primary and secondary particulate matter, classified according to the size of particles. PM$_{2.5}$ refers to particles with aerodynamic diameters under 2.5 micrograms (μm), while PM$_{10}$ refers to particles under 10 μm. Its main sources include the transport, power, residential, and industrial sectors as well as agricultural burning and forest fires. Long-term exposure to high concentrations of particulate matter increases the risk of cardiovascular disease, stroke, respiratory diseases and lung cancer, ischemic heart disease, and chronic obstructive pulmonary disease (Chen and Hoek 2020; Cohen et al. 2017). The toxicity of PM$_{2.5}$ varies, depending on its chemical composition—that is, its acidity (Thurston, Chen, and Campen 2022).
- *Ozone* is not emitted by primary sources; it is formed through a series of complex chemical reactions in the atmosphere, caused by energy transferred to nitrogen dioxide (NO$_2$) molecules when they absorb light (WHO 2021b). Methane is a precursor to ozone. Ozone has been associated with respiratory diseases, independent of other air pollutants. Each year, long-term exposure to ozone is responsible for up to 1 million deaths (Malley et al. 2017).
- *Nitrogen oxides* are a by-product of burning fossil fuels. They have been linked to chronic obstructive pulmonary disease, acute lower respiratory infections, and elevated levels of mortality (Huangfu and Atkinson 2020). In 2015 NO$_x$ emissions from diesel vehicles alone were responsible for 107,600 premature deaths in 11 regions (Anenberg et al. 2017).
- *Sulfur dioxide* (SO$_2$) is one of the main components of air pollution caused by the burning of fossil fuels, especially coal. It is linked to increased risk of asthma, lung cancer, respiratory disease, cardiovascular disease, and coronary heart disease (Kobayashi et al. 2020; Zheng et al. 2021). Global statistics for mortality from long-term exposure to it are not available (Orellano, Reynoso, and Quaranta 2021).
- *Black carbon and organic carbon.* Black carbon is a component of PM$_{2.5}$ and consists solely of carbon. Both black and organic carbon are formed through the incomplete combustion of fossil fuels, biofuels, and biomass. Both can exacerbate health risks and increase the risk of cardiovascular diseases, respiratory diseases, and cancer (Yang et al. 2021). In China in 2013, black carbon–related mortality was at least 265,000 (Wang et al. 2021).
- *Carbon monoxide* is a colorless, odorless toxic gas. It is a by-product of the burning of fossil fuels, including petrol, coal, natural gas, and kerosene and the incomplete combustion of carbon-based fuels like wood (WHO 2021b). Exposure to it increases the risk of cardiovascular diseases. Data from 337 cities in 18 countries show that a 1 μg per cubic meter increase in carbon monoxide concentration is associated with a 0.91 percent increase in total daily mortality (Chen et al. 2021).
- *Nonmethane volatile organic compounds* refer to a category of substances with different properties. They differ in their composition, but display similar behavior in the atmosphere (EEA 2015). They can affect people's cardiovascular and respiratory systems. Biomass and municipal solid waste combustion are significant sources of these compounds (Stewart et al. 2021).
- *Ammonia* emissions have been shown to be a precursor to secondary PM$_{2.5}$ emissions because they react with other pollutants such as SO$_2$ and NO$_x$ (Domingo et al. 2021). The agriculture sector—specifically fertilizers and livestock farming—are among the main sources of ammonia emissions. Estimates for the United States suggest that ammonia emissions cause 2,400 premature deaths each year (Domingo et al. 2021).

pollutants are gases or tiny particles that pass directly through the lungs and enter the bloodstream, affecting vital organs, such as the heart and lungs.

The use of polluting fossil fuels contributes substantially to a vast environmental and health crisis. Yet governments around the world explicitly subsidize the consumption of these fuels to the tune of around US$577 billion a year, entrenching their adverse effects. These subsidies perpetuate air pollution that aggravates a wide range of diseases, which, in turn, reduce quality of life, suppress productivity, increase health expenditures, and thus undermine countries' overall development prospects.

As this part of the report shows, 7.3 billion people, or 94 percent of the world's population, are exposed directly to *unsafe* average annual concentrations of fine particulate matter—one of the most pervasive air pollutants. And the scientific evidence is unequivocal that such air pollution has wide-ranging and profound impacts on human health and well-being. Based on the latest medical evidence, the World Health Organization (WHO) updated its air quality guidelines in 2021, significantly tightening the stringency of its 2005 guidelines (WHO 2021b). The revised threshold reflects a growing body of medical evidence on the wide-ranging global burden of disease associated with air pollution. There is strong evidence of the causal relationship between air pollution—especially particulate matter, ozone, NO_2, and SO_2—and all-cause mortality (WHO 2021b).

Estimates of the number of people affected by air pollution–related diseases vary, but they are uniformly staggering. For instance, the 2019 Global Burden of Disease report estimates that 4.5 million premature deaths are due to ambient air pollution, and another 2.3 million premature deaths are due to indoor air pollution each year (IHME 2020). $PM_{2.5}$ is responsible for the vast majority of air pollution–related deaths, and its impacts are on the rise. Between 2000 and 2019, $PM_{2.5}$-attributable deaths increased in all regions except Europe, Latin America, and North America (Southerland et al. 2022). This report assesses the extent to which subsidies may be responsible for this damage.

Poverty and exposure to air pollution

Poor people tend to be more exposed and more vulnerable to air pollution than rich people. A growing base of evidence shows that exposure to, and impact from, air pollution is not distributed equally and the feedback effects are discriminating. As health and productivity suffer, evidence from the United States has shown that air pollution reinforces socioeconomic inequalities. Poor and marginalized groups are often exposed to higher levels of pollution. In addition, these groups tend to be more vulnerable to the impacts of pollution, since they may have comorbidities and low-paying jobs that are more likely to require physical and outdoor labor and therefore entail heightened exposure. Pollution sources, such as industrial plants or transport corridors, are disproportionately located in low-income neighborhoods across the United States. The use of subsidized yet polluting biofuels for cooking, heating, and lighting further degrades air quality, especially for persons who cannot afford cleaner alternatives. And as air pollution increases, housing prices decline, which reinforces the low-income status of neighborhoods. Constraints regarding the availability and quality of health care further increase air pollution–related mortality among low-income groups.

In short, evidence from the United States indicates that social and economic marginalization makes people more exposed and more vulnerable to air pollution. In contrast, there is very limited evidence of the links between poverty, minority status, and exposure to harmful air pollution in low- and middle-income countries and at the

The unequal burden of air pollution on women, children, and ethnic minorities

Air pollution places an outsized burden on certain population groups. Women tend to have far higher exposure to indoor air pollution, especially in countries where the use of polluting cooking and heating fuels is common (Ali et al. 2021). Evidence from rural Xuanwei, China, shows that lifetime use of bituminous ("smoky") coal is associated with nearly a 100-fold higher risk of lung cancer mortality in women than the use of anthracite ("smokeless") coal (Bassig et al. 2020). The impacts of air pollution are also known to be particularly harmful for children in their early development stages. The World Health Organization (WHO 2018) estimates that each year, globally, around 600,000 children die from acute lower respiratory infections caused by polluted air. Air pollution can cause lifelong impacts by affecting neurological development and cognitive abilities, while increasing the risks of asthma and childhood cancer as well as lifelong chronic diseases such as cardiovascular disease (WHO 2018).

Ethnic minorities, too, bear a disproportionate burden. Polluting assets such as power plants or industrial zones have been shown to be located purposefully in low-income or ethnic minority neighborhoods. There is evidence for the United States that racial minority communities are significantly more likely to have new polluting manufacturing plants built in the vicinity; in contrast, communities with higher income levels are less likely to have new plants located nearby (Wolverton 2009). Similarly, ethnic composition drives environmental inequality in US cities—for instance, industrial air toxins affect black and Hispanic neighborhoods in the Midwest and South-Central regions the most (Zwickl, Ash, and Boyce 2014). In the period from 1995 to 2004, black communities were consistently more exposed to air pollution than white communities. Nonmonetary forms of marginalization can even dominate income levels in explaining exposure to air pollution: middle-class African Americans are exposed to more toxins than lower-income white communities.

global scale (box 2.2). A better understanding of the interplay between air pollution and poverty is crucial for several reasons. For one, correlations observed in the United States may not hold in low-income countries, where the nature of occupations and health care differs substantially. For another, if the implications of unsafe air pollution for health and productivity are significant, they may affect the development prospects and growth out of poverty, especially in low- and middle-income countries. This possibility is especially pertinent in low-income countries, which—as this study shows— tend to have relatively low levels of pollution compared to more industrial, middle-income countries. These countries may have an opportunity to grow wealthier without having to grow dirtier first.

Fossil fuels and air pollution

The particles that pollute the air are emitted from a wide range of sources. For instance, dust from vehicle brakes and industrial applications, wood-burning fireplaces, heating and cooking with charcoal, or burning practices in agriculture are significant sources of pollution. Natural sources—in particular, sandstorms and wildfires—also heighten the concentration of harmful particles. However, among all of the sources, one cause of air pollution stands out: the combustion of subsidized and underpriced fossil fuels. This report examines the contribution of subsidies to air pollution and explores the distributional effects of the pollution.

While some uncertainty remains on the exact contribution of fossil fuels to air pollution, studies agree that they have an outsized role. Fossil fuel combustion is a key contributor to particulate matter, SO_2, and NO_x (McDuffie et al. 2020). Fossil fuels contribute more $PM_{2.5}$ emissions globally than any other single source, like wildfires, dust, sea salt, or the burning of solid biofuels. As a result, fossil fuel–related air pollution is a major contributor to the worldwide burden of death and disease (box 2.3).

BOX 2.3

Indoor air pollution and the risks to human health

While outdoor air pollution is pervasive, poor air quality also poses severe health risks inside people's homes. Indoor air pollution results from the burning of solid fuels like coal, wood, dung, or crop waste for cooking, lighting, and heating (Ritchie and Roser 2022). The combustion of such substances releases various pollutants that are harmful to human health. As with ambient air pollution, long-term exposure to these pollutants increases the risk of life-threatening diseases, making indoor air pollution one of the leading global environmental risk factors (Murray et al. 2020).

Global estimates of premature deaths caused by indoor air pollution are similar in number to those caused by outdoor air pollution. The World Health Organization (WHO 2022) estimates that nearly 4 million people die annually from exposure to indoor air pollution, compared to the 4.2 million deaths from ambient air pollution in 2016 (WHO 2021a). The Global Burden of Disease study estimates that 2.3 million and 4.5 million premature deaths result from exposure to indoor and outdoor air pollution, respectively, each year (IHME 2020). Map B2.3.1 shows the global distribution of death rates from indoor air pollution in 2019.

MAP B2.3.1 **Death rates from indoor air pollution, 2019**

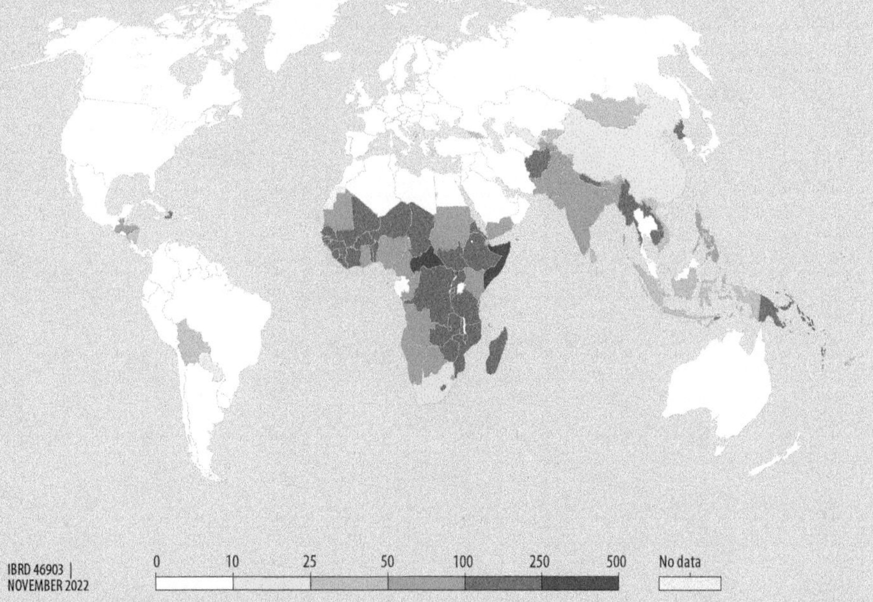

IBRD 46903 | NOVEMBER 2022

0   10   25   50   100   250   500   No data

Source: Adapted from Ritchie and Roser 2022.
Note: Units are the number of indoor air pollution–related deaths per 100,000 all-cause deaths.

(Continued)

BOX 2.3
Indoor air pollution and the risks to human health *(continued)*

Indoor air pollution poses particular health risks to people in low- and middle-income countries and vulnerable groups like children, women, the sick, and the elderly. The vast majority of deaths due to indoor air pollution—that is, 81 percent (or 1.8 million deaths)—occur in low- or lower-middle-income countries, especially in South Asia and Sub-Saharan Africa (IHME 2020). Globally, progress has been made in reducing deaths due to indoor air pollution, as the number of annual premature deaths was halved between 1990 and 2019. Yet improvements have been significantly smaller in Sub-Saharan Africa, where the annual number of deaths from indoor air pollution declined by only 15 percent in the same time frame (Ritchie and Roser 2022). Map B2.3.1 highlights the above-average mortality from air pollution in Sub-Saharan Africa and South Asia.

Indoor air pollution disproportionately affects poor people. Households that cannot afford clean cooking and heating fuels, like electricity or natural gas, are forced to use polluting solid fuels instead. Figure B2.3.1 summarizes the WHO's Energy Ladder, which highlights the higher dependency of poor households on toxic fuels. According to World Bank data, about 3 billion people—40 percent of the world's population—still lack access to modern energy sources for their home use (Ritchie, Roser, and

FIGURE B2.3.1 The Energy Ladder: The dominant energy sources for cooking and heating, by level of income

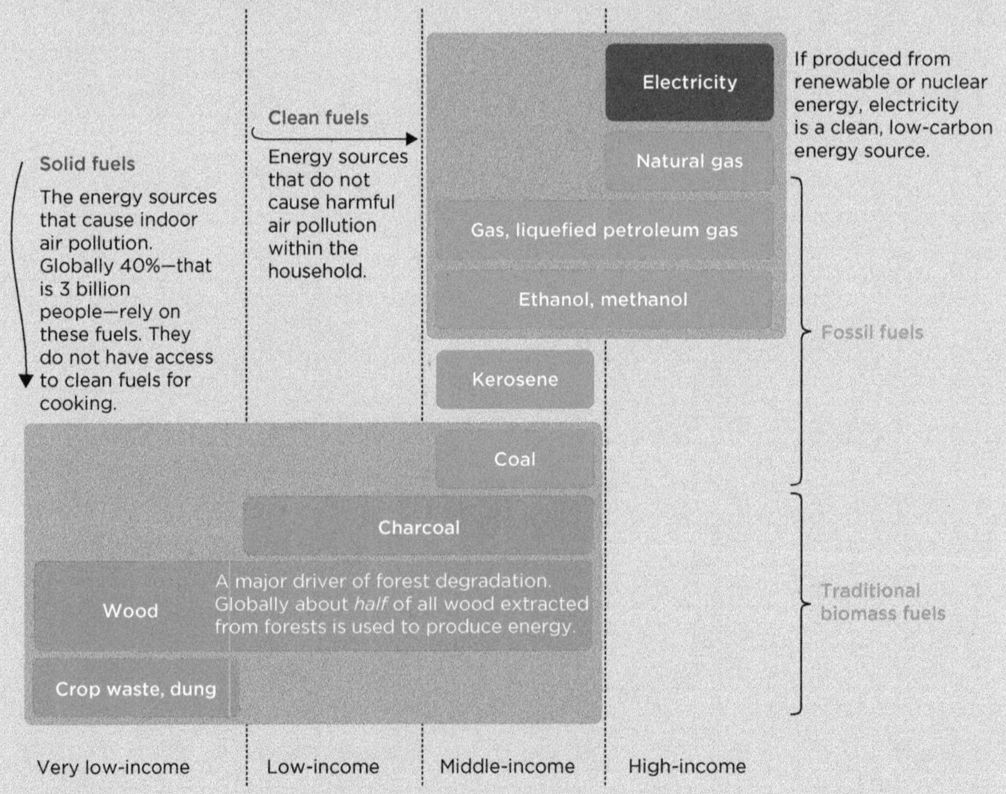

Source: Ritchie, Roser, and Rosado 2020, based on WHO 2006.

(Continued)

Underpricing of polluting activities

Air pollution is so pervasive because polluting activities are underpriced. Despite the well-documented detrimental impacts of fossil fuels, governments around the world provide explicit subsidies to lower the cost of using fossil fuels. Globally, explicit fossil fuel subsidies are estimated to have been around US$577 billion in 2021 (Parry, Black, and Vernon 2021). This amount is almost three times more than the global subsidies paid to the renewable energy sector (IRENA 2020) and almost six times more than what countries committed to raise in annual climate financing (US$100 billion) under the Paris Agreement on Climate Change.

Even when introduced with the best of intentions, these fossil fuel subsidies fail economically, fiscally, socially, and environmentally. By reducing the price of fossil fuels, subsidies incentivize the overconsumption of pollution-intensive fossil fuels in the short run. In the longer run, they aggravate fiscal imbalances, while entrenching pollution, inefficiency, inequality, corruption, and deep dependence on a highly damaging source of energy. Moreover, fossil fuel subsidies often consume large parts of public budgets, crowding out other productive public investments. In many countries, public spending on fossil fuel subsidies exceeds spending on the health or education sectors.

Even when fossil fuels are not subsidized explicitly, their prices do not fully reflect the damages that they cause, such as air pollution and climate change. The International Monetary Fund draws attention to the vast scale of underpricing—also known as implicit subsidies. The cost of implicit subsidies on fossil fuels is estimated at around 6 percent of global gross domestic product (GDP), or US$5.4 trillion in 2020. Climate change impacts and local air pollution constitute more than 75 percent of the total (Parry, Black, and Vernon 2021). At US$2.5 trillion a year, local air pollution is the single largest hidden cost of fossil fuels—far more than the size of explicit subsidies.

A case for action

The evidence presented in subsequent chapters explores the distributional incidence of fossil fuel subsidies and the distributional incidence of the air pollution to which they contribute. To this end, the report seeks to determine the extent to which fossil fuel subsidies are responsible for air pollution. This effort requires navigating through several complex causal links between explicit subsidies, implicit underpricing of polluting fuels and activities, and a range of moderating factors (box 2.4).

Subsidy reforms and the need for complementary policies to tackle pollution

Removing fossil fuel subsidies increases the cost of polluting fuels and thus should reduce fuel consumption, which, in turn, should reduce the emission of air pollutants. Put simply, prices drive pollution. This chain of argument follows the fundamentals of economic reasoning and opens several entry points for policy measures to mitigate air pollution. Thus the menu of policy options in figure B2.4.1 goes beyond fuel prices.

FIGURE B2.4.1 Entry points for antipollution policies

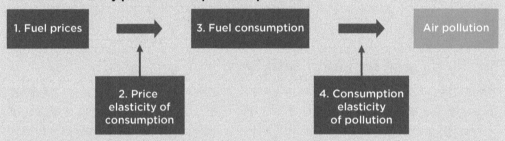

Source: World Bank.

1. *Fuel prices.* The price of fuel determines the cost of consumption. By increasing the price of polluting fuels, governments can incentivize consumers to use less—for instance, by driving less, investing in more fuel-efficient machinery, or switching to cleaner fuels. Removing explicit subsidies is the first step, but prices can be raised further to account for the adverse social costs of pollution—referred to as an environmental (externality) tax.
2. *The price elasticity of consumption.* The price elasticity of consumption measures how responsive consumers are to price changes. In theory, an increase in the unit cost of fuels would cause consumers to use less. In practice, consumers may face constraints (including those related to finance, information, technology, capacity, or behavioral biases), which mean that they cannot adjust their consumption in response to price changes. Especially in the short term, fuel consumption can be fairly unresponsive to price changes. To address this challenge, policy measures can be devised to increase the price elasticity of fuel consumption and improve the effectiveness of pricing policies. For instance, without affordable and efficient public transport systems, commuters may be unable to drive less—even when prices are increased. Investing in public transport or facilitating access to clean technologies can facilitate and accelerate the transition.
3. *Fuel consumption.* In some cases, policies that target fuel prices and consumption elasticities may still be insufficient or not feasible (for example, due to complex administration or public opposition). Policy measures can also directly target consumption to tackle air pollution—for example, by directly imposing a price on polluting activities (rather than on the underlying price of fuels). Inner-city congestion charges are a prime example of such a policy. Similarly, consumption can be regulated directly; for instance, by imposing driving bans for certain days of the week, certain parts of town, or certain types of high-polluting vehicles.
4. *The consumption elasticity of pollution.* Reductions in fossil fuel consumption do not translate one-to-one into lower air pollution. Technological standards differ substantially across countries, and air pollution regulation can have a substantial impact on reducing the pollution intensity of consumption. For instance, the same 10 liters of gasoline used for driving could result in much higher air pollution when the car is not equipped with a catalyzer or air filters or when the fuel is of lower quality.

(Continued)

Accordingly, chapter 3 explores the underpricing of fossil fuels through subsidies and assesses the extent to which price-based reforms can tackle the air pollution challenge. Chapter 4 offers a detailed account of the magnitude and distribution of the air pollution challenge, including the outsized burden that is borne by poorer people. Chapter 5 offers estimates of the distributional and health benefits of reforming explicit fossil fuel subsidies in 35 high-pollution high-subsidy countries. Overall, the new evidence presented shows that removing fossil fuel subsidies is not just a necessary first step toward addressing pervasive environmental externalities; it is an imperative for achieving healthier, more prosperous, and inclusive societies.

References

Ali, M., Y. Yu, B. Yousaf, M. A. Mujtaba Munir, S. Ullah, C. Zheng, X. Kuang, and M. H. Wong. 2021. "Health Impacts of Indoor Air Pollution from Household Solid Fuel on Children and Women." *Journal of Hazardous Materials* 416 (August 15): 126127.

Anenberg, S., J. Miller, R. Minjares, L. Du, D. Henze, F. Lacey, C. Malley, et al. 2017. "Impacts and Mitigation of Excess Diesel-Related NO$_x$ Emissions in 11 Major Vehicle Markets." *Nature* 545 (7655): 467–71.

Bassig, B., H. D. Hosgood, X.-O. Shu, R. Vermeulen, B. E. Chen, H. A. Katki, W. J. Seow, et al. 2020. "Ischaemic Heart Disease and Stroke Mortality by Specific Coal Type among Non-Smoking Women with Substantial Indoor Air Pollution Exposure in China." *International Journal of Epidemiology* 49 (1): 56–68.

Chen, J., and G. Hoek. 2020. "Long-Term Exposure to PM and All-Cause and Cause-Specific Mortality: A Systematic Review and Meta-Analysis." *Environment International* 143 (October): 105974.

Chen, K., S. Breitner, K. Wolf, M. Stafoggia, F. Sera, A. M. Vicedo-Cabrera, Y. Guo, et al. 2021. "Ambient Carbon Monoxide and Daily Mortality: A Global Time-Series Study in 337 Cities." *The Lancet Planetary Health* 5 (4): e191–e199.

Cohen, A. J., M. Brauer, R. Burnett, H. R. Anderson, J. Frostad, K. Estep, K. Balakrishn, et al. 2017. "Estimates and 25-Year Trends of the Global Burden of Disease Attributable to Ambient Air Pollution: An Analysis of Data from the Global Burden of Diseases Study 2015." *The Lancet* 389 (10082): 1907–18.

Domingo, N. G., S. Balasubramanian, S. K. Thakrar, M. A. Clark, P. J. Adams, J. D. Marshall, N. Z. Muller, et al. 2021. "Air Quality–Related Health Damages of Food." *Proceedings of the National Academy of Sciences* 118 (20): e2013637118.

EEA (European Economic Area). 2015. "Non-Methane Volatile Organic Compounds (NMVOC) Emissions." *EEA Indicator Specification,* September 4, 2015 (last update). https://www.eea.europa .eu/data-and-maps/indicators/eea-32-non-methane-volatile-1.

Huangfu, P., and R. Atkinson. 2020. "Long-Term Exposure to NO_2 and O_3 and All-Cause and Respiratory Mortality: A Systematic Review and Meta-Analysis." *Environment International* 144 (November): 105998.

IHME (Institute for Health Metrics and Evaluation). 2020. *Global Burden of Disease Study 2019 (GBD 2019) Results.* Seattle, WA: IHME. https://vizhub.healthdata.org/gbd-results/.

IRENA (International Renewable Energy Agency). 2020. "Renewable Capacity Statistics 2020." IRENA, Abu Dhabi.

Kobayashi, Y., J. M. Santos, J. G. Mill, N. C. Reis Junior, W. L. Andreao, T. T. de A. Albuquerque, and R. M. Stuetz. 2020. "Mortality Risks due to Long-Term Ambient Sulphur Dioxide Exposure: Large Variability of Relative Risk in the Literature." *Environmental Science and Pollution Research* 27 (29): 35908–17.

Malley, C. S., D. K. Henze, J. C. Kuylenstierna, H. W. Vallack, Y. Davila, S. C. Anenberg, M. C. Turner, et al. 2017. "Updated Global Estimates of Respiratory Mortality in Adults ≥ 30 Years of Age Attributable to Long-Term Ozone Exposure." *Environmental Health Perspectives* 125 (8): 087021.

McDuffie, E. E., S. J. Smith, P. O'Rourke, K. Tibrewal, C. Venkataraman, E. A. Marais, B. Zheng, et al. 2020. "A Global Anthropogenic Emission Inventory of Atmospheric Pollutants from Sector- and Fuel-Specific Sources (1970–2017): An Application of the Community Emissions Data System (CEDS)." *Earth System Science Data* 12 (4): 3413–42.

Murray, C. J. L., A. Y. Aravkin, P. Zheng, C. Abbafati, K. M. Abbas, M. Abbasi-Kangevari, F. Abd-Allah, et al. 2020. "Global Burden of 87 Risk Factors in 204 Countries and Territories, 1990–2019: A Systematic Analysis for the Global Burden of Disease Study 2019." *The Lancet* 396 (10258): 1223–49.

Orellano, P., J. Reynoso, and N. Quaranta. 2021. "Short-Term Exposure to Sulphur Dioxide (SO_2) and All-Cause and Respiratory Mortality: A Systematic Review and Meta-Analysis." *Environment International* 150 (May): 106434.

Parry, I., S. Black, and N. Vernon. 2021. "Still Not Getting Energy Prices Right: A Global and Country Update of Fossil Fuel Subsidies." Working Paper 2021/236, International Monetary Fund, Washington, DC.

Ritchie, H., and M. Roser. 2022. "Indoor Air Pollution." OurWorldInData.org, last revised January 2022. https://ourworldindata.org/indoor-air-pollution.

Ritchie, H., M. Roser, and P. Rosado. 2020. "Energy (Archive)." OurWorldInData.org. https:// ourworldindata.org/energy-archive.

Southerland, V. A., M. Brauer, A. Mohegh, M. S. Hammer, A. van Donkelaar, R. V. Martin, J. S. Apte, and S. C. Anenberg. 2022. "Global Urban Temporal Trends in Fine Particulate Matter ($PM_{2.5}$) and Attributable Health Burdens: Estimates from Global Datasets." *The Lancet Planet Health* 6 (2): e139–e146.

Stewart, G. J., B. S. Nelson, W. J. F. Acton, A. R. Vaughan, J. R. Hopkins, S. S. Yunus, C. N. Hewitt, et al. 2021. "Emission Estimates and Inventories of Non-Methane Volatile Organic Compounds from Anthropogenic Burning Sources in India." *Atmospheric Environment: X* 11 (October): 100115.

Taheripour, F., M. Chepeliev, R. Damania, and J. Russ. 2022. "Employment and Emission-Reduction Priorities: Europe and Central Asia." Background paper prepared for this report, World Bank, Washington, DC.

Thurston, G., L. C. Chen, and M. Campen. 2022. "Particle Toxicity's Role in Air Pollution." *Science* 375 (6580): 506.

Wang, Y., X. Li, Z. Shi, L. Huang, J. Li, H. Zhang, Q. Zhang, et al. 2021. "Premature Mortality Associated with Exposure to Outdoor Black Carbon and Its Source Contributions in China." *Resources, Conservation, and Recycling* 170 (July): 105620.

WHO (World Health Organization). 2006. *Fuel for Life: Household Energy and Health.* Geneva: WHO.

WHO (World Health Organization). 2018. *Air Pollution and Child Health: Prescribing Clean Air.* Geneva: WHO.

WHO (World Health Organization). 2021a. "Ambient (Outdoor) Air Pollution." WHO Fact Sheet, September 22, 2021. https://www.who.int/en/news-room/fact-sheets/detail/ambient-(outdoor)-air-quality-and-health.

WHO (World Health Organization). 2021b. *WHO Global Air Quality Guidelines: Particulate Matter (PM$_{2.5}$ and PM$_{10}$), Ozone, Nitrogen Dioxide, Sulfur Dioxide, and Carbon Monoxide.* Geneva: WHO.

WHO (World Health Organization). 2022. "Household Air Pollution and Health." WHO Fact Sheet, July 27, 2022. https://www.who.int/news-room/fact-sheets/detail/household-air-pollution-and-health.

Wolverton, A. 2009. "Effects of Socio-Economic and Input-Related Factors on Polluting Plants' Location Decisions." *BE Journal of Economic Analysis & Policy* 9 (1): 1–32.

Yang, J., M. J. Z. Sakhvidi, K. de Hoogh, D. Vienneau, J. Siemiatyck, M. Zins, M. Goldberg, et al. 2021. "Long-Term Exposure to Black Carbon and Mortality: A 28-Year Follow-Up of the GAZEL Cohort." *Environment International* 157 (December): 106805.

Zheng, X.-Y., P. Orellano, H. L. Lin, M. Jiang, and W. J. Guan. 2021. "Short-Term Exposure to Ozone, Nitrogen Dioxide, and Sulphur Dioxide and Emergency Department Visits and Hospital Admissions due to Asthma: A Systematic Review and Meta-Analysis." *Environment International* 150 (May): 106435.

Zwickl, K., M. Ash, and J. K. Boyce. 2014. "Regional Variation in Environmental Inequality: Industrial Air Toxics Exposure in US Cities." *Ecological Economics* 107 (November): 494–509.

CHAPTER 3

Subsidizing Toxic Air

The Vast Underpricing of Fossil Fuels and Their Use

"When we try to pick out anything by itself, we find it hitched to everything else in the universe."
—*John Muir*

CHAPTER AT A GLANCE

Air pollution originates from a wide range of sources:

- The fossil fuel–dependent power, transport, residential, and industrial sectors are major sources of ambient air pollution, including fine particulate matter ($PM_{2.5}$), sulfur dioxide (SO_2), nitrogen oxide (NO_x), and carbon monoxide. Fossil fuels are a leading driver for the emission and formation of air pollutants, although their exact contribution differs across sectors, types of air pollutant, countries, and studies.

Polluting activities are cheap because polluting fuels are underpriced:

- Globally, explicit fossil fuel subsidies were an estimated US$577 billion in 2021. Intended to improve energy affordability, these active subsidy programs have grown into major support schemes for the consumers and producers of oil, gas, and coal; as such, they impose a significant fiscal burden on governments and societies. About three-quarters of all energy subsidies go to fossil fuels.

- The scale of implicit underpricing of fossil fuels goes far beyond explicit subsidies—at US$5.4 trillion in 2020, implicit fossil fuel subsidies are equivalent to more than 6 percent of global gross domestic product (GDP). Implicit subsidies are an estimate of the negative externalities associated with fossil fuel consumption, including the social cost of carbon emissions, local air pollution, road congestion, and forgone tax revenues.

- Removing explicit fossil fuel subsidies alone is unlikely to bring fuel prices to their socially optimal level. At US$2.5 trillion a year, local air pollution was the single largest environmental cost of fossil fuels in 2020—far more than the size of explicit subsidies.

- Despite best intentions, fossil fuel subsidies are failing economically, fiscally, socially, and environmentally. Fossil fuel subsidies are typically implemented to alleviate poverty, redistribute national wealth, or promote economic development by supporting energy-intensive industries. Yet subsidies have proven ineffective in achieving these objectives, while at the same time entrenching pollution, inefficiency, inequality, corruption, and deep dependence on a highly damaging source of energy.

Fossil fuel prices matter for air pollution outcomes, but policies need to go further:

- When polluting fuels are expensive, people reduce their consumption—to some extent. Extensive evidence presented in this chapter suggests that, on average, a 10 percent increase in the unit price of energy results in a short-run reduction in consumption of around 2 percent. However, estimates vary across types of energy, sectors, and countries. The implication is that energy consumption is *inelastic*—that is, sluggishly responsive to prices, especially in the short term—*when cleaner alternatives are unavailable or unaffordable.*

- Raising prices alone may not be sufficient to tackle pollution. Prices are crucial in setting the incentives for consumption and the associated pollution, but their effectiveness in reducing air pollution depends on how responsive consumption choices are to prices and how responsive pollution levels are to consumption choices. To curb air pollution effectively, policies need to ensure that clean alternatives are available and affordable, address information and capacity constraints, and address individual biases.

Introduction

This chapter provides a nuanced account of how the underpricing of fossil fuels incentivizes overconsumption and contributes to anthropogenic air pollution. The chapter first makes the case that fossil fuels are a leading source of air pollutants around the world. It then explores the magnitude of "underpriced" fossil fuels, based on an extensive literature. For this purpose, it distinguishes between *explicit* and *implicit* subsidies. Explicit subsidies refer to active payments that lower the cost of polluting activities and inputs, while implicit subsidies refer to the unpriced external costs. Public support and subsidy programs in other sectors—for example, transport or agriculture—also contribute to air pollution. Finally, the chapter explores the power—and limits—of price-based instruments in triggering changes in polluting consumption choices.

Fossil fuels and air pollution

The use of fossil fuels in energy, industry, and transport is a leading source of air pollution. The particles that pollute the air are emitted from a wide range of sources. For instance, dust from vehicle brakes and industrial applications, wood-burning fireplaces, heating, cooking, or burning practices in agriculture have all been shown to be significant sources of pollution in certain localities. Natural sources—in particular, sandstorms and wildfires—also heighten the concentration of harmful particles. However, among all of the sources of air pollution, the combustion of fossil fuels stands out.

While some uncertainty remains on the exact contribution of fossil fuels to air pollution, studies agree that they play an outsized role. The Community Emissions Data System (CEDS) provides more specific evidence that fossil fuel combustion is a key contributor to air pollution (McDuffie et al. 2020).[1] Table 3.1 summarizes the share of different air pollutants that originate from fossil fuel combustion. It also lists the share of fossil fuel–related air pollutant emissions from three key polluting sectors. The numbers show

TABLE 3.1 Contribution of fossil fuels to major air pollutants and key emitting sectors

Pollutant	% of emissions caused by fossil fuels	Sectoral contribution (%)		
		Power	Industry	Transport
Nitrogen oxides	85	18	13	23
Sulfur dioxide	78	34	23	2
Carbon monoxide	48	2	6	32
Nonmethane volatile organic compounds	23	1	2	17
Black carbon	54	1	9	20
Organic carbon	19	1	2	7
Fine particulate matter ($PM_{2.5}$)[a]	43–58	13	15	9

Sources: Calculations based on supplemental materials by McDuffie et al. 2020. Estimates for share of anthropogenic $PM_{2.5}$ pollution based on McDuffie et al. 2021.
a. The sectoral emissions for $PM_{2.5}$ do not refer to the sectoral contribution of fossil fuels to anthropogenic $PM_{2.5}$ pollution, but rather to the overall sectoral contribution to anthropogenic $PM_{2.5}$ pollution.

that fossil fuel combustion makes up a significant part of most air pollutants that are dangerous to human health. Fossil fuels contribute more to $PM_{2.5}$ emissions globally than any other single source, like wildfires, dust, sea salt, or burning of solid biofuels, although the contribution of these sources can vary significantly across countries.

The *World Energy Outlook 2016,* published by the International Energy Agency (IEA), reported that in 2015 nearly 85 percent of anthropogenic airborne particle matter and more than 99 percent of SO_2 and NO_x particle emissions were related to energy production and use, mostly from burning fossil fuels. Among the types of fossil fuels, coal combustion takes the first place, accounting for 50 percent of all SO_2 emissions, followed by oil combustion, which accounts for 40 percent. For NO_x, which is increasing rapidly in low- and middle-income countries, more than 50 percent come from the transportation sector, 26 percent from industry, and 14 percent from power generation. In general, oil combustion accounts for more than 70 percent of NO_x emissions (IEA 2016). Further, more than half of all particulate matter pollution originates in the residential sector, especially in countries where fossil fuels like coal and kerosene and solid biofuels are used for heating and cooking (IEA 2016).

Fossil fuel–related air pollution is also a major contributor to the worldwide burden of death and disease, although estimates vary widely across studies. At the lower end of the estimates, the Global Burden of Disease study suggests that 4.5 million people died in 2019 from adverse health effects related to long-term exposure to ambient air pollution, with 4.1 million deaths caused by $PM_{2.5}$ (IHME 2020). Other estimates are more than twice as large. For instance, Lelieveld et al. (2020) put the deaths at 8.8 million worldwide, with a loss of life expectancy of 2.9 years. According to these estimates, life expectancy would be 1.1 years higher if fossil fuel combustion were ended or 1.7 years higher if all preventable

> The burning of coal, oil, and natural gas is a leading source of ambient $PM_{2.5}$— the pollutant responsible for the vast majority of air pollution–related deaths.

human-related emissions were ended. Vohra et al. (2021) provide an even higher estimate, arguing that ambient air pollution ($PM_{2.5}$)–related deaths from fossil fuel combustion are responsible for 10.2 million premature deaths globally each year. In particular, China and India, the countries that generate the most fossil fuel–driven $PM_{2.5}$ worldwide, account for 62 percent of total premature mortality, with 3.9 million and 2.5 million premature deaths, respectively.[2] While estimates differ across studies depending on the data sources and methodologies used, they all underscore the outsized contribution of fossil fuels. Box 3.1 outlines the other sources of pollution that receive public support.

BOX 3.1

Pollution sources other than fossil fuels that receive public support and subsidies

A significant proportion of air pollutants can be traced directly back to the combustion of fossil fuels: sulfur dioxide is emitted when burning coal for power generation, and particulate matter results from running a diesel generator or lighting a home with kerosene. It is well documented that burning fossil fuels is responsible for these harmful emissions. However, air pollutant emissions from other sources can also be driven by or linked partly to the continued dependence on fossil fuels.

Recent measurements have shown that particle pollution from tire wear of vehicles can be substantial even on modern vehicles with advanced pollution filtration mechanisms (Carrington 2022). Small particles of tire rubber pollute the air, water, and soils and contain toxic chemicals known to heighten the risk of cancer and other diseases. Similarly, the attrition of vehicle brakes and industrial machine parts has been shown to be a source of air pollutants. These pollutants are not necessarily a direct consequence of fossil fuel combustion, but underinvestment in public transport systems and underpricing of fossil fuels both contribute to people's dependence on private vehicles.

These sources of air pollution are not necessarily reflected in the data on air pollutants associated with fossil fuels. But several examples highlight how subsidies or public support programs in other sectors contribute to air pollution:

- *Transport sector spending.* Private vehicle travel—as opposed to public transport—is a major source of urban air pollution. Urban planning or infrastructure investment can incentivize private vehicle travel and aggravate air pollution. For instance, public spending to densify urban driving networks, extend lanes, increase the number of parking spaces, and crowd out investments from the public transport sector incentivize the use of polluting private vehicles, even if they do not directly lower the price of fossil fuels. Switching to electric vehicles cannot automatically address externalities such as air pollution from tire wear or congestion.

- *Agricultural subsidies.* In large parts of the world, farmers clear fields for the next season by burning crop residue. Studies have shown that income support for farming households (for example, through social protection programs) or subsidies for the adoption of mechanized harvesting technology can encourage agricultural burning and thus aggravate air pollution (Behrer 2019). Mechanized harvesting leaves more residue, which increases the use of burning. Reallocating subsidies to more efficient agricultural technology could help to reduce the need for burning and thus improve air quality (Shyamsundar et al. 2019). Moreover, fertilizer subsidies can drive up air pollution too—ammonia emissions associated with the use of nitrogen-based fertilizers are significant precursors to fine and ultrafine particles (box 8.3 in chapter 8 elaborates on this issue).

Explicit fossil fuel subsidies

Governments are actively subsidizing the overconsumption of polluting energy. As countries endeavor to industrialize and become more prosperous, their energy needs increase. This need has been a long-established hallmark of economic development, so it is no surprise that governments and the private sector around the world invest heavily in energy infrastructure. As a result, governments around the world have attempted to lower the cost of energy by means of "explicit" subsidization schemes. These schemes have grown into major support programs for the consumers and producers of oil, gas, and coal products. They have contributed to energy companies becoming a dominant force in the global economy, with revenues that far exceed the GDP of many countries. The world's largest company by market capitalization—an oil and gas company—reported revenues of about US$360 billion in 2021, which would make it the thirtieth largest economy in the world.

These subsidy programs not only cement the dominance of fossil fuels, but also come at huge societal costs, as they contribute not only to air pollution, but also to inequality and inefficiency. In effect, these subsidies underprice fossil fuels and the many polluting activities with which they are associated.

It is common practice to distinguish between *explicit* and *implicit* fossil fuel subsidies. Explicit subsidies represent the price difference between the supply cost and the price paid by the consumer, weighted by a country's aggregate fuel consumption (Parry, Black, and Vernon 2021). Implicit subsidies are defined as the price difference between the socially optimal price and the consumption price, thus also accounting for negative externalities and any suboptimally low taxes or regulations (see chapter 1 for further details). Regardless of their exact definition, fossil fuel subsidies rarely meet their original policy objective and instead entail a wide range of economic, fiscal, and social costs.

Explicit fossil fuel subsidies can take many different forms. The World Bank (Kojima and Koplow 2015, 4) defines explicit energy subsidies as

> a deliberate policy action by the government that specifically targets electricity, fuels, or heating and that results in one or more of the following effects:
> 1. It reduces the net cost of energy purchased.
> 2. It reduces the cost of energy production or delivery.
> 3. It increases the revenues retained by those engaged in energy production and delivery (energy suppliers).

Explicit subsidies can be categorized into two broad types: consumer subsidies and producer subsidies (IEA 2014; Parry, Black, and Vernon 2021; Whitley 2013). Both types effectively reinforce the underpricing of fossil fuels through active government interventions designed to reduce the cost of fossil fuel consumption or production.

Consumer subsidies are fiscal measures that lower the price of fossil fuel products below their market price (for example, the international market price or cost-recovery threshold), thus making them more affordable to end users. Consumer subsidies generally measure what governments spend on subsidies and do not reflect the wider societal externalities incurred by people (for example, due to carbon emissions or air pollution). As both market prices and domestic subsidized prices can be observed directly or estimated, consumer subsidies are easier to assess with available data than producer subsidies.

US$577 billion: The amount of global explicit subsidies for fossil fuels in 2021, according to the International Monetary Fund

Producer subsidies are more difficult to observe and quantify, as they may include any of a variety of concessions given to fossil fuel exploration, extraction, or processing firms or other energy-intensive companies, industries, or products (Bast et al. 2014, 2015; GSI 2010). Producer subsidies can be explicit transfers, such as grants, low-interest loans, direct payments (for example, upstream support for oil exploration), or tax exemptions; they also can be in-kind, such as credit subsidies, government guarantees to protect investments, derivatives, and subsidies through government procurement (guaranteed contracts), research, and public investment (OECD 2011; UNEP 2003; Whitley 2013).[3] The Global Subsidies Initiative has estimated producer subsidies for a series of countries, but estimates vary widely due to data issues (for example, GSI 2012). Overall, producer subsidies are thought to be in the range of US$80 billion to US$285 billion annually in low-income and in lower-middle-income countries and US$444 billion in G-20 countries (Bast et al. 2015; OECD 2013, 2015; Whitley 2013).

Based on these definitions, several attempts have been made to estimate the global magnitude of subsidies and their implications, most notably by the International Monetary Fund (IMF) and the IEA. The IMF estimates global explicit fossil fuel subsidies at around US$577 billion in 2021 (Parry, Black, and Vernon 2021), while the IEA's corresponding estimate is US$440 billion (IEA 2022). The range illustrates differences in the scope, definitions, and methodology of how explicit subsidies are measured and aggregated. Whatever the precise estimate may be, there is agreement that, around the world, substantial financial resources are being used to lower the price of polluting fossil fuels artificially.

26%: The share of global energy sector subsidies that goes to renewables

To understand the scale of fossil fuel subsidies being paid, it is helpful to compare these figures with international commitments to support clean technologies and decarbonization. For instance, under the Paris Agreement on Climate Change, governments committed to raise US$100 billion annually in climate financing—just a *fifth* of what they spend to prop up fossil fuels. Similarly, a study by the International Renewable Energy Agency (IRENA 2020) estimates that, out of all global energy sector subsidies, only 26 percent (about US$166 billion) went to support renewable energy in 2017 (figure 3.1). The vast majority, about 71 percent,

FIGURE 3.1 **Global energy sector subsidies, 2017**

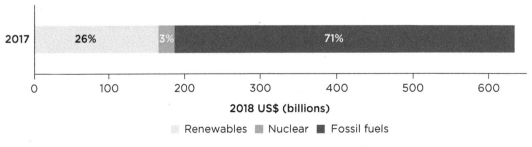

Source: IRENA 2020.

Detox Development

were explicit fossil fuel subsidies. In sum, the subsidies for fossil fuels dwarf those available for renewable energy.

The IMF's estimates of explicit fossil fuel subsidies offer a detailed account of where the largest subsidy programs are maintained and which types of fuel they support (figure 3.2). The vast majority of subsidies go to oil, natural gas, and electricity, while a much smaller share goes to diesel, gasoline, and coal (Parry, Black, and Vernon 2021). Figure 3.3 presents the 20 largest providers of fossil fuel consumption subsidies in 2020. The top five largest providers of subsidies—India, the Islamic Republic of Iran, the Russian Federation, Saudi Arabia, and República Bolivariana de Venezuela—together account for US$211 billion of subsidy payments in 2020, equivalent to almost half of global explicit subsidies.

> Fossil fuel subsidies tend to be especially high in resource-rich middle-income countries.

Perhaps paradoxically, fossil fuel subsidies tend to be especially high in resource-rich (middle-income) countries that are highly dependent on carbon-intensive economic activity. For instance, 32 percent (or US$147 billion) of global fossil fuel subsidies are paid in the Middle East and North Africa region. Under a business-as-usual scenario—that is, if current subsidy programs are not reformed—fossil fuel subsidies are expected to be around US$560 billion in 2025.

While consumer subsidies are paid mostly in resource-rich middle-income countries, numerous high-income economies have large producer subsidy schemes. The in-kind component of these subsidies is especially difficult to identify and measure, which is why there are so few comprehensive studies. A study of G-20 countries estimates that producer subsidies amounted to US$444 billion in 2014.[4] The largest share of production subsidies in 2014 came in the form of investments by state-owned

FIGURE 3.2 **Global explicit fossil fuel subsidies, 2015–25 (projected)**

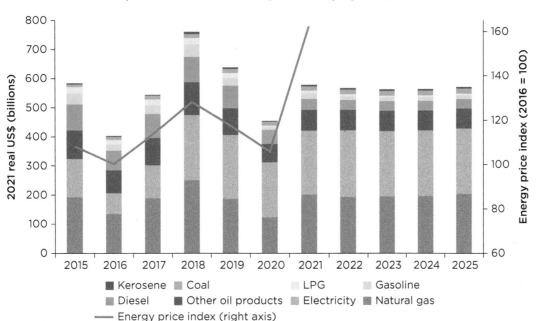

Source: Calculations based on data from Parry, Black, and Vernon 2021.
Note: The left axis depicts subsidies in 2021 real dollars; the right axis depicts a normalized energy price index.
LPG = liquefied petroleum gas.

FIGURE 3.3 Top 20 explicit fossil fuel subsidy programs, 2020

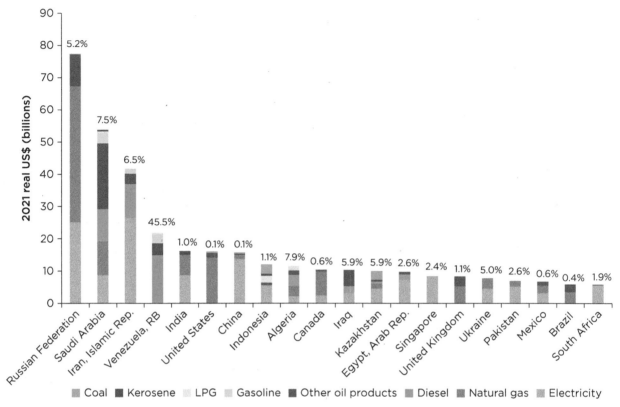

Source: Calculations based on data from Parry, Black, and Vernon 2021.
Note: Percentages denote the size of fossil fuel subsidies relative to GDP. LPG = liquefied petroleum gas.

enterprises and amounted to US$286 billion—64 percent of all producer subsidies for the G-20 (Bast et al. 2015).

The societal costs of air pollution

US$5.4 trillion: The scale of underpricing of fossil fuel-related externalities (implicit subsidies), according to the International Monetary Fund

"Implicit subsidies" reflect the underpricing of fossil fuels. Even when fossil fuels are not explicitly subsidized, their prices do not fully reflect the vast societal and environmental damages they cause, including air pollution and climate change. The underpricing of fossil fuels reinforces and incentivizes the polluting activities driving these externalities. The most commonly advocated policy measure to address such underpricing is to impose an externality tax on polluting fuels and activities. The failure to impose such a tax is, in principle, similar in consequence to an explicit subsidy: instead of pricing in societal costs, it makes fossil fuel consumption unsustainably cheap.

Apart from explicit subsidies, what is the magnitude of underpricing? The IMF refers to the failure to price in externalities as an "implicit subsidy" for fossil fuels (Coady et al. 2015, 2017; Parry, Black, and Vernon 2021). The measure of implicit subsidy is essentially an estimate of the negative externalities associated with fossil

FIGURE 3.4 Global sources of implicit fossil fuel subsidies and share of GDP, 2015–25 (projected)

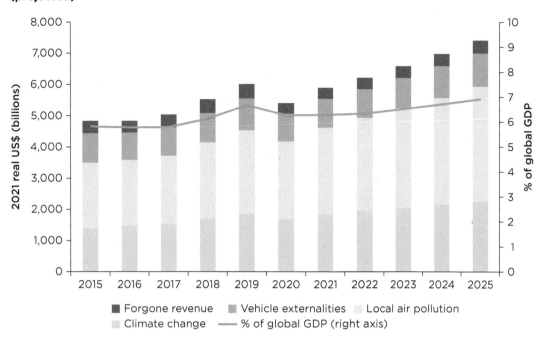

Source: Calculations based on data from Parry, Black, and Vernon 2021.

fuel consumption, including the social cost of carbon emissions, local pollution, road congestion, and forgone tax revenues (figures 3.4). By including negative externalities, the definition of "implicit subsidy" also includes lower energy tax rates. For instance, in the United Kingdom domestic energy use is taxed, albeit at a lower rate than other consumption goods—that is, 5 percent value added tax rather than 20 percent (HM Revenue and Customs 2022).

Explicit subsidies are dwarfed by the societal costs of air pollution. The IMF estimates the cost of implicit fossil fuel subsidies at US$5.4 trillion in 2020, with the impacts on local air pollution and global climate change constituting more than 75 percent of the total (Parry, Black, and Vernon 2021).[5] At US$2.5 trillion a year, local air pollution is the single largest unpriced environmental externality from fossil fuels. An important implication is that removing explicit subsidies alone is unlikely to bring fuel prices to their socially optimal level. The IMF's definition of implicit subsidies is thus particularly relevant from an environmental perspective, as it draws attention to the substantial external costs that result from fossil fuel use.

Figure 3.5 shows the 20 largest providers of implicit fossil fuel subsidies in 2020. Almost half of the global implicit subsidies stem from China, India, Russia, and the United States. The close correlation between the climate change externalities and the health costs of fossil fuels suggests that there is scope to tackle both problems simultaneously. In all 20 countries, the largest impacts are climate change and local air pollution. In a business-as-usual scenario, continued economic growth and its associated

US$2.65 trillion: The global societal cost associated with coal— larger than from any other fossil fuel

energy consumption could mean that implicit subsidies will reach more than US$7 trillion by 2025.

The measure of implicit subsidies also sheds light on the outsized environmental costs caused by certain fossil fuels. Coal, in particular, is the leading source of environmental costs, amounting to about US$2.65 trillion in 2021, which is more than the combined costs from gasoline, diesel, and natural gas. Thus coal makes an exceptionally large contribution to climate change and local air pollution (figure 3.6).

FIGURE 3.5 Top 20 providers of implicit fossil fuel subsidies, 2020

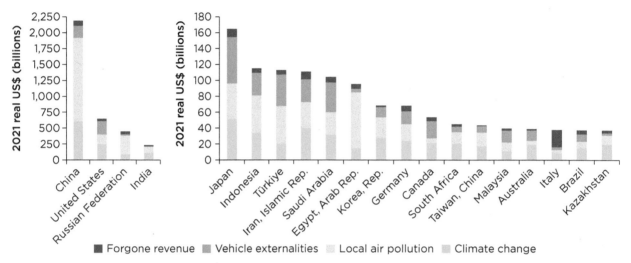

Source: Calculations based on data from Parry, Black, and Vernon 2021.

FIGURE 3.6 Global implicit fossil fuel subsidies, by type of fuel, 2015–25 (projected)

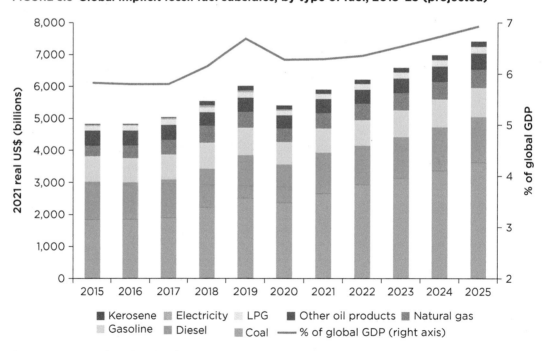

Source: Calculations based on data from Parry, Black, and Vernon 2021.
Note: LPG = liquefied petroleum gas.

Fossil fuel subsidies: Best intentions but detrimental outcomes

Fossil fuel subsidies are typically implemented with good intentions, such as to alleviate poverty, redistribute national wealth, or promote economic development by supporting energy-consuming industries (Rentschler and Bazilian 2017; Strand 2013). Especially in low-income countries, fossil fuel subsidies are established to improve the affordability of energy goods (box 3.2). Similarly, in many low-income economies, the objective of fuel subsidy policies has been to promote industrialization. Examples include Brazil and Nigeria, where the key objective in maintaining low energy prices has been to facilitate industrialization by conferring an advantage on domestic energy-intensive firms (de Oliveira and Laan 2010). However, evidence suggests that fossil fuel subsidies perform poorly at achieving these objectives and are generally detrimental to the economic, social, and environmental dimensions of sustainable development (Parry, Black, and Vernon 2021; Rentschler and Bazilian 2017).

While often well intentioned, these subsidy programs have proven to be ineffective in achieving their original objectives, while at the same time entrenching pollution and inefficiency. By artificially lowering the price of fossil fuels, governments remove incentives for investing in energy efficiency, modern electricity infrastructure, and low-carbon energy sources, including renewables (IEA 2014). As a consequence, subsidies cause a wide range of adverse unintended side effects that encompass

---

BOX 3.2

Subsidies: Intended for the poor, but benefiting the rich

Much of the public discourse on fuel subsidies in low- and middle-income countries has focused on the impact on the poor (Adam and Lestari 2008; Dube 2003; Gangopadhyay, Ramaswami, and Wadhwa 2005; IEA 2011; IMF 2013; Mourougane 2010; Rao 2012; Ruggeri Laderchi, Olivier, and Trimble 2013; World Bank 2010). While a common political justification for fossil fuel subsidies is to support the poor through subsidized energy supply, the literature shows unequivocally that most subsidies are regressive—that is, in absolute terms, most subsidies benefit the rich. Evidence from Nigeria, for instance, shows that certain energy sources are consumed disproportionately by the highest income quintile (Rentschler 2016). The top income quintile, for instance, accounts for 75 percent of overall petrol consumption, while the bottom income quintile accounts for just 1 percent. This stark ratio can be attributed to vehicle ownership and highlights how petrol subsidies disproportionately benefit the rich.

For 20 low- and middle-income countries, Arze del Granado, Coady, and Gillingham (2012) show that poorer households consume a disproportionately smaller fraction of total fuel and electricity supply. In fact, households in the top income quintile spend nearly 20 times more (per capita) on most energy goods than households in the bottom quintile. Kerosene is the only exception, with consumption distributed broadly and evenly across income quintiles, primarily because it is an inferior fuel that many low-income households use for cooking and lighting. Therefore, the bottom income quintile receives, on average, about 7 percent of the overall subsidy benefit. In contrast, the richest quintile alone receives, on average, almost 43 percent of the overall volume of subsidies.

economic inefficiencies, social injustices, and a host of environmental and health externalities (figure 3.7). Together, these unintended effects present a strong case in favor of subsidy reform.

Fossil fuel subsidies often consume a large part of public budgets, crowding out other productive public investments—for instance, in health, education, or transport infrastructure. In many countries, public spending on fossil fuel subsidies exceeds spending on health or education (figures 3.8 and 3.9). For instance, in Kazakhstan, public spending on fossil fuel subsidies is almost three times higher than spending on health or education, which is especially problematic, as investments in education and health are typically far less regressive and benefit lower-income groups more than fossil fuel subsidies. Thus if the goal is to alleviate poverty or stimulate the economy, investments in other public sectors are likely to be more effective than fossil fuel subsidies (Barbiero and Cournède 2013). Globally, 21 countries spent more than 5 percent of their annual budget on fossil fuel subsidies in 2019 (IMF 2022).

Subsidies also aggravate fiscal imbalances and environmental externalities. Fossil fuel subsidies have been shown to distort prices, aggravate fiscal imbalances, and reduce aggregate welfare (Plante 2014). They have particularly adverse effects on the balance of payments in oil-importing countries, as they exacerbate the difficulties in mitigating the effects of international energy prices (IMF 2008, 2013; Parry, Black, and Vernon 2021). Overall, soaring oil prices have turned fuel subsidies into an unsustainable financial burden for governments. In 2020 República Bolivariana de Venezuela allocated about 45 percent of its annual GDP to explicit fossil fuel subsidies, while Algeria, Saudi Arabia, Suriname, and Tajikistan allocated around 8–10 percent. The Islamic Republic of Iran, Iraq, Kazakhstan, Libya, and Sudan all allocated roughly 6 percent of their respective

FIGURE 3.7 Best intentions and detrimental effects of fossil fuel subsidies

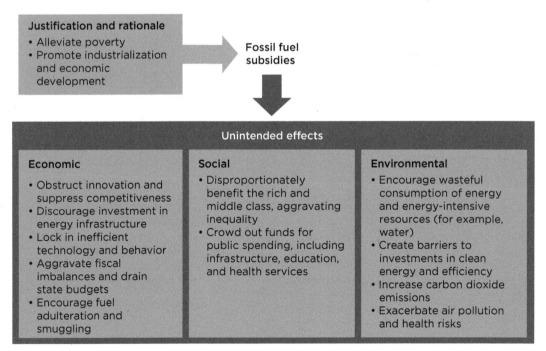

Source: Based on Rentschler and Bazilian 2017.

Detox Development

FIGURE 3.8 Fossil fuel subsidies and health expenditures as a share of GDP in select countries, 2019

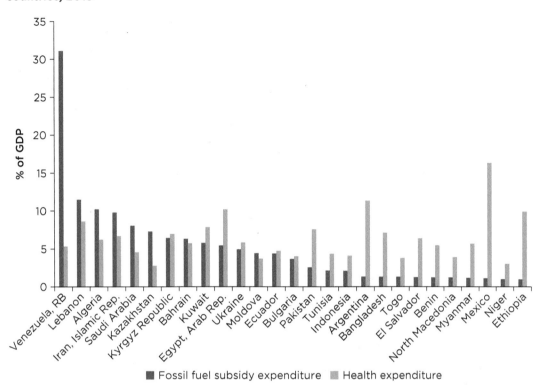

Sources: Calculations based on 2019 data from Parry, Black, and Vernon 2021 and World Bank BOOST data (https://www.worldbank.org/en/programs/boost-portal).

FIGURE 3.9 Fossil fuel subsidies and education expenditures as a share of GDP in select countries, 2019

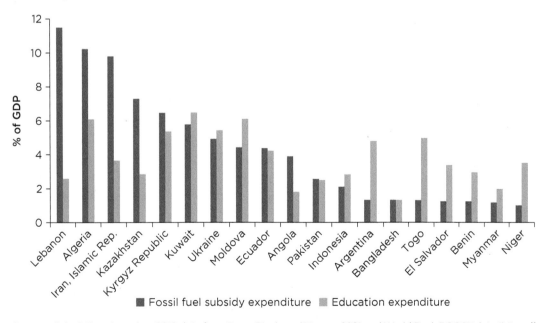

Sources: Calculations based on 2019 data from Parry, Black, and Vernon 2021 and World Bank BOOST data (https://www.worldbank.org/en/programs/boost-portal).

GDP to fossil fuel subsidies (Parry, Black, and Vernon 2021). Considering a surge in debt levels across low- and middle-income countries—aggravated by the COVID-19 pandemic—inefficient fuel subsidy programs add to fiscal pressures. As of May 2022, 38 low-income countries were considered to be at high risk or already in acute debt distress. Repurposing expenditures and spending more efficiently would yield a triple dividend—economic, environmental, and distributional.[6]

As chapter 4 highlights for the case of air pollution, fossil fuel subsidies contribute to severe environmental externalities. By underpricing fossil fuel consumption, subsidies undermine international efforts to curb climate change. The IMF's recent estimates show that removing explicit fossil fuel subsidies is necessary—albeit insufficient—to put the world on track to achieve the ambitious target of 1.5° global warming (Parry, Black, and Vernon 2021). By one estimate, removing explicit fuel subsidies globally would result in an up to 4 percent net reduction in global greenhouse gases by 2030 (Jewell et al. 2018). Given this limited effect, the removal of explicit fossil fuel subsidies is just a first step toward addressing the underpricing of fossil fuels. In addition to reforming explicit subsidy programs, the wider societal costs of fossil fuel consumption need to be reflected in fossil fuel prices—for instance, through externality taxes and complementary policies.

Fossil fuel subsidies have also been shown to have a wide range of unintended consequences, including fuel adulteration and smuggling (Calvo-Gonzales, Cunha, and Trezzi 2015; Rentschler and Hosoe 2022; Victor 2009). Evidence from the Arab Republic of Egypt and the Republic of Yemen shows that fuel subsidies impose substantial external costs in the form of traffic congestion, local pollution, and associated health impacts, but also deplete scarce water resources in the agriculture sector by incentivizing the excessive pumping of groundwater (Coady et al. 2017; Commander 2012).

Polluting fossil fuels and the role of price signals

The price of fossil fuels is crucial in setting the incentives for consumption and the associated pollution. However, the effectiveness of price-based policies, such as subsidy reform, in reducing air pollution depends on how responsive consumption choices are to prices and how responsive pollution levels are to consumption choices.

A fundamental principle of economics is that lowering the price of a good tends to increase people's demand for it. The extent to which consumption choices are sensitive to price changes can be measured empirically in the form of *price elasticities* of demand. In the context of air pollution, price elasticities are a crucial component in understanding the underlying forces that drive pollution outcomes. For instance, lower gasoline prices would result in higher consumption, which, intuitively, would increase the level of pollution from fine particulate matter.

The relationship between fossil fuel prices and consumption has been studied extensively. A large-scale review by Labandeira, Labeaga, and López-Otero (2017) documents more than 400 empirical studies that provide almost 2,000 separate elasticity estimates covering different countries, time frames, users, and types of fossil fuels. While the existing literature focuses heavily on high-income countries, a review conducted for this report suggests that elasticity estimates in low- and middle-income countries broadly align with estimates from higher-income countries (box 3.3).

Price elasticities of demand have consistently been shown to be negative for fossil fuels, as expected—that is, higher prices lead to lower consumption—although responsiveness may vary significantly across countries and types of fuel (Dahl 2012; Labandeira, Labeaga, and López-Otero 2017). Figure 3.10 provides a summary of average elasticity estimates for different types of energy and groupings.

Across different types of energy and different users, a 10 percent increase in the unit price of energy results in a short-run reduction in consumption of 2 percent. The reduction in consumption is relatively low, reflecting the fact that consumers adjust their choices, but are likely to face constraints to doing so in the short run. For instance, they may try to avoid driving for leisure trips, but cannot stop commuting to work altogether.

In the long run, however, consumption choices are more responsive to price changes. The average long-run price elasticity estimates in the literature suggest that a sustained

---

BOX 3.3
Technical spotlight: A meta-analysis of price elasticities in low- and middle-income countries

How much is demand for fossil fuels likely to change if their prices are raised? While existing empirical studies have answered this question for thousands of specific cases—specific types of energy, different users, or different countries—the vast majority of these studies have been done in high-income economies. The reason is largely because of data availability issues, as estimating price elasticities requires detailed information on price patterns and consumption decisions by users.

In a dedicated technical study prepared for this report, Triyana et al. (2022) conducted a meta-analysis to unearth recent elasticity estimates specifically for low- and middle-income countries. The meta-analysis began with a systematic screening of empirical studies and identification of sources of convergence and divergence between studies. Three common scientific databases that act as repositories of academic articles were queried, and a systematic keyword search was used to identify relevant studies. The keyword search yielded 1,351 potentially relevant studies, which were screened first based on their abstracts and then through a full-text review.

The meta-analysis identified a range of recent empirical studies for Brazil, Cameroon, China, Ghana, Indonesia, the Islamic Republic of Iran, Lebanon, Mexico, Mozambique, and Pakistan. Most studies in these countries focus on gasoline, while very few studies also consider diesel, natural gas, or liquefied petroleum gas (LPG). The average effect across these studies suggests an elasticity of −0.28. In other words, a 10 percent increase in gasoline prices leads to an estimated 2.8 percent reduction in consumption (figure B3.3.1). This estimate from the meta-analysis broadly aligns with elasticity estimates constructed by Dahl (2012).

(Continued)

BOX 3.3
Technical spotlight: A meta-analysis of price elasticities in low- and middle-income countries (continued)

FIGURE B3.3.1 Price elasticities of gasoline demand

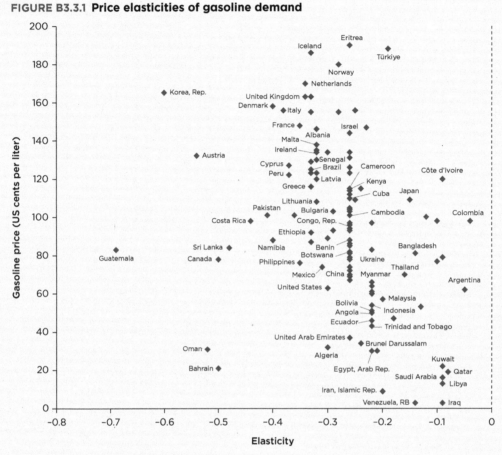

Source: Based on data in Dahl 2012.

The meta-analysis confirmed that even the recent empirical literature continues to focus heavily on high-income countries. The evidence base for robust elasticity estimates in low- and middle-income countries continues to be patchy. Addressing this shortfall is crucial for data-driven and evidence-based policy making.

10 percent increase in energy prices leads to a drop in consumption of between 4 percent and 7 percent, depending on the type of fuel and user (figure 3.10). However, demand remains inelastic (sluggish response to price changes). A *relatively* more elastic demand in the long run implies that consumers will shift away from fossil fuels if given time to adjust (Sterner 2007). For instance, in the case of transport fuels, consumers could choose to replace fuel-inefficient vehicles or reduce travel (for example, by moving closer to their workplace); likewise, governments could facilitate such consumer choices by investing in public transit systems.

FIGURE 3.10 Average price elasticities of energy demand

a. By type of energy

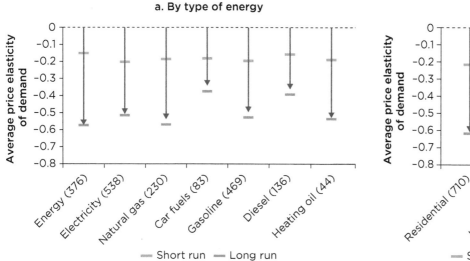

b. By type of use

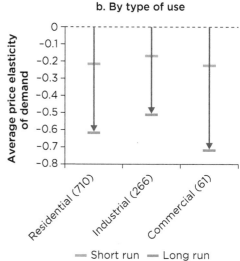

Source: Calculations based on Labandeira, Labeaga, and López-Otero 2017.
Note: Energy refers to a pooled basket of energy goods. Car fuels refer to a pooled basket of vehicle fuels. Numbers in brackets indicate the number of observations (that is, studies) on which these averages are based.

While most studies confirm that fossil fuel consumption responds to price changes and that demand is inelastic, the magnitude of the effect can vary widely across countries and types of fuel. For instance, in the short run, a 10 percent rise in gasoline prices leads to an estimated 2.1 percent reduction in consumption in Türkiye, compared to a 1.5 percent reduction in Cameroon; the long-run effect, however, is only 4.8 percent in Türkiye, compared to a (more elastic) 14.3 percent reduction in Cameroon (Erdogu 2014 and Sapnken et al. 2018, respectively). These elasticity values indicate that the response can be significantly larger in the long run than in the short run. Therefore, increasing fuel prices can have slow but lasting positive effects on air quality.

However, as figure 3.10 suggests, energy consumption is, on average, "inelastic"—in other words, the change in consumption in response to a change in price is relatively small. By implication, achieving meaningful reductions in fossil fuel consumption may require substantial increases in fuel prices or complementary policy measures.

> The consumption of polluting fuels tends to be price "inelastic": The change in consumption in response to a change in price is smaller than the price change itself.

Air pollution and price signals

To what extent do changes in energy prices affect air pollution outcomes? Perhaps surprisingly, the answer is not straightforward, as air pollution is driven by a wide range of factors, including the affordability and availability of clean alternative technologies, the quality of fuels, and regulatory requirements for public good technologies like air filters.

In addition to targeting price levels, policy measures also should improve the availability and affordability of clean alternatives, address information constraints and behavioral biases, and mandate the use of public good technologies such as air filters.

Case studies offer mixed evidence of the direct link between fuel prices and air pollution (box 3.4). In principle, higher gasoline prices incentivize people to reduce the distance traveled, to improve fuel efficiency, or to find alternative means of transportation. Thus by reducing the quantity of gasoline used, price signals should lower the level of air pollution. This channel between fuel prices and air quality—that is, changes in fuel prices affecting consumption and hence air quality—has been studied in urban settings as far-ranging as Australia, China, and the Islamic Republic of Iran:

- A study on fuel prices and air quality in Tehran shows that a 10 percent increase in gasoline prices reduces carbon dioxide and PM_{10} concentrations, though only by 0.2 percent and 0.12 percent, respectively (Raeissi, Khalilabad, and Hadian 2022). However, due to a substitution away from gasoline and toward other fuels, the same increase in gasoline prices results in an increase in concentrations of NO_2 of 0.11 percent in the short run and 0.2 percent in the long run. Overall, the authors conclude that, while increasing fuel prices improves air quality, the effect is rather small, and other measures might prove more effective—for instance, more rigorous regulation of the pollution intensity of vehicles.

- Similar effects have been documented for the city of Brisbane, Australia, with respect to the impact of a change in fuel prices on air quality (Barnett and Knibbs 2014). Their results indicate that higher fuel prices do not lead to a significant change in concentrations of particulate matter ($PM_{2.5}$ and PM_{10}); however, increasing diesel prices leads to substantial decreases in concentrations of NO_x. The study estimates that raising the cost

---

BOX 3.4

Technical spotlight: A global empirical analysis of the relationship between energy prices and air pollution

How far can fossil fuel subsidy reform and price-based instruments go in tackling urban air pollution? In a technical background paper prepared for this report, Mayr and Rentschler (2022) analyze the nature of the relationship between fossil fuel prices and air pollution outcomes. The study uses panel data on the unit price of gasoline, diesel, coal, and liquefied petroleum gas (LPG), covering 133 countries over a 19-year period from 2000 to 2019 totaling 2,465 observations. The study constructed a novel panel data set of annual average urban $PM_{2.5}$ concentrations from van Donkelaar et al. (2021), containing particle concentrations at the center of the capital city of each country. This information is combined with a data set of average annual energy prices for 133 countries, based on Parry, Black, and Vernon (2021). The study then uses the common correlated effects estimator (Pesaran 2006) to understand the role of energy prices in determining air pollution outcomes in 133 cities. The role of energy consumption and income per capita is also investigated.

Overall, the study finds a robust negative relationship between transport fuel prices and particle concentrations. A US$1 increase in the average annual retail price of common transport fuels (diesel, gasoline) is associated with a decline in annual average $PM_{2.5}$ concentrations in capital cities of around 22 micrograms per cubic meter ($\mu g/m^3$). However, the same effect is not found in the case of coal and LPG. The lack of reliable higher-frequency energy price data also poses a challenge for analyzing the price-pollution relationship. The annual frequency removes finer variations that could be explored in a more detailed follow-up study.

of diesel by $A 0.15 per liter would result in significant health benefits, decreasing daily emergency hospital admissions by 2 percent or approximately 215 people per year.

- For 11 regions in China over the 2006 to 2015 period, a study found that vehicle volumes, energy consumption, and GDP per capita are all significantly correlated with higher concentrations of $PM_{2.5}$, PM_{10}, and ozone (Xu et al. 2019). Vehicle volumes were consistently found to increase pollution levels. In other words, more driving means more pollution. However, data on 21 cities in Northern China show that more than half of the major air pollutants (that is, $PM_{2.5}$, PM_{10}, and SO_2) originate from coal-fired heating in autumn and winter (Lin and Ling 2021). Still, changes in coal prices do not affect the level of air pollution, as the government controls heating prices.

Overall, these case studies suggest that the price of polluting fuels may affect air pollution outcomes in some cases, but less so in others. In an exploratory study conducted for this report, Mayr and Rentschler (2022) build on these case studies and statistically explore the relationship between fossil fuel prices and air pollution levels for 133 countries using 20-year-long panel data. The results suggest that a US$1 per liter increase in the average annual retail price of common transport fuels (diesel, gasoline) is associated with a decrease in the annual average concentrations of $PM_{2.5}$ in capital cities of at least 22.2 micrograms per cubic meter ($\mu g/m^3$)—a substantial improvement in air quality that could reduce all-cause mortality by around 16 percent. This association suggests that price hikes, on average, indeed reduce pollution levels. However, the results also suggest strong heterogeneity across countries. In other words, the price-pollution relationship is not as pronounced in many countries.

Moreover, no significant relationship could be established for coal or liquefied petroleum gas (LPG) prices and air pollution. Several factors may explain this finding. Coal is a heavily polluting fuel used especially for power generation and industrial applications. Such uses tend to be unresponsive to shorter-term price variations—for instance, coal plants have long life spans and provide a consistent baseload to the power sector. LPG, by contrast, is a relatively clean-burning fuel that makes a small contribution to urban air pollution, which is why higher consumption of LPG may not degrade air quality outcomes as much as higher consumption of other fossil fuels.

Prices matter, but antipollution measures need to go further

The relationship between gasoline prices and air pollution may not be as pronounced as the relationship between gasoline prices and consumption for several reasons. For one, the availability and affordability of clean technologies will affect the impact of a change in prices. For example, vehicle fleets with better fuel efficiency and exhaust filters would register a smaller change in the fine particulate concentrations for the same reduction in distance traveled than a "dirtier" fleet. Similarly, depending on the quality of gasoline, the same amount of fuel consumed for driving can have different impacts on air pollution. Therefore, depending on the composition of the fleet or the quality of fuel, $PM_{2.5}$ concentrations in some countries might respond more or less strongly to a change in prices.

Another factor could be that the price elasticity of gasoline demand varies significantly across and even within countries. For instance, consumers can more easily substitute away from driving in a city with a well-developed public transit system. Efficient public transit systems are not available or affordable in all countries, and hence behavioral responses to changes in fuel prices differ from place to place. Lastly, some high-pollution fossil fuels, such as coal, are used for baseload power generation in many countries, thus making its demand extremely inelastic. For example, responses to a price change might take time to manifest. In short, besides price levels, several other entry points exist for policy measures to mitigate air pollution, which explains the heterogeneity in the impact of prices on $PM_{2.5}$ concentrations (see box 2.4 in chapter 2).

While the results of this report provide evidence of the efficacy of price changes, multiple other avenues exist for reducing air pollution. Nonprice measures, such as technology mandates and regulations on catalyzers, are underutilized and could improve air quality without diminishing consumption (UNEP 2021). Ambient air quality standards can be found in the legislative instruments of 64 percent of countries, although they are much more common in high-income countries. Thus encouraging the implementation of air quality legislation, such as mandatory filters and fuel mileage requirements for cars, could improve urban air quality. Ultimately, curbing air pollution is not a goal that can be accomplished through a single price adjustment or a single instrument. Instead, it requires a combination of measures working in tandem. Consumer resistance to price increases for fossil fuels may be another reason for complementary instruments to control pollution levels in a second-best world beset with political economy constraints. As chapter 5 shows, air quality can be improved by reforming explicit fossil fuel subsidies, but more ambitious and comprehensive measures are needed to bring air pollution down to safe levels.

Notes

1. Hoesly et al. (2018) introduce an open-source annual historical (1750–2014) CEDS inventory, which contains the emissions of seven key atmospheric pollutants: nitrogen oxide, carbon monoxide, sulfur dioxide, ammonia, nonmethane volatile organic compounds, black carbon, and organic carbon. McDuffie et al. (2020) update the CEDS and report contemporary estimates of annual country-level emissions for four categories of fuel (coal, biofuel, oil, and natural gas) and remaining process-level emissions from 1970 to 2017.

2. Lelieveld et al. (2020) base their estimates on when fossil fuel emissions peaked in China, before the dramatic reduction in emissions caused by the country's recent stringent mitigation measures. Thus the measures in China might have reduced annual mean $PM_{2.5}$ across the country by between 30 percent and 50 percent from 2013 to 2018 as well as mortality, depending on the circumstances of each region (Zhai et al. 2019).

3. Such in-kind fossil fuel subsidies have also been labeled "implicit subsidies." This definition differs from the International Monetary Fund definition of "implicit subsidies," which reserves this term for the environmental externalities associated with fossil fuel use.

4. This amount included US$97 billion in China, US$79 billion in Russia, US$52 billion in Saudi Arabia, US$50 billion in Brazil, US$21 billion in the United States, US$9 billion in the United Kingdom, and US$5 billion in Australia.

5. These estimates are sensitive to the assumed social cost of carbon. Parry, Black, and Vernon (2021) assume a carbon price of US$60 per ton of CO_2 (tCO_2) in 2020, which is consistent with lower-bound estimates in line with a 2° warming scenario. To monetize the costs of air pollution, Parry, Black, and Vernon (2021) rely on an inflation and real GDP growth–adjusted valuation approach from the Organisation for Economic Co-operation and Development (OECD) for avoided deaths (OECD 2012). For the average OECD country in 2020, this method implies a value of US$4.6 million per death avoided. This number is adjusted with relative income per capita to extrapolate it to other countries, assuming a unitary elasticity for the mortality value.

6. For the IMF debt sustainability analysis for countries eligible for funding through the Poverty Reduction and Growth Trust as of May 31, 2022, see https://www.imf.org/external/Pubs/ft/dsa/DSAlist.pdf.

References

Adam, L., and E. Lestari. 2008. "Ten Years of Reforms: The Impact of an Increase in the Price of Oil on Welfare." *Journal of Indonesian Social Sciences and Humanities* 1: 121–39.

Arze del Granado, F. J., D. Coady, and R. Gillingham. 2012. "The Unequal Benefits of Fuel Subsidies: A Review of Evidence for Developing Countries." *World Development* 40 (11): 2234–48.

Barbiero, O., and B. Cournède. 2013. "New Econometric Estimates of Long-Term Growth Effects of Different Areas of Public Spending." Economics Department Working Paper 1100, Organisation for Economic Co-operation and Development, Paris.

Barnett, A. G., and L. D. Knibbs. 2014. "Higher Fuel Prices Are Associated with Lower Air Pollution Levels." *Environment International* 66 (May): 88–91.

Bast, E., A. Doukas, S. Pickard, L. Van Der Burg, and S. Whitley. 2015. *Empty Promises: G20 Subsidies to Oil, Gas, and Coal Production.* London: Overseas Development Institute; Washington, DC: Oil Change International. https://odi.org/en/publications/empty-promises-g20-subsidies-to-oil-gas-and-coal-production/.

Bast, E., S. Makhijani, S. Pickard, and S. Whitley. 2014. *The Fossil Fuel Bailout: G20 Subsidies for Oil, Gas, and Coal Exploration.* London: Overseas Development Institute; Washington, DC: Oil Change International.

Behrer, P. 2019. "Earth, Wind, and Fire: The Impact of Anti-Poverty Efforts on Indian Agriculture and Air Pollution." Working paper, Harvard University, Cambridge, MA.

Calvo-Gonzales, O., B. Cunha, and R. Trezzi. 2015. "When Winners Feel Like Losers: Evidence from an Energy Subsidy Reform." Policy Research Working Paper 7227, World Bank, Washington, DC.

Carrington, D. 2022. "Car Tyres Produce Vastly More Particle Pollution Than Exhausts, Tests Show." *The Guardian*, June 3, 2022. https://www.theguardian.com/environment/2022/jun/03/car-tyres-produce-more-particle-pollution-than-exhausts-tests-show.

Coady, D., I. Parry, L. Sears, and B. Shang. 2015. "How Large Are Global Energy Subsidies?" Working Paper 15/105, International Monetary Fund, Washington, DC.

Coady, D., I. Parry, L. Sears, and B. Shang. 2017. "How Large Are Global Fossil Fuel Subsidies?" *World Development* 91 (March): 11–27.

Commander, S. 2012. "A Guide to the Political Economy of Reforming Energy Subsidies." IZA Policy Paper 52, Institute for the Study of Labor, Bonn.

Dahl, C. A. 2012. "Measuring Global Gasoline and Diesel Price and Income Elasticities." *Energy Policy* 41 (February): 2–13.

de Oliveira, A., and T. Laan. 2010. *Lessons Learned from Brazil's Experience with Fossil-Fuel Subsidies and Their Reform.* Geneva: International Institute for Sustainable Development.

Dube, I. 2003. "Impact of Energy Subsidies on Energy Consumption and Supply in Zimbabwe: Do the Urban Poor Really Benefit?" *Energy Policy* 31 (15): 1635–45.

Erdogdu, E. 2014. "Motor Fuel Prices in Turkey." *Energy Policy* 69 (June): 143–53.

Gangopadhyay, S., B. Ramaswami, and W. Wadhwa. 2005. "Reducing Subsidies on Household Fuels in India: How Will It Affect the Poor?" *Energy Policy* 31 (2): 125–37.

GSI (Global Subsidies Initiative). 2010. "A How-to Guide: Measuring Subsidies to Fossil Fuel Producers." GSI, Geneva.

GSI (Global Subsidies Initiative). 2012. "Fossil Fuels—At What Cost? Government Support for Upstream Oil and Gas Activities in Russia." GSI, Geneva.

HM Revenue and Customs. 2022. "VAT Rates on Different Goods and Services." HM Revenue and Customs, London. https://www.gov.uk/guidance/vat-on-fuel-and-power-notice-70119.

Hoesly, R. M., S. J. Smith, L. Feng, Z. Klimont, G. Janssens-Maenhout, T. Pitkanen, J. J. Seibert, et al. 2018. "Historical (1750–2014) Anthropogenic Emissions of Reactive Gases and Aerosols from the Community Emissions Data System (CEDS)." *Geoscientific Model Development* 11 (1): 369–408.

IEA (International Energy Agency). 2011. *World Energy Outlook 2011.* Paris: IEA.

IEA (International Energy Agency). 2014. *World Energy Outlook 2014.* Paris: IEA.

IEA (International Energy Agency). 2016. *World Energy Outlook 2016—Special Report: Energy and Air Pollution*. Paris: IEA.

IEA (International Energy Agency). 2022. "Energy Subsidies—Tracking the Impact of Fossil-Fuel Subsidies." IEA, Paris. https://www.iea.org/topics/energy-subsidies.

IHME (Institute for Health Metrics and Evaluation). 2020. *Global Burden of Disease Study 2019 (GBD 2019) Results*. Seattle, WA: IHME. https://vizhub.healthdata.org/gbd-results/.

IMF (International Monetary Fund). 2008. *Fuel and Food Price Subsidies: Issues and Reform Options*. Washington, DC: IMF.

IMF (International Monetary Fund). 2013. *Case Studies on Energy Subsidy Reform: Lessons and Implications*. Washington, DC: IMF.

IMF (International Monetary Fund). 2022. Fossil Fuel Subsidies by Country and Fuel Database (2021). Washington, DC: IMF. https://www.imf.org/en/Topics/climate-change/energy-subsidies.

IRENA (International Renewable Energy Agency). 2020. "Renewable Capacity Statistics 2020." IRENA, Abu Dhabi.

Jewell, J., D. McCollum, J. Emmerling, C. Bertram, D. E. Gernaat, V. Krey, L. Paroussos, et al. 2018. "Limited Emission Reductions from Fuel Subsidy Removal Except in Energy-Exporting Regions." *Nature* 554 (7691): 229–33.

Kojima, M., and D. Koplow. 2015. "Fossil Fuel Subsidies: Approaches and Valuation." Policy Research Working Paper 7220, World Bank, Washington, DC. https://openknowledge.worldbank.org/bitstream /handle/10986/21659/WPS7220.pdf?sequence=1&isAllowed=y.

Labandeira, X., J. M. Labeaga, and X. López-Otero. 2017. "A Meta-Analysis on the Price Elasticity of Energy Demand." *Energy Policy* 102 (March): 549–68.

Lelieveld, J., A. Pozzer, U. Pöschl, M. Fnais, A. Haines, and T. Münzel. 2020. "Loss of Life Expectancy from Air Pollution Compared to Other Risk Factors: A Worldwide Perspective." *Cardiovascular Research* 116 (11): 1910–17.

Lin, B., and C. Ling. 2021. "Heating Price Control and Air Pollution in China: Evidence from Heating Daily Data in Autumn and Winter." *Energy and Buildings* 250 (November 1): 111262.

Mayr, K., and J. Rentschler. 2022. "Fossil Fuel Prices and Air Pollution: Evidence from a Panel of 133 Countries." Policy Research Working Paper, World Bank, Washington, DC.

McDuffie, E. E., R. V. Martin, J. V. Spadaro, R. Burnett, S. J. Smith, P. O'Rourke, M. S. Hammer, et al. 2021. "Source Sector and Fuel Contributions to Ambient $PM_{2.5}$ and Attributable Mortality across Multiple Spatial Scales." *Nature Communications* 12 (1): 1–12.

McDuffie, E. E., S. J. Smith, P. O'Rourke, K. Tibrewal, C. Venkataraman, E. A. Marais, B. Zheng, et al. 2020. "A Global Anthropogenic Emission Inventory of Atmospheric Pollutants from Sector- and Fuel-Specific Sources (1970–2017): An Application of the Community Emissions Data System (CEDS)." *Earth System Science Data* 12 (4): 3413–42.

Mourougane, A. 2010. "Phasing Out Energy Subsidies in Indonesia." Economics Department Working Paper 808, Organisation for Economic Co-operation and Development, Paris.

OECD (Organisation for Economic Co-operation and Development). 2011. "Fossil Fuel Support." OECD Secretariat Background Report for G20 Meeting of Finance Ministers. OECD, Paris.

OECD (Organisation for Economic Co-operation and Development). 2012. "Mortality Risk Valuation in Environment, Health and Transport Policies." OECD Publishing, Paris.

OECD (Organisation for Economic Co-operation and Development). 2013. "Inventory of Estimated Budgetary Support and Tax Expenditures for Fossil Fuels 2013." OECD, Paris. https://www.oecd .org/publications/inventory-of-estimated-budgetary-support-and-tax-expenditures-for-fossil -fuels-2013-9789264187610-en.htm.

OECD (Organisation for Economic Co-operation and Development). 2015. "OECD Companion to the Inventory of Support Measures for Fossil Fuels 2015." OECD, Paris.

Parry, I. W. H., S. Black, and N. Vernon. 2021. "Still Not Getting Energy Prices Right: A Global and Country Update of Fossil Fuel Subsidies." Working Paper 2021/236, International Monetary Fund, Washington, DC.

Pesaran, M. H. 2006. "Estimation and Inference in Large Heterogeneous Panels with a Multifactor Error Structure." *Econometrica* 74 (4): 967–1012.

Plante, M. 2014. "The Long-Run Macroeconomic Impacts of Fuel Subsidies." *Journal of Development Economics* 107: 129–43.

Raeissi, P., T. H. Khalilabad, and M. Hadian. 2022. "The Impacts of Fuel Price Policies on Air Pollution: Case Study of Tehran." *Environmental Science and Pollution Research International* 29: 11780–89.

Rao, N. D. 2012. "Kerosene Subsidies in India: When Energy Policy Fails as Social Policy." *Energy for Sustainable Development* 16 (1): 35–43.

Rentschler, J. 2016. "Incidence and Impact: The Regional Variation of Poverty Effects due to Fossil Fuel Subsidy Reform." *Energy Policy* 96: 491–503.

Rentschler, J., and M. Bazilian. 2017. "Reforming Fossil Fuel Subsidies: Drivers, Barriers, and the State of Progress." *Climate Policy* 17 (7): 891–914.

Rentschler, J., and N. Hosoe. 2022. "Illicit Schemes: Fossil Fuel Subsidy Reforms and the Role of Tax Evasion and Smuggling." Policy Research Working Paper 9907, World Bank, Washington, DC.

Ruggeri Laderchi, C., A. Olivier, and C. Trimble. 2013. *Balancing Act: Cutting Energy Subsidies While Protecting Affordability*. Report 76820. Washington, DC: World Bank.

Sapnken, F., T. G. Tamba, S. N. Essiane, F. D. Koffi, and D. Njomo. 2018. "Modeling and Forecasting Gasoline Consumption in Cameroon Using Linear Regression Models." *International Journal of Energy Economics and Policy* 8: 111–20.

Shyamsundar, P. N. P. Springer, H. Tallis, S. Polasky, M. L. Jat, H. S. Sidhu, P. P. Krishnapriya, et al. 2019. "Alternatives to Crop Residue Burning in India." *Science* 365 (6453): 536–38.

Sterner, T. 2007. "Fuel Taxes: An Important Instrument for Climate Policy." *Energy Policy* 35 (6): 3194–202.

Strand, J. 2013. "Political Economy Aspects of Fuel Subsidies: A Conceptual Framework." Policy Research Working Paper 6392, World Bank, Washington, DC.

Triyana, M., L. Nguyen, C. Donnelley, and K. Mayr. 2022. "The Effect of Agricultural Input Subsidies on Productivity: A Meta-Analysis." Background paper prepared for this report, World Bank, Washington, DC.

UNEP (United Nations Environment Programme). 2003. "Energy Subsidies: Lessons Learnt in Assessing Their Impact and Designing Policy Reforms." United Nations, Geneva.

UNEP (United Nations Environment Programme). 2021. "Regulating Air Quality: The First Global Assessment of Air Pollution Legislation." United Nations, Geneva.

van Donkelaar, A., M. S. Hammer, L. Bindle, M. Brauer, J. R. Brook, M. J. Garay, N. C. Hsu, et al. 2021. "Monthly Global Estimates of Fine Particulate Matter and Their Uncertainty." *Environmental Science & Technology* 55 (22): 15287–300.

Victor, D. 2009. *The Politics of Fossil-Fuel Subsidies*. Geneva: Global Subsidies Initiative.

Vohra, K., A. Vodonos, J. Schwartz, E. A. Marais, M. P. Sulprizio, and L. J. Mickley. 2021. "Global Mortality from Outdoor Fine Particle Pollution Generated by Fossil Fuel Combustion: Results from GEOS-Chem." *Environmental Research* 195 (April): 110754.

Whitley, S. 2013. "Time to Change the Game: Fossil Fuel Subsidies and Climate." Overseas Development Institute, London.

World Bank. 2010. "Subsidies in the Energy Sector: An Overview." Background paper for the World Bank Group Energy Sector Strategy, July 2010. World Bank, Washington, DC.

Xu, W., J. Sun, Y. Liu, Y. Xiao, Y. Tian, B. Zhao, and X. Zhang. 2019. "Spatiotemporal Variation and Socioeconomic Drivers of Air Pollution in China during 2005–2016." *Journal of Environmental Management* 245 (September 1): 66–75.

Zhai, S., D. J. Jacob, X. Wang, L. Shen, K. Li, Y. Zhang, K. Gui, T. Zhao, and H. Liao. 2019. "Fine Particulate Matter ($PM_{2.5}$) Trends in China, 2013–2018: Separating Contributions from Anthropogenic Emissions and Meteorology." *Atmospheric Chemistry and Physics* 19 (16): 11031–41.

CHAPTER 4

Virtually Inescapable

The Scale and Distribution of Toxic Air Pollution

"Water and air, the two essential fluids on which all
life depends, have become global garbage cans."
—*Jacques-Yves Cousteau*

CHAPTER AT A GLANCE

The burden of air pollution is vast:

- Globally 7.3 billion people, or 94 percent of the world's population, face air pollution levels considered unsafe by the World Health Organization (WHO). For 2.8 billion people, pollution levels are hazardous—with concentrations of particulate matter ($PM_{2.5}$) greater than 35 micrograms per cubic meter ($\mu g/m^3$). The all-cause mortality rate is more than 24 percent higher in such areas than in safe areas.

- Almost a third of the world's population—2.32 billion people—are exposed to toxic sulfur dioxide (SO_2) emissions from coal-fired power plants. Expensive to build, coal power plants tend to be located in richer countries and in richer regions within countries. Globally, the burden of air pollution from coal plants falls on higher-income countries, but this pattern reverses locally.

The burden is also distributed unequally:

- In the proximity of coal plants, downwind areas tend to be poorer and more polluted than upwind areas. Compared to upwind areas, downwind areas witness 13 percent higher SO_2 concentrations, 2.9 percent lower gross domestic product (GDP) per capita, and 2.5 percent lower GDP on average. The regressive environmental burden of air pollution may reinforce the social marginalization and low-income status of affected communities. This pattern holds in rich and poor countries alike.

- Approximately 1 in 10 people exposed to unsafe levels of air pollution lives in extreme poverty. For the extreme poor, the same level of air pollution likely implies more severe health risks compared to higher-income households, as the effects of air pollution are compounded by vulnerabilities associated with poverty, including inequitable access to affordable health care.

- Air pollution is highest in middle-income countries. About 64.5 percent of people in lower-middle-income countries are exposed to hazardous levels of $PM_{2.5}$, compared to just 4.4 percent in low-income countries and 0.9 percent in high-income countries. Low-income countries have an opportunity to ensure that development does not come hand-in-hand with intensifying air pollution and the associated detrimental effects.

Introduction

Air pollution is one of the leading causes of death worldwide, especially for poorer people, who tend to be more exposed and vulnerable. For instance, new evidence produced for this report shows that globally 2.32 billion people—almost a third of the world's population—are exposed to SO_2 emissions from coal-fired power plants. But even though the burden of air pollution may appear ubiquitous, it is not distributed equally. This chapter presents new evidence from about 70 countries showing that, in the proximity of coal plants, downwind areas are systematically poorer and more polluted than upwind areas.

To document the scale and distribution of the challenge, the chapter begins by offering global estimates of population exposure to the most prevalent pollutant: $PM_{2.5}$. It also offers new estimates of the global number of poor people exposed to $PM_{2.5}$ concentrations considered unsafe by the WHO, highlighting the burden of pollution on the most vulnerable. It then presents novel analyses and results on the extent and distribution of one of the most toxic sources of air pollution: SO_2 emissions from the world's coal-fired power plants.

The global burden of air pollution

This section presents new research showing that 7.3 billion people are directly exposed to unsafe average annual concentrations of $PM_{2.5}$ (box 4.1). Low- and middle-income countries account for 80 percent of people exposed to unsafe levels of $PM_{2.5}$. Moreover, 716 million poor people (living on under US$1.90 per day) live in areas with unsafe levels of air pollution.

BOX 4.1

Technical spotlight: New evidence on global air pollution exposure and poverty

While most studies have focused on air pollution in rich countries, a better understanding of the interplay between air pollution and poverty is crucial for several reasons. Studies from high-income countries on the health risks associated with air pollution may not be directly transferable to low-income communities, where the nature of occupations and health care differs substantially. The effects of air pollution for health and productivity will be felt the strongest in low-income countries, which tend to have relatively low levels of anthropogenic air pollution compared to more industrial middle-income countries. These countries have an opportunity to ensure that future development does not come hand-in-hand with intensifying air pollution and the associated detrimental effects on health and well-being.

A new study prepared for this report provides a comprehensive account of the relationship between exposure to ambient (outdoor) air pollution, economic development, and poverty in 211 countries and territories (Rentschler and Leonova 2022). It presents global estimates of exposure using the World Health Organization's (WHO's) 2021 revised thresholds for fine particulate matter. In addition, it provides estimates of the number of poor people exposed to unsafe concentrations of $PM_{2.5}$. These findings are based on high-resolution global data on ambient air pollution (concentrations of $PM_{2.5}$). The study uses the gridded data set of annual mean ground-level concentrations of fine particulate matter provided by van Donkelaar et al. (2021), offering global coverage and a resolution of 0.01 degrees (map B4.1.1). This air pollution data set is overlaid with the

(Continued)

BOX 4.1
Technical spotlight: New evidence on global air pollution exposure and poverty
(continued)

WorldPop Global High Resolution Population data set to capture the spatial distribution of population as well as subnational poverty estimates from the World Bank's Global Subnational Atlas of Poverty, which is based on harmonized household surveys. Exposure headcount estimates are computed for each grid cell, by categorizing different classes of air pollution hazard—5, 10, 15, 25, and 35 micrograms per cubic meter ($\mu g/m^3$)—in line with WHO guidelines (WHO 2021).

As a global modeled data set, some uncertainty is to be expected in air pollution measurements (Alvarado et al. 2019). However, sensitivity tests suggest good agreement with ground measurements (van Donkelaar et al. 2021). For more spatially nuanced analyses—for instance, at the neighborhood or street level—alternative data based on local measures would be required; however, in low-income countries, these data are often unavailable with wide spatial coverage. The global $PM_{2.5}$ data set provides information on the total concentration of particles, but not on the spatial variation in the chemical composition (that is, acidity) of particles (Thurston, Chen, and Campen 2022).

MAP B4.1.1 $PM_{2.5}$ concentrations in Southeast Asia

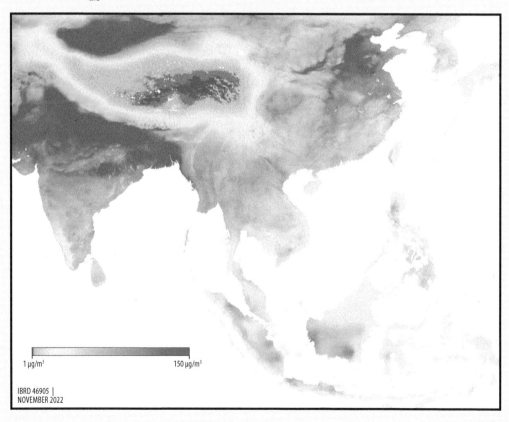

1 µg/m³   150 µg/m³

IBRD 46905 |
NOVEMBER 2022

Source: Data by van Donkelaar et al. 2021.
Note: PM$_{2.5}$ = fine particulate matter; µg/m³ = micrograms per cubic meter.

Around half of them are located in just three countries: the Democratic Republic of Congo, India, and Nigeria. Air pollution levels are particularly high in middle-income countries, where economies tend to rely heavily on polluting industries and technologies.

The distribution of unsafe and hazardous pollution

A global study conducted for this report shows that globally 7.3 billion people, or 94 percent of humanity, face air pollution levels considered unsafe by the WHO (that is, annual average concentration over 5 μg/m³ [figure 4.1]), while 2.8 billion are exposed to "hazardous" concentrations of $PM_{2.5}$ over 35 μg/m³ (Rentschler and Leonova 2022). These areas have a 24 percent higher mortality risk than safe areas. While the problem is global, it is especially prevalent in certain regions. At 2.2 billion people, the East Asia and Pacific region has the highest number of people exposed to unsafe $PM_{2.5}$ concentrations—corresponding to about 95 percent of the region's total population. In the South Asia region, about 1.8 billion people are exposed to unsafe levels of air pollution—about 99 percent of the region's population. In sum, almost the entire population of both East Asia and Pacific and South Asia endures unsafe levels of $PM_{2.5}$.

In other regions, the proportion of the population exposed to dangerous levels of air pollution is slightly lower, but still constitutes the majority of the population. In Europe and Central Asia, the Middle East and North Africa, Sub-Saharan Africa, and the United States and Canada, between 95 percent and 92 percent of the respective regional population is exposed to unsafe concentrations of $PM_{2.5}$. In Latin America and the Caribbean, exposure as a share of population is slightly lower (84 percent). Figure 4.2 breaks down estimated exposure by region.

Several countries have particularly large populations directly exposed to unsafe levels of air pollution (figure 4.3). The two most populous countries, China and India, also have the highest absolute numbers of population exposed to unsafe air pollution. About 36 percent of all people exposed to unsafe concentrations of $PM_{2.5}$ air pollution globally reside in China or India. In India, 1.36 billion people (99 percent of the population) are exposed to unsafe $PM_{2.5}$ concentrations (that is, over 5 μg/m³), of which 1.33 billion (96 percent) face hazardous levels (over 35 μg/m³). In China, 1.41 billion people (99 percent of the population) face unsafe levels of $PM_{2.5}$ concentrations, of which 0.765 billion (53 percent) face hazardous levels. Measured by WHO standards, exposure to air pollution is exceptionally severe in India.

Map 4.1 presents the relative exposure for all countries. It demonstrates that, in large parts of the world and across all regions, the vast majority of the population is exposed to unsafe levels of $PM_{2.5}$ (over 5 μg/m³; map 4.1, panel a). Unlike, for example, flood hazards, which are highly localized, air pollution tends to cover and move across large

FIGURE 4.1 Global population exposed to different levels of air pollution risk

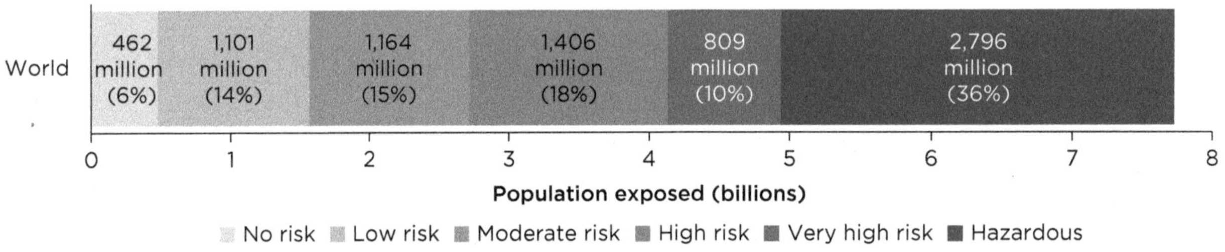

| World | 462 million (6%) | 1,101 million (14%) | 1,164 million (15%) | 1,406 million (18%) | 809 million (10%) | 2,796 million (36%) |

Population exposed (billions)

□ No risk ▨ Low risk ▦ Moderate risk ▪ High risk ■ Very high risk ■ Hazardous

Source: Rentschler and Leonova 2022, 8.
Note: In line with World Health Organization guidelines, "hazardous" means air pollution concentrations over 35 μg/m³; "very high risk" means 25–35 μg/m³; "high risk" means 15–25 μg/m³; "moderate risk" means 10–15 μg/m³; "low risk" means 5–10 μg/m³; and "no risk" means under 5 μg/m³. μg/m³ = micrograms per cubic meter.

FIGURE 4.2 **Population exposed to air pollution, by region and as a share of total regional population**

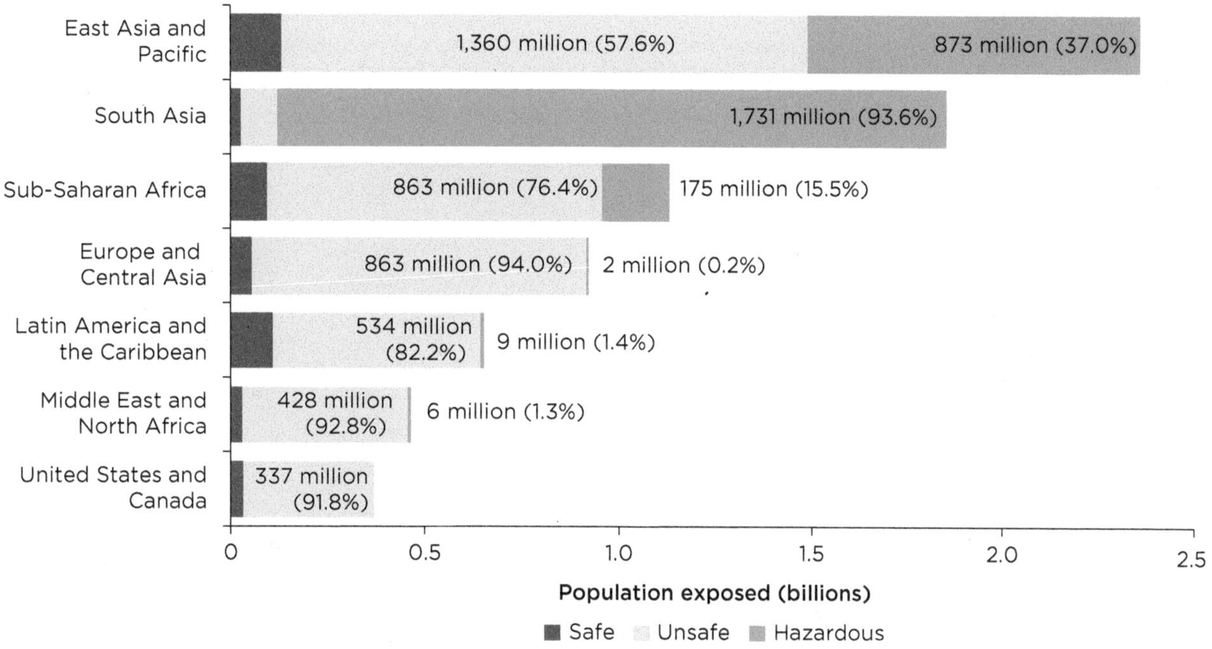

East Asia and Pacific		1,360 million (57.6%)	873 million (37.0%)
South Asia		1,731 million (93.6%)	
Sub-Saharan Africa		863 million (76.4%)	175 million (15.5%)
Europe and Central Asia		863 million (94.0%)	2 million (0.2%)
Latin America and the Caribbean		534 million (82.2%)	9 million (1.4%)
Middle East and North Africa		428 million (92.8%)	6 million (1.3%)
United States and Canada		337 million (91.8%)	

Population exposed (billions)

■ Safe ░ Unsafe ■ Hazardous

Source: Rentschler and Leonova 2022, 9.
Note: Safe levels are defined as being below 5 µg/m³, and unsafe levels are defined as being 5 to 35 µg/m³. Hazardous levels are defined as being over 35 µg/m³. µg/m³ = micrograms per cubic meter.

areas—often blanketing entire cities or regions. As a consequence, if large proportions of a population live in densely populated areas, they tend to be exposed collectively to unsafe levels of pollution. Driven by inefficient transport systems, inferior fuel quality, and polluting technologies in industry, heating, cooking, and power generation, people in low- and middle-income countries face particularly high levels of pollution (over 15 µg/m³, map 4.1, panel b); exposure levels are especially high in Central America, Western and Middle Africa, Eastern Europe, the Middle East, as well as in Central, South, and East Asia. The regions where large parts of the population face hazardous concentrations of $PM_{2.5}$ (over 35 µg/m³) are Eastern China, the Indian subcontinent, and parts of Western Africa (map 4.1, panel c).

Inequality in air pollution exposure

This subsection explores the links between pollution exposure and income level. Considering the income level of the population exposed to pollution is crucial because poorer people tend to be more vulnerable to serious health impacts. Poorer people tend to have higher exposure to air pollution because they are often employed in occupations that require outdoor physical labor. Once affected by pollution-related diseases, they also tend to have more limited access to adequate and affordable health care, thus increasing mortality rates. In addition, poorer countries tend to have less developed health care systems. In short, considering the interplay between pollution exposure, income level, and poverty can shed light on the vulnerability of affected populations.

Members of marginalized groups like ethnic and religious minorities, indigenous populations, and others are also more exposed and vulnerable to air pollution for these same reasons, but data limitations prevent this connection from being explored in depth.

FIGURE 4.3 **Top 10 countries with the highest number and proportion of population exposed to unsafe and hazardous levels of PM$_{2.5}$**

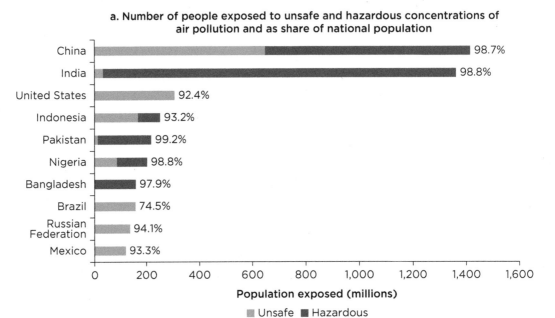

a. Number of people exposed to unsafe and hazardous concentrations of air pollution and as share of national population

Population exposed (millions)

■ Unsafe ■ Hazardous

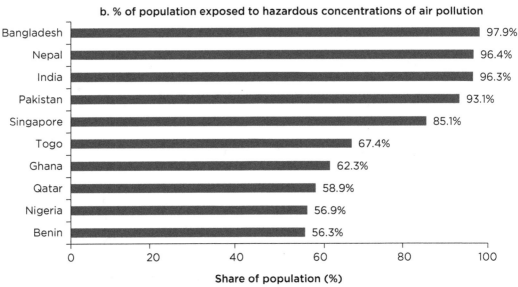

b. % of population exposed to hazardous concentrations of air pollution

Share of population (%)

Source: Rentschler and Leonova 2022, 10.
Note: Unsafe levels are defined as being 5 to 35 µg/m³. Hazardous levels are defined as being over 35 µg/m³. PM$_{2.5}$ = fine particulate matter; µg/m³ = micrograms per cubic meter.

Health impacts from air pollution in low-income countries and communities

People in low- and middle-income countries are especially prone to experiencing adverse health impacts. The scientific evidence is in strong agreement that air pollution is one of the leading causes of death (Lelieveld et al. 2020). Due to compounding factors, this burden is borne disproportionately by low- and middle-income countries. Less stringent air quality regulations, prevalence of older polluting machinery and vehicles, subsidized fossil fuels,

MAP 4.1 **Share of population exposed to unsafe levels of PM$_{2.5}$**

a. % of population exposed to PM$_{2.5}$ over 5 µg/m³

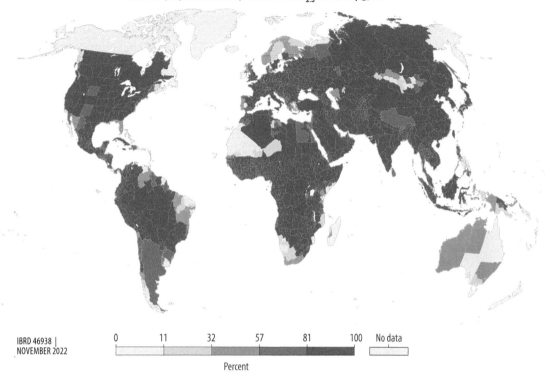

IBRD 46938 |
NOVEMBER 2022

0   11   32   57   81   100   No data

Percent

b. % of population exposed to PM$_{2.5}$ over 15 µg/m³

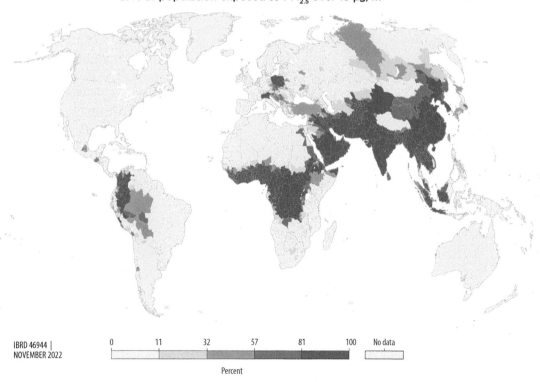

IBRD 46944 |
NOVEMBER 2022

0   11   32   57   81   100   No data

Percent

(Continued)

MAP 4.1 **Share of population exposed to unsafe levels of PM$_{2.5}$** *(continued)*

c. % of population exposed to PM$_{2.5}$ over 35 µg/m³

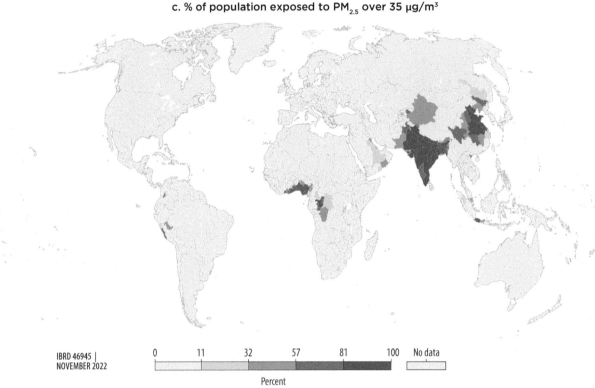

IBRD 46945 |
NOVEMBER 2022

0   11   32   57   81   100   No data

Percent

Source: Rentschler and Leonova 2022.
Note: PM$_{2.5}$ = fine particulate matter; µg/m³ = micrograms per cubic meter.

congested urban transport systems, rapidly developing industrial sectors, and cut-and-burn practices in agriculture contribute to high levels of concentration.

In addition, high proportions of physical and outdoor labor mean that a larger proportion of the population faces greater exposure to pollution. Constraints in terms of accessibility, availability, and quality of health care provision further exacerbate air pollution–related mortality in low- and middle-income countries (Lelieveld et al. 2020). Evidence on air pollution in India suggests that a 1 gigawatt increase in coal-fired capacity corresponds to a 14 percent increase in infant mortality rates in districts near versus far from the plant site (Barrows, Garg, and Jha 2019). This effect is two to three times larger than estimates for the high-income world.

Poverty and pollution interact in various ways (box 4.1). The burden of air pollution falls disproportionately on low- and middle-income countries. However, even in richer countries, poorer, more marginalized communities tend to be exposed to higher pollution levels (Bell, Zanobetti, and Dominici 2013; Jbaily et al. 2022). This inequality in exposure is well documented in the United States, where data on socioeconomic characteristics are available with high spatial disaggregation. Studies have shown that industrial planning policies have disproportionately placed polluting industries in areas with large ethnic minorities or low-income populations.[1] Strikingly, the disparity in exposure has been increasing over time (Chay and Greenstone 2005).

Since environmental amenities such as air quality influence real estate prices, a self-enforcing cycle may emerge such that lower-income households locate in cheaper, more polluted neighborhoods. And since exposure to high levels of pollution affects health, labor productivity, and human capital, people become trapped in poverty, resulting in a low-level separating equilibrium.

Outside of the United States, evidence on the inequality in air pollution exposure is more limited—often due to a lack of socioeconomic data with high spatial disaggregation. The few studies for African and Asian countries tend to confirm the presence of inequality in exposure found in the United States. For instance, Rao et al. (2021) show that in India the mortality risk due to $PM_{2.5}$ falls disproportionately on low-income households. Overall, there is a significant gap in the literature on the extent to which the poor in low- and middle-income countries are affected by air pollution globally.

> Countries' pollution intensities along their economic development path are not set in stone.

Air pollution is most pervasive in middle-income countries

Pollution levels differ according to region and, by implication, the stage of economic development and industrialization of a country. Indeed, the vast majority of people breathing unsafe air are located in middle-income countries (map 4.2). Of the 7.3 billion people exposed to unsafe concentrations of $PM_{2.5}$ (over 5 $\mu g/m^3$), around half (3.4 billion) live in low- or lower-middle-income countries. In middle-income countries, 2.8 billion people are exposed to hazardous levels of $PM_{2.5}$ (over 35 $\mu g/m^3$), compared to just 40.5 million in low- and high-income countries combined.

In relative terms—that is, as a share of the overall population—$PM_{2.5}$ exposure is also highest in middle-income countries. About 65 percent of people in lower-middle-income countries are exposed to levels of $PM_{2.5}$ over 35 $\mu g/m^3$, compared to just 4.4 percent in

MAP 4.2 **Mean concentrations of PM$_{2.5}$**

IBRD 46950 | NOVEMBER 2022

Mean concentrations of PM$_{2.5}$ ($\mu g/m^3$): 0 5 10 15 25 35 145 No data

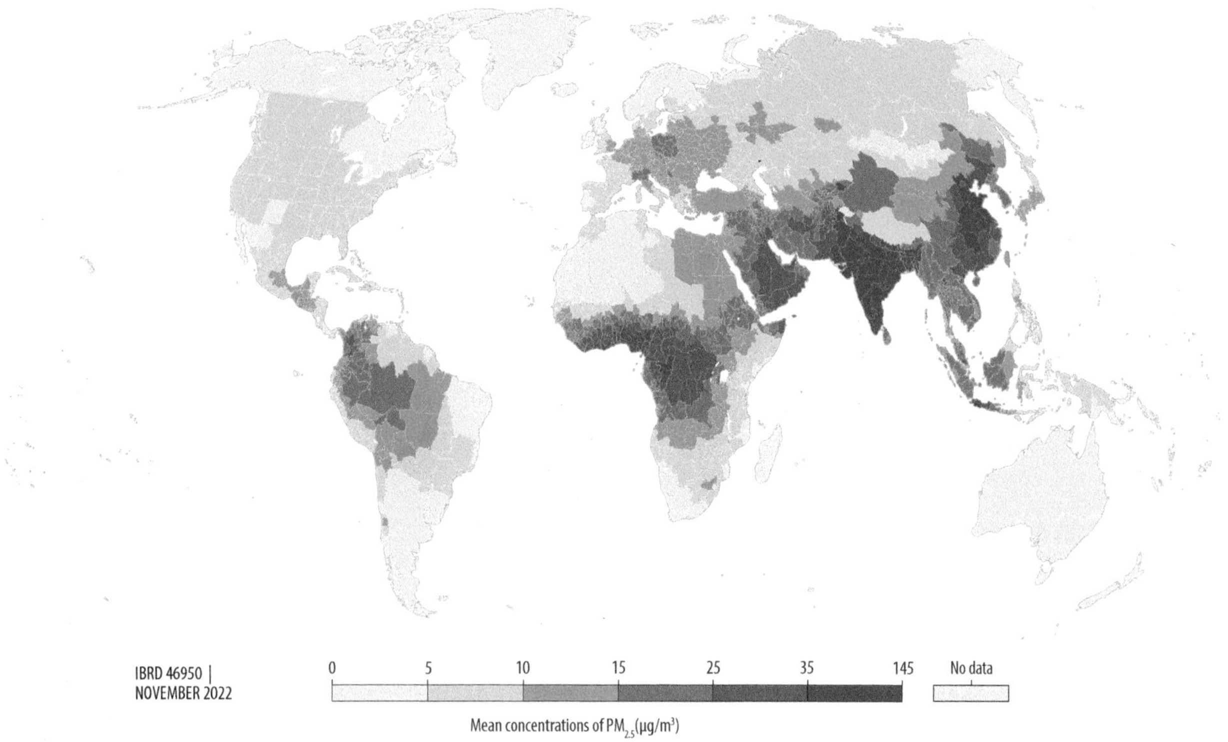

Source: Rentschler and Leonova 2022.
Note: PM$_{2.5}$ = fine particulate matter; $\mu g/m^3$ = micrograms per cubic meter.

low-income countries and 0.9 percent in high-income countries. The pattern holds regardless of which threshold of concentration is considered: exposure to more than 5 μg/m³ (the safe threshold recommended by the WHO) or 10 μg/m³ or 35 μg/m³ (figure 4.4). The high levels of ambient air pollution in middle-income countries are due in large part to the rapid economic growth and industrialization of South and East Asia.

FIGURE 4.4 Global population exposure to PM$_{2.5}$ concentrations, by headcount and percentage of population exposed

a. Headcount of population exposed, by exposure level and country income group

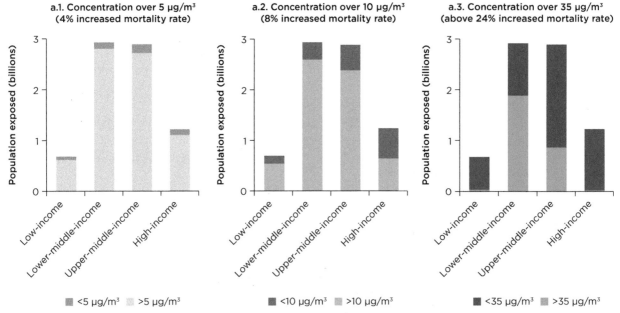

b. % of population exposed, by exposure level and country income group

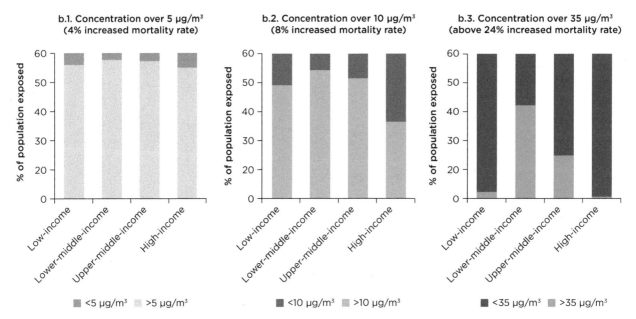

Source: Rentschler and Leonova 2022, 17.
Note: Green bars represent the population exposed to levels of PM$_{2.5}$ concentration over 5, 10, and 35 μg/m³, respectively. Blue bars represent the population exposed to levels below the threshold. PM$_{2.5}$ = fine particulate matter; μg/m³ = micrograms per cubic meter.

The reasons for this pattern are well established. Air pollution levels are highest in middle-income countries, where polluting industries and technologies dominate the economy and regulations do not prioritize air quality. With some exceptions, concentrations of air pollution are relatively low in the lowest-income countries, as economic activities (for example, agriculture) tend to rely less on fossil fuels, and the consumption of polluting goods (for example, high electricity use or private car ownership) is limited to small population groups. In high-income countries, pollution is low, as economic activity tends to be focused on less-polluting sectors (for example, services), polluting activities tend to be off-shored, while clean technologies are widely available and mandated by regulation.

Therefore, the pollution intensity along the economic development path is not set in stone. Whether today's low-income countries will face intensifying pollution as a by-product of development depends on the availability and affordability of clean technologies as well as the incentive structure for their adoption. The provision of subsidies for fossil fuel consumption directly undermines the uptake of such clean technologies, entrenching high levels of pollution in low- and middle-income countries, where such subsidies are particularly common (chapter 3).

Poverty and air pollution exposure

Outside of the United States, evidence on the inequality in exposure to air pollution is more limited, with very little evidence from low- and middle-income countries, often due to a lack of socioeconomic data at high spatial disaggregation.[2] This section fills this gap in the literature.

Air pollution exposure among the world's poor is estimated by combining estimates of air pollution exposure with survey-based subnational data on poverty (table 4.1). Approximately 1 out of 10 people exposed to unsafe levels of air pollution lives in extreme poverty. The estimates show that 716 million people living in extreme poverty (that is, living on less than US$1.90 per day) are directly exposed to unsafe concentrations of $PM_{2.5}$; of these, 405 million (or 57 percent) are in Sub-Saharan Africa (figure 4.5). Further, 275 million people living in extreme poverty are exposed to hazardous concentrations of $PM_{2.5}$.

275 million: The number of people living in extreme poverty who are exposed to hazardous concentrations of $PM_{2.5}$

TABLE 4.1 **Population exposed to $PM_{2.5}$, by poverty threshold**

Indicator	Poverty threshold (consumption, US$ per day)		
	US$1.90	US$3.20	US$5.50
Number of poor (millions)	768	1,853	3,034
% of global population who are poor	9.9	23.9	39.2
Number of poor who are exposed to *unsafe* $PM_{2.5}$ levels (millions)	716	1,752	2,870
% of population who are poor and exposed to *unsafe* $PM_{2.5}$ levels	9.3	22.7	37.1
Number of poor who are exposed to *hazardous* $PM_{2.5}$ levels (millions)	275	938	1,573
% of population who are poor and exposed to *hazardous* $PM_{2.5}$ levels	3.5	12.1	20.3

Source: Rentschler and Leonova 2022, 13.
Note: Unsafe levels are defined as being over 5 μg/m³. Hazardous levels are defined as being over 35 μg/m³. $PM_{2.5}$ = fine particulate matter; μg/m³ = micrograms per cubic meter.

FIGURE 4.5 Air pollution, poverty, and quality of health care

a. Number of poor people exposed to unsafe air pollution, by poverty threshold and region

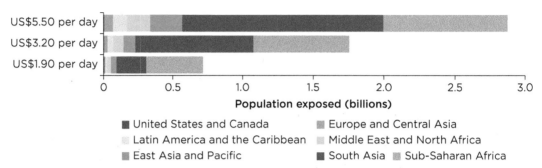

b. % of population who are poor (below US$1.90 per day) and exposed to unsafe air pollution, compared to country's Healthcare Access and Quality Index

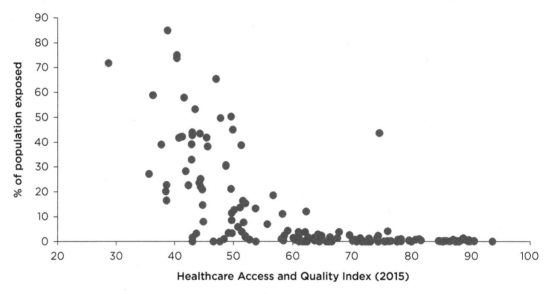

Sources: Panel a: Rentschler and Leonova 2022, 13. Panel b: GBD Healthcare Access and Quality Collaborators 2015.
Note: Each data point represents one country. Unsafe $PM_{2.5}$ levels are defined as being over 5 μg/m³. $PM_{2.5}$ = fine particulate matter; μg/m³ = micrograms per cubic meter. The Healthcare Access and Quality Index is measured on a scale from 0 (worst) to 100 (best), based on GBD Healthcare Access and Quality Collaborators (2015).

When poverty is defined using less extreme (that is, higher) income thresholds, the number of air pollution–exposed poor people increases significantly. Overall, 4 out of every 10 people exposed to unsafe levels of $PM_{2.5}$ globally live on less than US$5.50 a day. Around 1.8 billion people live in areas with unsafe levels of air pollution, while also living on less than US$3.20 a day. The number increases to 2.9 billion when considering incomes below US$5.50 a day. Increasing the poverty threshold from US$1.90 to US$5.50 doubles the number of poor people exposed to unsafe $PM_{2.5}$ levels in Sub-Saharan Africa from 405 million to 877 million.[3] In South Asia, the number of the poor and pollution-exposed population increases more than sixfold, from 220 million to 1.43 billion; in East Asia, the increase is also sixfold, from 38 million to 229 million. These large population groups are particularly vulnerable to high exposure to air pollution, because accessibility and quality of health care are especially low in places where high air pollution coincides with high poverty rates.

Country-level poverty and pollution exposure

The distribution of poor people exposed to unsafe or hazardous levels of pollution is highly skewed: almost half (48.6 percent) are located in just three countries. India, with a population of more than 202 million, has the highest absolute number of extreme poor exposed to unsafe levels of $PM_{2.5}$, corresponding to 14.7 percent of India's overall population (figure 4.6). Nigeria is a distant second, followed by the Democratic Republic of Congo. Overall, the top 10 countries account for 67.8 percent of the total incidence of all of the extreme poor who are exposed to unsafe concentrations of $PM_{2.5}$ globally (figure 4.7).

FIGURE 4.6 Top 10 countries with the largest number of poor people (at US\$1.90 per day) exposed to unsafe and hazardous levels of $PM_{2.5}$

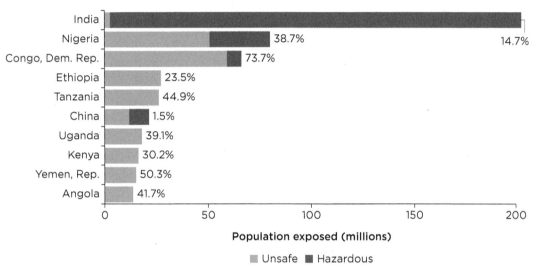

Source: Rentschler and Leonova 2022, 14.
Note: Numbers following bar labels are the percentage of the population who are poor and exposed to unsafe or hazardous levels of $PM_{2.5}$. Unsafe levels are defined as being 5 to 35 μg/m³. Hazardous levels are defined as being over 35 μg/m³. $PM_{2.5}$ = fine particulate matter; μg/m³ = micrograms per cubic meter.

FIGURE 4.7 Top 10 countries with the largest percentage of poor people (at US\$1.90 per day) exposed to hazardous levels of $PM_{2.5}$

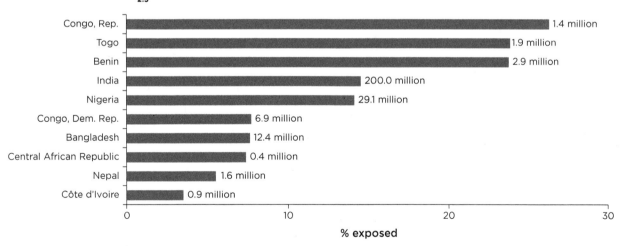

Source: Rentschler and Leonova 2022, 14.
Note: Numbers following bar labels are the number of the population who are poor and exposed to hazardous levels of $PM_{2.5}$. Hazardous levels are defined as being over 35 μg/m³. $PM_{2.5}$ = fine particulate matter; μg/m³ = micrograms per cubic meter.

Notably, 7 of the top 10 countries where the extreme poor are exposed to high levels of pollution are located in Sub-Saharan Africa.[4] As is evident on map 4.3, extreme poverty and exposure to unsafe concentrations of $PM_{2.5}$ coincide most acutely in Sub-Saharan Africa. However, when considering higher poverty thresholds, exposure also becomes

MAP 4.3 Regional distribution of air pollution and poverty

a. % of the population exposed to unsafe levels of $PM_{2.5}$ and living in poverty at US$1.90 per day

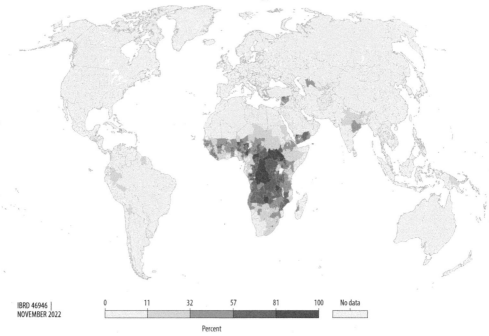

IBRD 46946 |
NOVEMBER 2022

| 0 | 11 | 32 | 57 | 81 | 100 | No data |

Percent

b. % of the population exposed to unsafe levels of $PM_{2.5}$ and living in poverty at US$3.20 per day

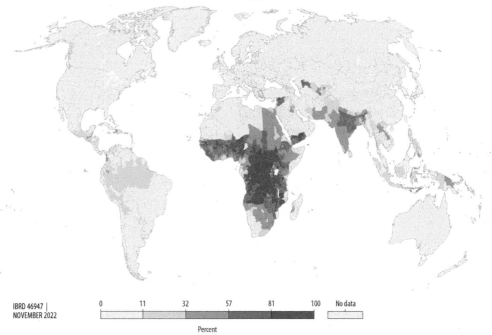

IBRD 46947 |
NOVEMBER 2022

| 0 | 11 | 32 | 57 | 81 | 100 | No data |

Percent

(Continued)

Detox Development

MAP 4.3 Regional distribution of air pollution and poverty *(continued)*

c. % of the population exposed to unsafe levels of PM$_{2.5}$ and living in poverty at US$5.50 per day

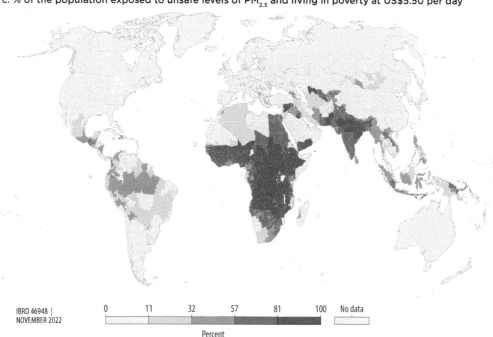

| IBRD 46948 | NOVEMBER 2022 | 0 | 11 | 32 | 57 | 81 | 100 | No data |

Percent

Source: Rentschler and Leonova 2022.
Note: Unsafe levels are defined as being over 5 μg/m³. PM$_{2.5}$ = fine particulate matter; μg/m³ = micrograms per cubic meter.

apparent in areas of the Middle East, South and East Asia, and Latin America. Klaiber (2022) conducts a case study for Vietnam showing that nuance is crucial when assessing whether poor people are more exposed to air pollution than nonpoor people (box 4.2).

BOX 4.2
Nuances matter: Air pollution and poverty in Vietnam

Klaiber (2022) analyzes the exposure of Vietnam's population to different thresholds of fine particulate matter (PM$_{2.5}$) pollution for the years 2014 to 2019 and uses a spatial econometric model to analyze the connection between income levels and air pollution in 678 districts. While Vietnam has made great progress in fighting poverty in urban areas, poverty rates remain high in rural and minority populations. Like in most countries, the vast majority of the population—about 93 percent—lives above the World Health Organization's recommended level of annual air quality: PM$_{2.5}$ of 5 micrograms per cubic meter (μg/m³).

At a national level, ambient PM$_{2.5}$ pollution is higher in wealthier urban districts. The metropolitan areas around Hanoi and Ho Chi Min City have significantly higher levels of air pollution than the surrounding areas (figure B4.2.1). The presence of industrial zones is also associated with higher district-level air pollution. In other words, areas with high economic activity are more polluted but also wealthier.

Yet this is only part of the story. While exposure rates are similar for poor and nonpoor people in areas of very low and very high air pollution (5 μg/m³ and more than 35 μg/m³), poor people are significantly overrepresented in areas where annual average concentrations of PM$_{2.5}$ are between 25 μg/m³ and 35 μg/m³, suggesting that the risk of being exposed to at least 25 μg/m³ of annual average PM$_{2.5}$ is 28 percent higher for a person who is poor.

(Continued)

BOX 4.2
Nuances matter: Air pollution and poverty in Vietnam *(continued)*

This pattern may be due to the larger number of poor people living in periurban and adjacent rural areas who still face dangerously high pollution levels, albeit not as high as in urban cores. Focusing solely on urban areas risks overlooking low-income and ethnic minority groups who lack access to the same advanced health care and economic opportunities as urban groups. National averages can hide such spatial nuances, which nevertheless need to be considered when making policy choices.

MAP B4.2.1 Bivariate distribution of poverty and PM$_{2.5}$ pollution (over 15 µg/m³) in Vietnam

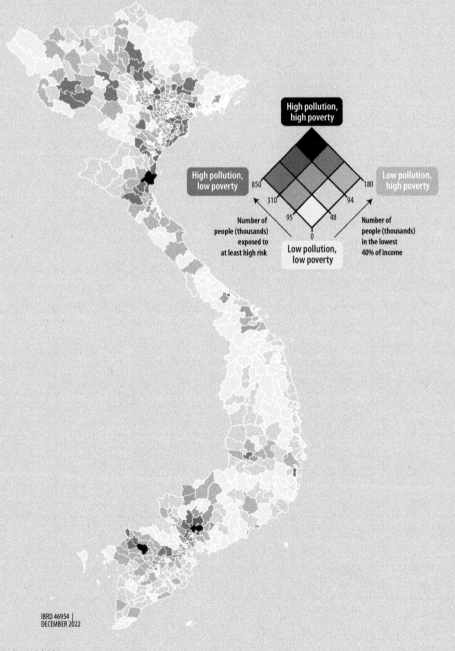

Source: Klaiber 2022.
Note: PM$_{2.5}$ = fine particulate matter; µg/m³ = micrograms per cubic meter.

An unequal burden: New evidence on air pollution from the world's coal-fired power plants

While fossil fuels are collectively responsible for the lion's share of anthropogenic air pollution, coal is particularly harmful. Indeed, not all $PM_{2.5}$ particles are created equal—the chemical composition and hence toxicity of $PM_{2.5}$ particles differ by the source of pollution (Thurston, Chen, and Campen 2022). $PM_{2.5}$ particles associated with the combustion of fossil fuels have been shown to be more toxic due to their higher levels of acidity (for example, sulfuric particulate matter from coal burning).

This section presents new evidence from a global analysis of coal plants conducted for this report (Du, Rentschler, and Russ 2022). It documents the regressive nature of air pollution and the health burden associated with the operation of coal power plants. It finds that air pollution from coal plants increases with income level. In other words, coal power plants tend to be located in richer countries and richer regions within countries. Globally, 2.32 billion people are exposed to air pollution originating from coal power plants.

However, in the proximity of plants, areas that are downwind are associated with higher pollution and lower income levels than areas that are upwind. Thus in countries rich or poor, lower-income groups are disproportionately affected by air pollution. This finding suggests that coal-fired power plants may be strategically located in a way that reinforces the spatial sorting of low- and high-income communities. It may be indicative of systematic environmental injustice concerns at a global scale.

Unprofitable and toxic, yet coal power persists

Coal power is associated with tremendous societal costs, including air pollution, climate change, acid rain, wastewater discharge, and wastage of financial resources. In the United States alone, coal-fired power plants release more than 3.1 million tons of SO_2, 1.5 million tons of nitrogen oxides (NO_x), and 0.2 million tons of particulate matter each year.[5] They also release substantial lead, mercury, volatile organic compounds, arsenic, and other toxic chemicals. The evidence is unequivocal that these air pollutants are linked with heightened risks of asthma, cancer, heart and lung ailments, neurological diseases, and other adverse impacts on public health (Manisalidis et al. 2020). The societal impact of coal plants goes beyond local air pollution. In 2020 the burning of coal was responsible for around 40 percent of global carbon dioxide emissions—more than any other fossil fuel—and more than 70 percent of greenhouse gas emissions from the electricity sector.

Nevertheless, coal remains a widely used source of energy supply around the world. The Global Coal Plant Tracker estimates that about 2,500 coal power plants were in active operation in January 2022.[6] Almost 500 more were in the planning or construction stage, while more than 1,000 coal plants were retired between 2000 and 2021. Based on a technology invented in the 1880s, coal-fired power generation provides some 2 terawatts of capacity today—about half of which is in China.

In addition to their high environmental and health costs, coal-fired power plants operate with low efficiency and a lack of profitability. More than 40 percent of the world's coal plants are operating at a loss, not least due to high fuel costs, and this share could rise to 75 percent by 2040 (Gray et al. 2018). However, power plants are still expanding in number and capacity, as their economic viability is propped up by fossil fuel subsidies. In 2021, China, India,

92%: The share of planned coal power plants in East Asia that are forecasted to operate at a financial loss

Indonesia, Japan, and Vietnam had plans to build more than 600 new coal-fired power plants, even though 92 percent of them are expected to be unprofitable (Carbon Tracker 2021). Each year governments around the world spend around US$13.6 billion dollars to lower the price of coal artificially, thus perpetuating the operation of unprofitable coal-fired power plants and the associated air pollution (Parry, Black, and Vernon 2021).

Contribution of coal to air pollution

Coal-fired power plants are a major source of toxic air pollution. The Global Coal Plant Tracker documents more than 6,000 coal-fired power generation units, sited in almost 4,000 unique locations. These plants are distributed widely across the world, but they are especially concentrated in Europe, South and East Asia, and the United States (map 4.4). In a technical background study for this report, Du, Rentschler, and Russ (2022) use wind trajectory models to simulate how prevailing wind patterns disperse particle emissions from power plants to surrounding areas (box 4.3). For all of the world's coal-fired plants, these models yield simulated SO_2 pollution maps that enable pollution exposure analyses. For instance, figure B4.3.1 shows SO_2 pollution patterns in China, Europe, and India, documenting pollution hotspots near power plants and their surrounding areas, with concentrations decreasing with the distance to plants.

New evidence produced for this report shows that globally 2.32 billion people—almost a third of the world's population—are exposed to emissions from coal-fired power plants; 1.29 billion of whom are in China and India alone (Du, Rentschler, and Russ 2022). Overall, 13 countries have more than half of their population exposed to air pollution from coal plants. Figure 4.8 shows the top 10 countries with the largest number and highest proportion of population affected by pollution from coal plants.

MAP 4.4 Global distribution of coal-fired power plants

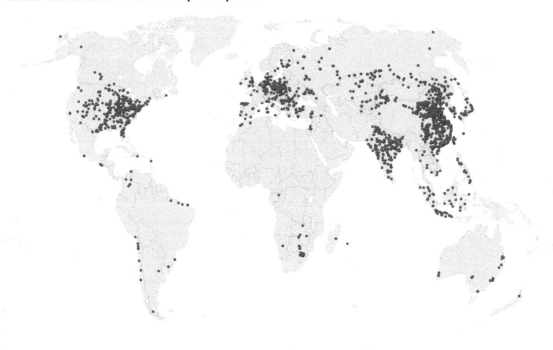

IBRD 46904 |
NOVEMBER 2022

Source: Global Coal Plant Tracker, https://globalenergymonitor.org/projects/global-coal-plant-tracker/.
Note: This map shows the spatial distribution of coal-fired power plants as of January 2013. Each dot denotes an operating unit.

Detox Development

BOX 4.3
Technical spotlight: New evidence on air pollution from the world's coal-fired power plants

In a study conducted for this report, Du, Rentschler, and Russ (2022) develop a new high-resolution global map of air pollution associated with coal-fired power plants and analyze patterns of exposure and economic development. The study analyzes air pollution originating from around 4,000 coal-fired power plants that were operational in the period up to 2013.

The study uses data on the location and emission intensity of coal-fired power plants to simulate the continuous output of air pollutants for each plant. It then models the movement and dispersion behavior of pollutants at each power plant location, based on prevailing wind conditions. For this purpose, it uses the Hybrid Single-Particle Lagrangian Integrated Trajectory (HYSPLIT) model of the National Oceanic and Atmospheric Administration. HYSPLIT is designed to compute air parcel trajectories to determine how far and in what direction a parcel of air, and subsequently air pollutants,

FIGURE B4.3.1 Simulated SO$_2$ pollution from coal-fired power plants dispersed by prevailing wind patterns in select countries

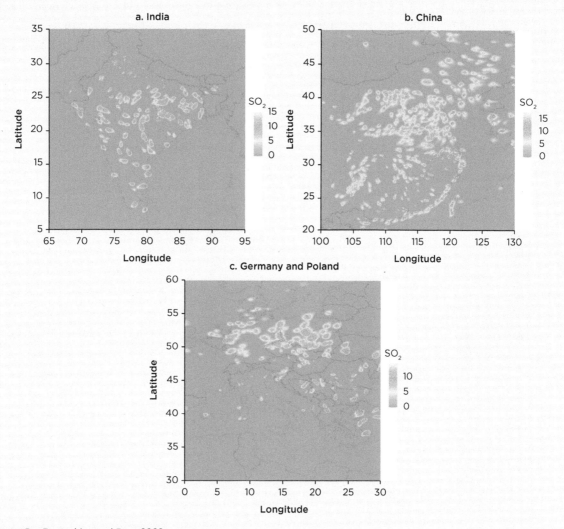

Source: Du, Rentschler, and Russ 2022.
Note: Units are ln(μg/m³) at a spatial resolution of 0.01 arc-degrees. SO$_2$ = sulfur dioxide; μg/m³ = micrograms per cubic meter.

(Continued)

Technical spotlight: New evidence on air pollution from the world's coal-fired power plants
(continued)

will travel. Du, Rentschler, and Russ (2022) then use three-dimensional Gaussian dispersion modeling to simulate how pollutants disperse along wind trajectories. This approach yields a novel high-resolution map of sulfur dioxide (SO_2) particle pollution for coal-fired power plants with global coverage (figure B4.3.1).

The study then compares this spatial data set on pollution concentrations with a gridded data set on economic activity (GDP) and income (GDP per capita). The high spatial resolution enables a systematic comparison of incomes in the upwind and downwind areas of power plants. The study also explores the relationship between the income level of countries or communities and their exposure to toxic air pollution from coal plants. The use of high-resolution gridded data on GDP per capita allows the study to consider up to about 35 million individual locations (that is, mapped pixels) with separate observations.

This large-scale analysis unearthed a key finding that previous studies have mostly shown only for specific case studies in high-income countries. In a globally prevailing pattern, there is systematic inequality in the exposure to coal plant pollution. In the proximity of coal plants, downwind areas are associated with higher pollution and lower income levels than upwind areas—indicative of strategic location choices and spatial sorting of communities. In addition, the study enabled global estimates of exposure to toxic SO_2 pollution from coal plants, which are presented in this report.

FIGURE 4.8 Top 10 countries with the greatest population exposure to coal power plant pollution

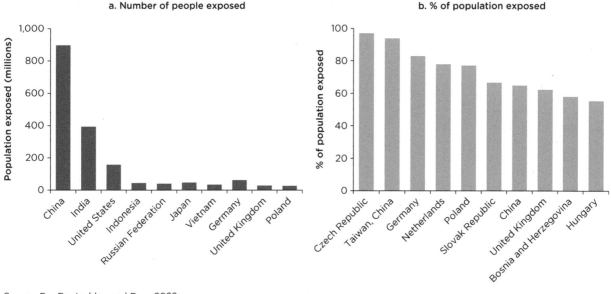

a. Number of people exposed

b. % of population exposed

Source: Du, Rentschler, and Russ 2022.

China, India, and the United States have 896 million, 392 million, and 157 million people exposed, respectively, due to their high concentration of plants and high population densities. In the Czech Republic, 97 percent of the population is exposed to pollution from coal plants, including high transboundary pollution from foreign plants.

FIGURE 4.9 Top 10 countries with the greatest population exposure to transboundary pollution

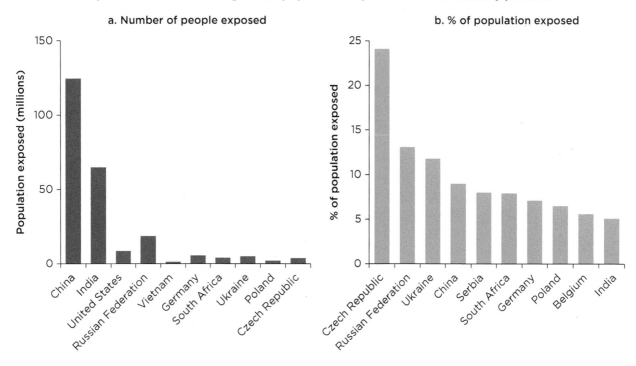

a. Number of people exposed

b. % of population exposed

Source: Du, Rentschler, and Russ 2022.

Air pollution knows no borders, as it is carried by prevailing winds. Some 247.5 million people worldwide are exposed to transboundary air pollution from foreign coal plants. The top affected countries are shown in figure 4.9. Transboundary pollution affects 24.1 percent of the population in the Czech Republic, where Germany and Poland are the primary sources of transboundary flows.

Studies have shown that polluting activities display a systematic tendency to be located near administrative borders. This location reduces the environmental and health costs for domestic residents, while creating transboundary externalities. Monogan, Konisky, and Woods (2017) show that major air polluters in the United States are more likely to be located near the downwind borders of states. Hence in the United States, 57 percent of coal-fired electricity generators are located within 5 kilometers of a county border and 25 percent are within 5 kilometers of a state border (Morehouse and Rubin 2021). In comparison, natural gas plants, which have substantially lower emissions and face lower emission-based environmental regulatory pressures, do not exhibit this pattern of spatial sorting. The presence of environmental free-riding has also been detected where water pollution is disproportionately borne by downstream jurisdictions (Lipscomb and Mobarak 2016; Sigman 2005) and in toxic emissions into the air and water near state borders relative to same-state regions (Helland and Whitford 2003).

The global analysis conducted for this report shows that toxic air pollution and the associated health impacts from coal-fired power plants are substantial. Globally, 160 million people face SO_2 pollution from coal plants that exceeds the WHO guideline of

> Coal plants tend to be located near administrative borders— thus avoiding air quality regulation and offloading pollution across jurisdictions.

40 μg/m³—and this does not account for pollution from other sources. These figures translate to substantial health impacts and excess mortality that are in line with estimates from earlier studies on the health effects of coal-fired power generation.[7]

Disproportionate burden on poorer communities

In the proximity of coal power plants, poorer communities face a disproportionate burden. While coal power plants tend to be located in richer countries and regions, the air pollution they cause may still disproportionately affect poorer communities. Case study evidence, especially from the United States, has documented that low-income and ethnic minority communities are disproportionately exposed to air pollution. In other words, the economic activities that contribute to growth and prosperity are occurring at the expense of socially marginalized groups. But is the pattern of unequal exposure in the United States also found in other countries around world, in particular, in low- and middle-income countries? So far, no systematic evidence has been available.

> New evidence shows that, in the proximity of coal plants, downwind areas are systematically poorer and more polluted than upwind areas.

Du, Rentschler, and Russ (2022) provide the first large-scale evaluation of the unequal burden of air pollution associated with coal-fired power plants. The study analyzes the relationship between income and air pollution for more than 1.3 million unique locations (pixels) in the vicinity of more than 3,800 coal plants located in 71 countries. It finds that power plant locations are indeed distributed unevenly across income groups. A key reason for this pattern is that prevailing winds carry pollution emissions in a certain direction—instead of affecting all surrounding neighborhoods evenly (figure 4.10). In the proximity of coal plants, downwind areas are systematically poorer and more polluted than upwind areas. Specifically, compared to upwind areas,

FIGURE 4.10 Modeled wind dispersion of air pollutants at the Mauda power plant in India

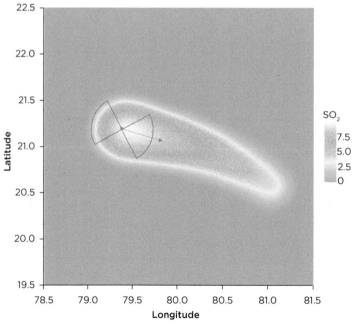

Source: Du, Rentschler, and Russ 2022.
Note: Units are ln(μg/m³) at a spatial resolution of 0.01 arc-degrees. SO₂ = sulfur dioxide.

downwind areas witness around 50 percent higher SO_2 concentrations, 2.9 percent lower GDP per capita, and 2.5 percent lower GDP.

These estimates confirm the hypothesis that more polluted areas downwind tend to host lower-income, socially marginalized communities. Two key mechanisms can explain this pattern: strategic location choices and spatial sorting (box 4.4). When choosing the location of new coal-fired power plants, operators may strategically select spots where prevailing winds carry air pollutants to lower-income, socially marginalized communities. These communities are likely to have weaker bargaining power to oppose such siting decisions. Once the plant assumes operations, the regressive environmental burden of air pollution may be aggravated by spatial sorting that reinforces the causes of social marginalization. As lower environmental quality (including air quality) is priced into land values, high-income households are able to sort into cleaner upwind areas, while low-income households are more likely to move into less expensive, more polluted areas. In addition, affected communities experience heightened risks of adverse health impacts, which further degrade productivity and incomes.

> Socially marginalized communities often are exposed to higher levels of pollution.

In short, these results for the world's 3,800 coal power plants offer the first global evidence consistent with the notion that the choice of plant location is influenced by the income profile of surrounding areas and that air pollution may reinforce the low-income status of downwind areas through spatial sorting effects. Evidence from previous case studies can shed further light on the functioning of the underlying mechanisms.

Country and within-country income level and air pollution from coal plants

Richer countries tend to have more coal power and associated pollution. Not all countries can afford to invest in coal-fired power plants—at the same time, not all countries are prepared to put up with the environmental pollution and financial inefficiencies of coal plants. As coal-fired power plants are increasingly operating at a loss, maintaining them requires large coal or power sector subsidies that may be unaffordable for many governments, especially in low-income countries.

To assess the relationship between income levels and air pollution from coal plants, Du, Rentschler, and Russ (2022) examine this issue across 71 countries for which data are available (box 4.3). Results show that the exposure to air pollution from coal power plants varies significantly by income level. In general, countries with higher national income tend to have more coal-fired power plants.

The relationship between SO_2 air pollution and income levels is also shown to be consistent with the distribution of coal plants. Countries with higher incomes tend to face more pollution from coal power plants. The correlation is found to be positive and concave, suggesting that the increase in SO_2 levels is slower at higher levels of GDP per capita. This finding is consistent with the notion that, as incomes rise, there is a tendency to slow the investment in coal and move to cleaner sources.

Within countries, coal-fired plants tend to be located in richer regions with high economic activity. New high-resolution data sets on the spatial distribution of economic

FIGURE 4.11 National income and coal-fired power plants

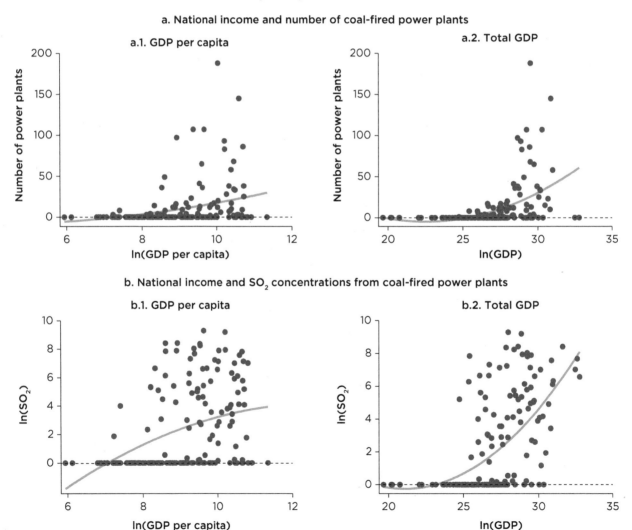

a. National income and number of coal-fired power plants

Source: Du, Rentschler, and Russ 2022.
Note: Each data point represents one country. SO_2 = sulfur dioxide.

activity, incomes, air pollution, and coal plant locations also enable large-scale granular assessments of the relationship between income levels and pollution exposure.

The patterns suggest a relationship between income and pollution that is consistent across countries. Around the world, higher-income areas tend to be exposed to higher SO_2 pollution from coal-fired power plants (figure 4.11). Every additional 10 percentage point increase in GDP per capita is associated with a 1.6 percent higher level of SO_2 particle concentration. In short, regions with higher incomes and more economic activity tend to have poorer air quality. This is no coincidence, as power plants tend to be located in the vicinity of large power consumers, such as cities and industrial areas, in regions with higher economic activity. The twist is that even in the richer regions communities that are relatively poor are disproportionately exposed to pollution. As the next chapter shows, reforming fossil fuel subsidies can be pro-poor in terms of both fiscal and health benefits.

Notes

1. For more details on air pollution disparities in the United States, refer to Bell and Ebisu (2012); Bell, Zanobetti, and Dominici (2013); Colmer et al. (2020); Fann et al. (2018); Kioumourtzoglou et al. (2016); Mikati et al. (2018); Patel et al. (2021); and Tessum et al. (2021).

2. The few studies for African and Asian countries tend to confirm the presence of inequality in exposure found in the United States. For instance, Rao et al. (2021) show that in India the mortality risk due to $PM_{2.5}$ falls disproportionately on low-income households. Hajat, Hsia, and O'Neill (2015) offer a systematic review of the literature on the economic disparities and confirm the US result. Overall, there is a significant gap in the literature on the extent to which the poor in low- and middle-income countries are affected by air pollution globally.

3. In Sub-Saharan Africa, 39.3 percent of the region's total population lives in extreme poverty (US$1.90), and 91.8 percent of the region's total population is exposed to unsafe levels of $PM_{2.5}$.

4. In the 10 countries with the highest share of the population who are poor and exposed to high concentrations of air pollution, about 50 percent to 85 percent of the population is estimated to be exposed, most of whom are in Sub-Saharan Africa (9 out of 10).

5. Data from the 2014 National Emissions Inventory compiled by the US Environmental Protection Agency.

6. The Global Coal Plant Tracker can be found at https://globalenergymonitor.org/projects/global-coal-plant-tracker/.

7. Vohra et al. (2021) document 10.2 million global excess deaths per year due to $PM_{2.5}$ from fossil fuel combustion. In the United States, 350,000 premature deaths are attributed to emissions from the fossil sector. The number in India is 2.5 million people per year, representing more than 30 percent of all-cause deaths. Cropper et al. (2021) conclude that 112,000 deaths are attributable annually to coal-fired power plants in India. Kushta et al. (2021) estimate that 18,000 to 106,000 deaths in Europe could be avoided by phasing out emissions from coal power plants.

References

Alvarado, A., A. McVey, J. Hegarty, E. Cross, C. Hasenkopf, R. Lynch, E. Kennelly, et al. 2019. "Evaluating the Use of Satellite Observations to Supplement Ground-Level Air Quality Data in Selected Cities in Low- and Middle-Income Countries." *Atmospheric Environment* 218 (December 1): 117016.

Barrows, G., T. Garg, and A. Jha. 2019. *The Health Costs of Coal-Fired Power Plants in India*. IZA Discussion Paper 12838. Bonn: Institute for the Study of Labor.

Bell, M., and K. Ebisu. 2012. "Environmental Inequality in Exposures to Airborne Particulate Matter Components in the United States." *Environmental Health Perspectives* 120 (12): 1699–704.

Bell, M., A. Zanobetti, and F. Dominici. 2013. "Evidence on Vulnerability and Susceptibility to Health Risks Associated with Short-Term Exposure to Particulate Matter: A Systematic Review and Meta-Analysis." *American Journal of Epidemiology* 178 (6): 865–76.

Carbon Tracker. 2021. "Paris Target at Risk as Five Countries Plan 80% of World's New Coal Power." Carbon Tracker, June 30, 2021. https://carbontracker.org/paris-target-at-risk-as-five -countries-plan-80-of-worlds-new-coal-power/.

Chay, K., and M. Greenstone. 2005. "Does Air Quality Matter? Evidence from the Housing Market." *Journal of Political Economy* 113 (2): 376–424.

Colmer, J., I. Hardman, J. Shimshack, and J. Voorheis. 2020. "Disparities in PM$_{2.5}$ Air Pollution in the United States." *Science* 369 (6503): 575–78.

Cropper, M., R. Cui, S. Guttikunda, N. Hultman, P. Jawahar, Y. Park, X. Yao, and X. P. Song. 2021. "The Mortality Impacts of Current and Planned Coal-Fired Power Plants in India." *Proceedings of the National Academy of Sciences* 118 (5): e2017936118.

Du, X., J. Rentschler, and J. Russ. 2022. "People's Unequal Exposure to Air Pollution: Evidence for the World's Coal-Fired Power Plants." Background paper prepared for this report, World Bank, Washington, DC.

Fann, N., E. Coffman, B. Timin, and J. T. Kelly. 2018. "The Estimated Change in the Level and Distribution of PM$_{2.5}$-Attributable Health Impacts in the United States: 2005–2014." *Environmental Research* 167 (November): 506–14.

GBD Healthcare Access and Quality Collaborators. 2015. "Healthcare Access and Quality Index Based on Amenable Mortality 1990–2015: A Novel Analysis from the Global Burden of Disease Study 2015." *The Lancet* 390 (10091): 231–66.

Gray, M., S. Ljungwaldh, L. Watson, and I. Kok. 2018. "Powering Down Coal: Navigating the Economic and Financial Risks in the Last Years of Coal Power." Carbon Tracker Initiative, London.

Hajat, A., C. Hsia, and M. O'Neill. 2015. "Socioeconomic Disparities and Air Pollution Exposure: A Global Review." *Current Environmental Health Reports* 2 (4): 440–50.

Hanlon, W. W. 2020. "Coal Smoke, City Growth, and the Costs of the Industrial Revolution." *Economic Journal* 130 (626): 462–88.

Heblich, S., A. Trew, and Y. Zylberberg. 2021. "East-Side Story: Historical Pollution and Persistent Neighborhood Sorting." *Journal of Political Economy* 129 (5): 1508–52.

Helland, E., and A. B. Whitford. 2003. "Pollution Incidence and Political Jurisdiction: Evidence from the TRI." *Journal of Environmental Economics and Management* 46 (3): 403–24.

Jbaily, A., X. Zhou, J. Liu, T. Lee, L. Kamareddine, S. Verguet, and F. Dominici. 2022. "Air Pollution Exposure Disparities across US Population and Income Groups." *Nature* 601: 228–33.

Khanna, G., W. Liang, A. M. Mobarak, and R. Song. 2021. "The Productivity Consequences of Pollution-Induced Migration in China." NBER Working Paper w28401, National Bureau of Economic Research, Cambridge, MA.

Kioumourtzoglou, M. A., J. Schwartz, P. James, F. Dominici, and A. Zanobetti. 2016. "PM$_{2.5}$ and Mortality in 207 US Cities: Modification by Temperature and City." *Epidemiology* 27 (2): 221–27.

Klaiber, C. 2022. "Air Pollution and Poverty in Vietnam: A Geospatial Analysis." Background paper prepared for this report, World Bank, Washington, DC.

Kushta, J., N. Paisi, H. D. Van Der Gon, and J. Lelieveld. 2021. "Disease Burden and Excess Mortality from Coal-Fired Power Plant Emissions in Europe." *Environmental Research Letters* 16 (4): 045010.

Lelieveld, J., A. Pozzer, U. Pöschl, M. Fnais, A. Haines, and T. Münzel. 2020. "Loss of Life Expectancy from Air Pollution Compared to Other Risk Factors: A Worldwide Perspective." *Cardiovascular Research* 116 (11): 1910–17.

Lipscomb, M., and A. M. Mobarak. 2016. "Decentralization and Pollution Spillovers: Evidence from the Re-drawing of County Borders in Brazil." *Review of Economic Studies* 84 (1): 464–502.

Manisalidis, I., E. Stavropoulou, A. Stavropoulos, and E. Bezirtzoglou. 2020. "Environmental and Health Impacts of Air Pollution: A Review." *Frontiers in Public Health* 8: 14.

Mikati, I., A. F. Benson, T. J. Luben, J. D. Sacks, and J. Richmond-Bryant. 2018. "Disparities in Distribution of Particulate Matter Emission Sources by Race and Poverty Status." *American Journal of Public Health* 108 (4): 480–85.

Monogan, J. E., III, D. M. Konisky, and N. D. Woods. 2017. "Gone with the Wind: Federalism and the Strategic Location of Air Polluters." *American Journal of Political Science* 61 (2): 257–70.

Morehouse, J., and E. Rubin. 2021. "Downwind and Out: The Strategic Dispersion of Power Plants and Their Pollution." Working paper, Center for Growth and Opportunity, Utah State University, Logan.

Parry, I., S. Black, and N. Vernon. 2021. "Still Not Getting Energy Prices Right: A Global and Country Update of Fossil Fuel Subsidies." IMF Working Paper WP/21/236, International Monetary Fund, Washington, DC.

Patel, L., E. Friedman, S. Johannes, S. Lee, H. O'Brien, and S. Schear. 2021. "Air Pollution as a Social and Structural Determinant of Health." *Journal of Climate Change and Health* 3 (August): 100035.

Rao, N. D., G. Kiesewetter, J. Min, S. Pachauri, and F. Wagner. 2021. "Household Contributions to and Impacts from Air Pollution in India." *Nature Sustainability* 4 (10): 859–67.

Rentschler, J., and N. Leonova. 2022. "Air Pollution and Poverty: $PM_{2.5}$ Exposure in 211 Countries and Territories." Policy Research Working Paper 10005, World Bank, Washington, DC.

Sigman, H. 2005. "Transboundary Spillovers and Decentralization of Environmental Policies." *Journal of Environmental Economics and Management* 50 (1): 82–101.

Tessum, C., D. Paolella, S. Chambliss, J. Apte, J. Hill, and J. Marshall. 2021. "$PM_{2.5}$ Polluters Disproportionately and Systemically Affect People of Color in the United States." *Science Advances* 7 (18): eabf4491.

Thurston, G., L. C. Chen, and M. Campen. 2022. "Particle Toxicity's Role in Air Pollution." *Science* 375 (6580): 506.

van Donkelaar, A., M. Hammer, L. Bindle, M. Brauer, J. Brook, M. Garay, C. Hsu, et al. 2021. "Monthly Global Estimates of Fine Particulate Matter and Their Uncertainty." *Environmental Science & Technology* 55 (22): 15287–300.

Vohra, K., A. Vodonos, J. Schwartz, E. A. Marais, M. P. Sulprizio, and L. J. Mickley. 2021. "Global Mortality from Outdoor Fine Particle Pollution Generated by Fossil Fuel Combustion: Results from GEOS-Chem." *Environmental Research* 195 (April): 110754.

WHO (World Health Organization). 2021. *WHO Global Air Quality Guidelines: Particulate Matter ($PM_{2.5}$ and PM_{10}), Ozone, Nitrogen Dioxide, Sulfur Dioxide and Carbon Monoxide.* Geneva: World Health Organization. https://apps.who.int/iris/handle/10665/345329.

CHAPTER 5

Pro-Poor and Pro-Health
The Benefits of Reforming Subsidies

"As long as poverty, injustice and gross inequality
persist in our world, none of us can truly rest."

—Nelson Mandela

CHAPTER AT A GLANCE

Fossil fuel subsidy reform is pro-poor:

- *In absolute terms, richer households consume significantly more energy than poorer ones and thus lose more when subsidies are removed.* By owning more cars and by heating and lighting bigger houses, richer people consume more energy and thus benefit more from energy subsidy schemes. Evidence suggests that the richest income group always loses more from the removal of subsidies than the poorest—on average, 13 times more in subsidy reform simulations for 19 countries.

- *As a share of income, poor people are not necessarily hit harder by subsidy reform; it depends on the country context.* Common convention holds that, as a share of income, poorer households incur a larger loss from subsidy reform, but the data offer mixed evidence. As a share of total spending, energy consumption is similar in size across income groups and so are the relative impacts of subsidy removal. In subsidy reform simulations for 19 countries, the richest income group lost, on average, 10 percent more than the poorest, in relative income terms.

Fossil fuel subsidy reforms can reduce air pollution and save lives:

- *Removing fossil fuel subsidies could reduce concentrations of particulate matter ($PM_{2.5}$) by between 2 percent and 40 percent, depending on the country considered.* These air quality benefits are largest in countries with large subsidy programs for the most polluting fuels and where fossil fuel consumption is highly responsive to price changes.

- *Reforming explicit fossil fuel subsidies in 25 high-pollution, high-subsidy countries could save about 360,000 air pollution–related deaths between 2022 and 2035.* While saving 360,000 lives is significant, it is but a fraction of the 4.5 million annual deaths associated with outdoor ambient air pollution around the world. This fraction is small because fossil fuel subsidy reform does not automatically yield large environmental and health benefits in every case. The size and types of fuel in the subsidy program matter and, in a few cases, can even lead to detrimental substitution effects.

- *Removing explicit fossil fuel subsidies is the first step, but much more is needed.* The removal of explicit (that is, active) fossil fuel subsidies is just a first step toward addressing the underpricing of fossil fuels. Explicit subsidies are dwarfed by the magnitude of the social costs of fossil fuels, so removing explicit subsidies alone will not fix climate change or air pollution. In addition to reforming explicit subsidy programs, the wider societal costs of fossil fuel consumption need to be reflected in the price of fossil fuels and supported by complementary policies.

The distributional implications of fossil fuel subsidy reform

This chapter examines the main economic consequences of subsidy removal or reform. It begins by exploring the distributional impacts in both absolute and relative terms. The results shed light on the extent to which richer people benefit from fossil fuel subsidies compared to the poor, but also on how the extent of this inequality can differ across countries.

While a common political justification for fossil fuel subsidies is to support the poor by subsidizing the energy supply, the literature clearly shows that most subsidies are regressive—that is, in absolute terms, most subsidies benefit the rich. Nevertheless, relative to income, the adverse effects of subsidy removal have been suggested to be greatest for the poor in some countries (Arze del Granado, Coady, and Gillingham 2012; IEA et al. 2010; Ruggeri Laderchi, Olivier, and Trimble 2013; World Bank 2010).

This section assesses the distributional impact of subsidy removal by estimating the mean effect on the consumption of different income groups. The scenario analyzed simulates the phasing out of all explicit fossil fuel subsidies in a country within one year—all subsidies for coal, natural gas, liquefied petroleum gas (LPG), gasoline, diesel, and kerosene. The revenues from reforming the fossil fuel subsidy—that is, the government expenses saved—are redistributed as a direct lump-sum cash transfer to households in the bottom 40 percent of the income distribution.

Depending on the features of the existing fossil fuel subsidy program, the reform scenario differs between countries (box 5.1). For instance, the economic impact and changes in consumer prices associated with reform will be larger in countries with high levels of fossil fuel subsidies than in countries with lower subsidies. Consequently, the effect on the level of consumption and income will be larger in those countries. Similarly, the distributional impact of reform depends heavily on people's consumption of energy and energy-intensive goods. These patterns of consumption differ significantly across countries, in line with preferences, the sectoral composition of economic activities, the prevailing mix of energy and technology, and other factors. As this section demonstrates, the wide range of country-specific drivers of energy consumption also means that the potential distributional impacts of subsidy reform will vary significantly across countries.

Household income level and the benefits of subsidy reform

In absolute terms, richer households consume significantly more energy than poorer ones. Higher-income households in most countries consume more energy than lower-income households. By owning more cars and by heating and lighting bigger houses, rich people consume more energy and thus benefit more from energy subsidy schemes. In turn, when energy subsidies are removed, the richest households tend to incur the largest loss in absolute monetary terms. For this reason, fossil fuel subsidy reforms are considered to be a progressive—that is, pro-poor—reform measure. However, the degree of progressiveness can differ from country to country, depending on the pattern of household consumption at different income levels.

To illustrate distributional impacts, this section first presents in-depth results for 4 out the 19 countries for which distributional impacts were simulated: Brazil, China, Indonesia, and Mexico. These countries were chosen based on the availability of household-level data and the large size of their fossil fuel subsidy programs. Figure 5.1 presents the absolute mean consumption effects from removing existing fossil fuel

BOX 5.1
Technical spotlight: Assessing the distributional and health benefits of fossil fuel subsidy reform for 35 countries using the World Bank's Carbon Pricing Assessment Tool

In a study conducted for this report, Klaiber, Rentschler, and Dorband (2022) use the World Bank's Climate Policy Assessment Tool (CPAT) to analyze the distributional impact of fossil fuel subsidy reform in 19 countries as well as the health benefits of subsidy reform in 25 countries. Together, the analysis covers 35 unique countries with high levels of air pollution and (or) large energy subsidy programs. The subsidy reforms simulated in the analysis entail a complete phaseout of all fossil fuel subsidies within one year. The revenues from this reform—that is, the government savings—are transferred to households at the lowest 40 percent of income as a direct lump-sum cash transfer.

The analysis of the distributional impacts on household consumption is based on household-level surveys in 19 countries, as implemented in CPAT. The changes are expressed as shares of disposable income before reform and measured in absolute per capita monetary terms at the income decile level. Where data are available, differences between urban and rural populations are estimated (Klaiber, Rentschler, and Dorband 2022). The changes in mean consumption are based on the estimated changes in energy prices resulting from the subsidy reform modeled by CPAT. The estimated energy prices are combined with household budget information and linked to input-output data to estimate changes in consumption. The resulting estimates assume a full cost pass-through with no behavioral adjustment.

The analysis of health impacts is available for more countries. Klaiber, Rentschler, and Dorband (2022) limit the analysis to countries with high levels of air pollution and countries with large fossil fuel subsidy programs. Avoided deaths are the key measure of health impacts. Using CPAT, they are estimated by changes in energy prices, which change the level of fuel consumption. These changes in fuel consumption are translated into changes in emitted air pollutants via fuel-specific emissions factors using the GAINS model. These changes, in turn, affect aggregate concentrations of air pollutants, which are estimated using source receptor matrixes (TM5-FASST). Finally, relative risk functions for different pollutants are used to estimate the health impacts (averted deaths) resulting from the changed level of emissions, relative to a no-reform business-as-usual scenario.

The analysis shows that fossil fuel subsidy reforms save lives and that the health effects are greatest in countries with high levels of fossil fuel subsidies. For the 25 countries in the analysis, the results suggest that removing the fossil fuel subsidy could reduce the number of avoidable deaths from air pollution exposure by at least 360,000 between 2022 and 2035. Regarding the distributional impacts of a fossil fuel subsidy reform, the analysis suggests that country-specific factors play a large role; in absolute terms, richer people benefit more from fossil fuel subsidies, as they consume more energy-intensive goods. Thus the results support evidence from the literature that fossil fuel subsidies are regressive and are unsuited to reducing poverty.

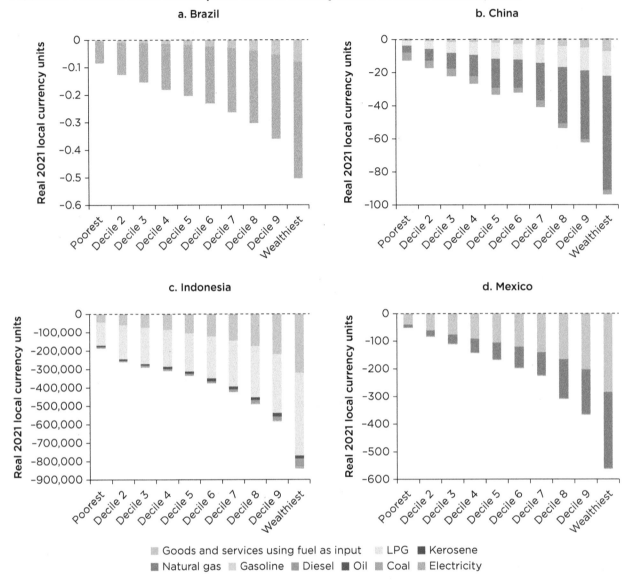

FIGURE 5.1 **Absolute mean consumption effect of subsidy reform in select countries, 2030**

a. Brazil

b. China

c. Indonesia

d. Mexico

Goods and services using fuel as input ▪ LPG ▪ Kerosene ▪ Natural gas ▪ Gasoline ▪ Diesel ▪ Oil ▪ Coal ▪ Electricity

Source: Klaiber, Rentschler, and Dorband 2022, based on the World Bank CPAT (Climate Policy Assessment Tool).
Note: LPG = liquefied petroleum gas.

subsidies for each of the countries by income decile and type of fuel. The consumption effect captures the change in disposable income in local currency units associated with the increase in price as a result of phasing out fossil fuel subsidies.

In all four countries, the overall effect on consumption of phasing out fossil fuel subsidies is negative. This finding is representative of all 19 countries analyzed. As the subsidies are removed, the unit price increases for energy goods and for goods that use energy as an input. The absolute size of the effect measured in local currency units increases by income decile—for richer households with higher baseline spending, the absolute monetary loss is also higher. In the status quo, the richer income deciles profit most from fossil fuel subsidies and lose most in absolute terms when they are phased out.

The consumption effects resulting from fossil fuel subsidy reform can be both direct and indirect. When subsidies are phased out, the price of fossil fuel increases, making consumption of fossil fuels more expensive, which has an immediate effect on energy costs—for instance, as households purchase gas for heating and cooking or gasoline for transportation. However, there is also a substantial indirect effect, as unit costs also increase for all goods and services that use fossil fuels as an input. These indirect supply chain effects are often much larger than the direct effects and affect a wide variety of consumption goods, as energy is an input to almost everything that is consumed in an economy. However, the way in which these effects play out depends crucially on which types of energy are subsidized and who consumes them. If subsidies target the consumers of industrial energy (that is, firms), then large indirect effects can be expected. If subsidies target the types of energy consumed by end users, then direct effects may dominate.

Hence, in some cases, indirect effects due to subsidy removal can outweigh direct ones. New results suggest that, in Mexico, the strongest effect on people's consumption is due to a change in the price of goods and services that use fuel as an input rather than the direct effects of higher energy prices. In Brazil, in contrast, price changes for electricity consumed by households have the strongest effect on mean consumption. In Indonesia, two-thirds of the negative consumption effect is driven by price changes for LPG, although indirect price effects from energy-intensive goods and services are also significant. In China, the largest impact on mean consumption stems from higher prices for natural gas, while the indirect effects on consumption are relatively limited. This situation may occur when the subsidized energy good is used primarily by households, rather than firms, so that the price increase is not passed along the value chain. Overall, the effect, size, and drivers of the consumption effect depend on a variety of factors, such as whether fuels were subsidized before the reform, the availability of energy sources, price elasticities, consumption patterns, and sectoral structure of the economy. Therefore, the overall size and composition of the mean consumption effect differ between countries.

Energy consumption differs not only across income groups, but also across the urban-rural divide. In many countries, direct and indirect energy consumption by urban households is significantly higher than energy consumption by rural ones, even at the same income level. This urban-rural divide reflects different patterns of consumption of energy-intensive goods and services, such as a higher dependence on urban commuting. Consequently, urban populations may benefit more from fossil fuel subsidy programs than rural populations (Rentschler 2016). Likewise, urban populations stand to lose more in monetary terms when subsidies are phased out. This finding is consistent with the observations in some studies that public opposition to the removal of energy subsidies has been particularly fierce in more prosperous urban areas, such as the large-scale protests in Abuja and Lagos in response to Nigeria's 2012 announced plan to remove fuel subsidies (Rentschler 2016).

Direct and indirect effects: Subsidy reforms affect people directly through changes in the price of fuel, and indirectly through changes in the price of goods and services that use fuels as inputs.

Urban-rural divide: Urban populations are more affected by fossil fuel subsidy reforms. The urban-rural divide of the distributional effects is significantly larger for richer deciles.

The urban-rural divide can be illustrated by disaggregating the simulated impacts of energy subsidy reforms on consumption. Figure 5.2 shows the absolute mean consumption effect, measured in local currency units, associated with the full removal of energy subsidies. It shows that the mean consumption effect is consistently larger for urban populations than for rural ones; this pattern holds for all income groups and all four countries. The differences between urban and rural populations are quite small for the poorer income deciles but grow in size and become substantial for the richer ones. This finding

FIGURE 5.2 **Absolute mean consumption effect of subsidy reform, urban versus rural, in select countries, 2030**

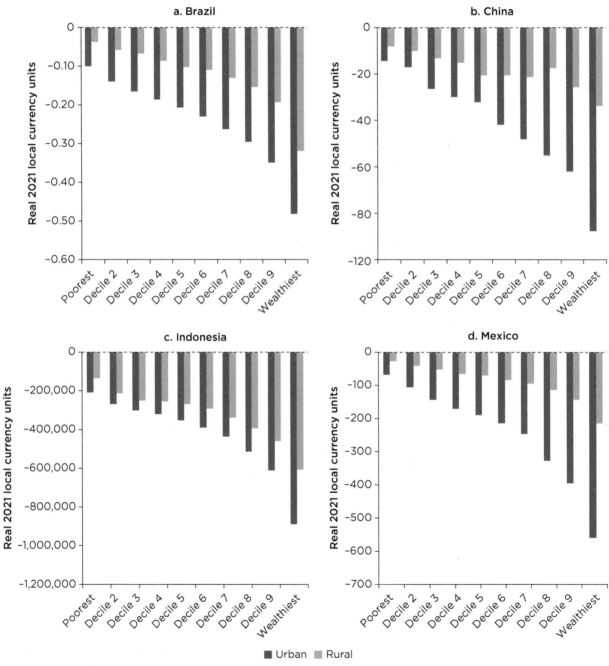

Source: Klaiber, Rentschler, and Dorband 2022, based on the World Bank Climate Policy Assessment Tool (CPAT).

underlines the findings in figure 5.1: richer income deciles profit the most from fossil fuel subsidies. Yet the degree of the urban-rural divide differs across countries—for instance, it is more pronounced in Brazil than in Indonesia (figure 5.2).

While energy subsidy reforms are almost always progressive in absolute terms, the degree of progressivity can vary significantly. This variation is illustrated by the results of subsidy reform simulations for 19 countries for which the distributional consumption impacts are assessed and mean consumption effects on the richest and the poorest income deciles are compared. Figure 5.3 summarizes these results. The bars show the ratio of consumption effects between the richest and poorest income deciles. For instance, households in the richest income decile in Rwanda are estimated to lose 80 times more in absolute monetary terms than households in the poorest decile, reflecting the fact that Rwanda's poorest households, often in rural agricultural communities, consume very limited amounts of energy. On average, across all 19 countries, the mean consumption effect is 12.6 times larger for the richest income decile than for the poorest.

12.6: On average, people in the richest income decile lose 12.6 times more from the removal of subsidies than the poorest.

FIGURE 5.3 Ratio of mean consumption effects of subsidy reform for the richest to the poorest income deciles in select countries

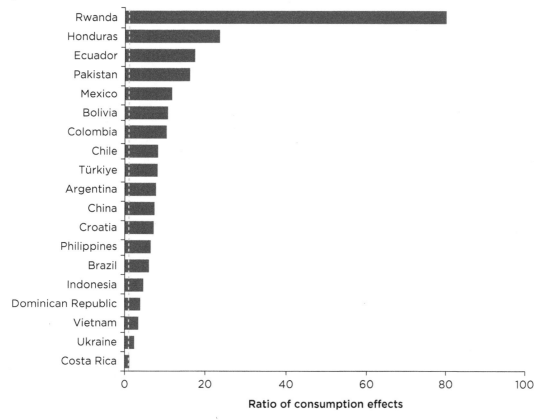

Source: Klaiber, Rentschler, and Dorband 2022, based on the World Bank CPAT (Climate Policy Assessment Tool).
Note: The dashed vertical line indicates where the impact ratio between the richest and poorest is equal to 1—that is, where the mean changes in consumption (measured in absolute terms) resulting from the simulated subsidy reform are equal for the richest and poorest income deciles.

These results confirm that the regressive nature of fossil fuel subsidies is consistent across countries. In figure 5.3 the dashed vertical line indicates where the impact ratio between the richest and poorest is equal to 1—that is, where the mean changes in consumption (measured in absolute terms) resulting from the simulated subsidy reform are equal for the richest and poorest income deciles. Only in Costa Rica is the absolute monetary effect of subsidy reform estimated to be nearly equal for the richest and poorest income deciles. For 14 out of 19 countries, the ratio of consumption losses is greater than 5. These results confirm once again that, in virtually all countries considered, the richest income decile benefits the most from fossil fuel subsidies. Yet the results also confirm that the degree of the distributive consumption effect varies widely between countries and depends on a multitude of local economic factors.

Effect of reform on energy consumption as a share of income

Relative to income levels, the progressiveness of subsidy reforms is less pronounced than in absolute monetary terms. Common convention holds that poorer households may incur a larger loss from subsidy reform, when considered as a share of income, but the data offer mixed evidence. There is no doubt that, on average, energy consumption constitutes a significant share of household spending across all income groups in virtually all countries. Some country case studies suggest that subsidy removal may hit poor households the hardest because they spend a larger share of their income on energy than rich households (see, for instance, Arze del Granado, Coady, and Gillingham 2012; IEA et al. 2010; Ruggeri Laderchi, Olivier, and Trimble 2013; World Bank 2010). This section systematically examines the change in disposable income associated with fossil fuel subsidy removal as a share of total income.

As the previous sections show, the absolute monetary loss of consumption due to subsidy removal is usually far larger for rich households than for poor ones. But when examined in relation to income, this difference disappears in most of the 19 countries considered. As a share of total spending, energy consumption tends to be of similar size across income groups in the countries covered in this analysis. As a consequence, the consumption impacts of subsidy removal represent a similar share of people's income, regardless of income level. Estimated relative consumption effects for four sample countries are presented in figure 5.4; a summary for all 19 country simulations is presented in figure 5.5. The results suggest that—as a share of income—the consumption effect associated with fuel subsidy reform is similar for households in the poorest and richest income deciles.

In Mexico, for instance, the relative consumption effect is similar for the richest and the poorest income deciles. In Brazil, China, and Indonesia, the relative mean consumption effect is larger for the lowest income decile than for the highest, implying that in these countries the richest income deciles are more affected in absolute terms, but poor people are more affected in relative terms. The reason is that the poorer income deciles spend a larger part of their income on energy and energy-intensive goods and are therefore relatively more affected. Just like for absolute consumption effects, the indirect price effects on goods and services using fuel as an input can be a significant part of people's consumption losses, as seen in Indonesia and Mexico.

Overall, the size of effect depends heavily on the consumption patterns in a country, the energy intensity of consumption, and the size and nature of the subsidization program. For example, in Indonesia, fossil fuel subsidies contribute between 1.2 percent and 1.4 percent to people's consumption expenditures. In China, the size of effect is much smaller, at only 0.08 percent to 0.019 percent of consumption expenditure. Out of

FIGURE 5.4 **Relative mean consumption effect of subsidy reform in select countries, 2030**

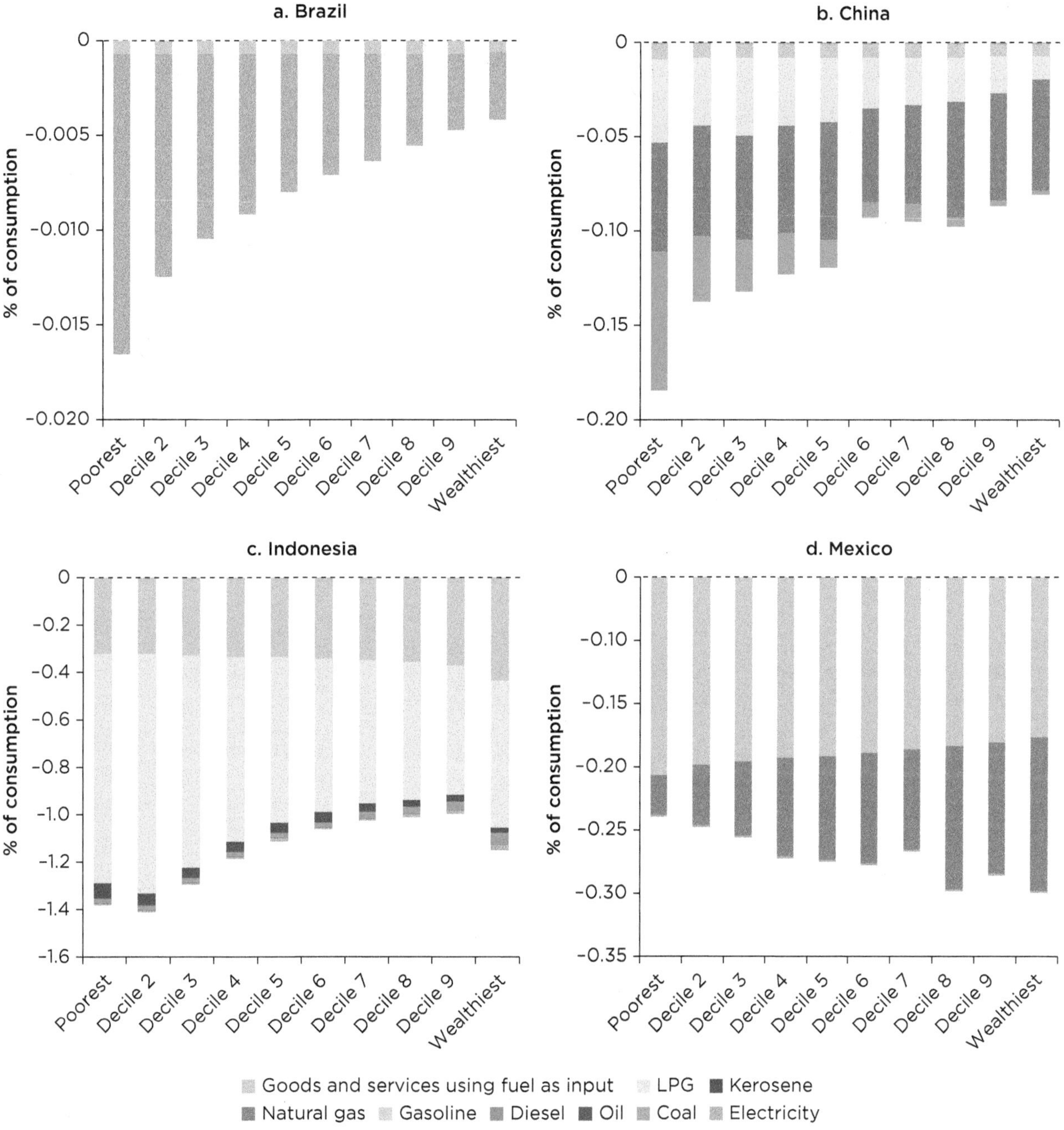

Source: Klaiber, Rentschler, and Dorband 2022, based on the World Bank CPAT (Climate Policy Assessment Tool).
Note: LPG = liquefied petroleum gas.

the four countries, results suggest that the Indonesian population would be affected the most by fossil fuel subsidy reform. In the wider sample of 19 countries, the relative size of the mean consumption effect across income deciles is smaller than 1 percent for most countries (12 out of 19). In these countries, the price effects due to subsidy reform cause less than a 1 percent loss of disposable income. Therefore, the relative cost of fossil fuel subsidy reforms for individuals can be quite small.

1.1: Relative to their income, people in the richest income decile only lose 1.1 times more from subsidy removal than the poorest.

Relative to income, the consumption effects of subsidy reform are similar for the poorest and richest households, although there are exceptions. The consumption effect relative to income levels is estimated for the same sample of 19 countries. Figure 5.5 displays the ratio of the relative mean consumption effects of the richest to the poorest income deciles. Three rough categories of countries are evident: countries where the richest income decile is affected more strongly in relative terms than the poorest, countries where the two groups are affected roughly equally, and countries where the poorest income decile is affected more than the richest. In this sample, 7 out of 19 countries fall into the second group, where the relative consumption effect is similar for the poorest and richest deciles (that is, the ratio is between 0.8 and 1.2). Five countries are in the first category, and seven are in the third, where the poorest income decile loses more relative to their income.

Overall, these estimates do not confirm the commonly held notion that poor people always spend a larger share of their income on energy than richer people—and thus incur larger relative losses from subsidy removal. In this sample of countries, energy expenditure tends to account for a similar share of total expenditure across income groups. In the majority of sample countries, the average relative impact of fossil fuel subsidy reform tends to be similar across income groups. However, there are notable exceptions, which

FIGURE 5.5 Ratio of the relative mean consumption effect of subsidy reform for the richest and poorest income deciles and the difference between them in select countries

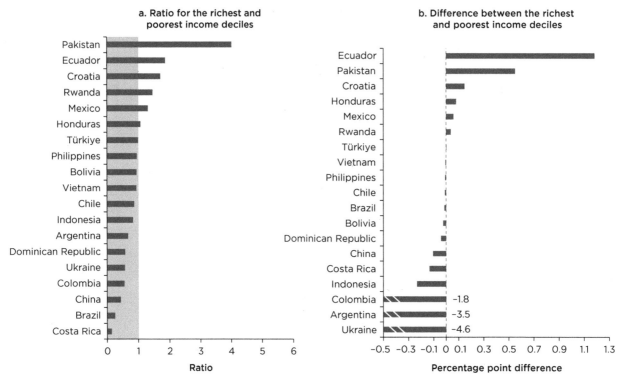

Source: Klaiber, Rentschler, and Dorband 2022, based on the World Bank CPAT (Climate Policy Assessment Tool).
Note: Ratio represents the consumption loss incurred by the richest income decile divided by the loss incurred by the poorest. The shaded area in panel a represents countries where the poorest income decile loses more relative to their income than the richest decile.

Detox Development

are crucial to consider for the design of effective reform strategies with well-planned compensation and social protection components.

Fossil fuel subsidy reforms save lives

Insofar as subsidy reform reduces the consumption of harmful fossil fuels, health outcomes will improve. However, the air pollution and health benefits of fossil fuel subsidy reform can differ substantially between countries and need to be assessed carefully (Enriquez, Larsen, and Sanchez-Triyana 2019). A fossil fuel subsidy reform does not automatically yield large environmental and health benefits in every case. Several factors are crucial in determining the extent to which energy subsidy reforms can indeed induce the behavioral and technological change required to reduce the consumption of polluting fuels. These factors include, among others:

- *The magnitude of explicit fossil fuel subsidies.* Depending on the primary policy objective for which fossil fuel subsidy schemes were introduced in the first place, the magnitude of subsidies can differ substantially (chapter 3). For instance, resource-rich fossil fuel–exporting low- and middle-income countries tend to have particularly large consumption subsidy programs. When substantial shares of the public budget are spent to lower fossil fuel prices artificially, they are likely to induce substantial overconsumption and thus pollution. In contrast, other countries maintain relatively small and targeted subsidy programs that have a limited impact on overall levels of energy consumption. Naturally, the pollution and health benefits of subsidy removal depend substantially on the size of subsidies to begin with and the number of people who experience adverse health effects.

- *The type of subsidized fuels and their relative prices.* In addition, the pollution and health benefits of subsidy removal are prone to be larger in countries that subsidize highly polluting types of fuel or industries, such as coal for the power sector or gasoline for transportation. In contrast, when existing subsidy programs target fossil fuels with relatively low air pollution footprints, such as LPG, the pollution and health benefits of subsidy removal are naturally more limited. In fact, subsidy removal for cleaner fuels such as LPG can even result in heightened air pollution. For instance, when subsidy removal increases the relative price of LPG as a clean cooking fuel, low-income households may switch to less expensive but more polluting alternatives such as charcoal or kerosene. Similarly, the removal of natural gas subsidies in the power sector may make coal more competitive, especially if it continues to be subsidized. Relative prices between more and less polluting types of fuel are crucial determinants of the fuel-switching responses to subsidy removal (Rentschler and Kornejew 2017).

- *The responsiveness of energy consumption to prices.* In theory, an increase in the unit cost of fuels would cause consumers to use less. In practice, consumers face a wide range of constraints, including constraints related to finance, information, technology, capacity, or behavioral biases. These constraints mean that consumers may be unable or unwilling to adjust their consumption in response to price changes. Especially in the short term, fuel consumption can be fairly unresponsive to price changes, as noted in chapter 3. In the longer term, people may choose to move closer to their workplace to reduce their commute time or to benefit from better public transit infrastructure. In addition, an emerging literature describes how behavioral biases such as habit and status quo can influence consumer decisions.

This section presents estimates from simulations of the air pollution and health benefits of fossil fuel subsidy reform in 25 countries. To illustrate the mechanisms of subsidy reforms, it presents more detailed results for 4 of the 25 countries: Algeria, China, Indonesia, and the Islamic Republic of Iran. Since the overall country effects of phasing out fossil fuel subsidies depend on the existing policies, the types of subsidies in these four countries are discussed briefly. This issue is important because the resulting improvements in air quality and health depend on substitution effects, which differ between countries depending on which fuels are subsidized and country-specific fuel price elasticities (chapter 3). These countries are selected because they have large fossil fuel subsidy programs and face substantial air pollution challenges. While all four countries have high levels of fossil fuel subsidies, the types of subsidized fuels differ across countries, providing interesting insights into the pollution and health effects of fossil fuels (box 5.2).

BOX 5.2

No two subsidy schemes are the same: Subsidies in Algeria, China, Indonesia, and the Islamic Republic of Iran

Algeria maintains one of the largest fossil fuel subsidy programs worldwide. Explicit subsidies on fossil fuels in 2020, according to the International Monetary Fund (IMF 2022), were US$11.54 billion, equivalent to 8 percent of Algeria's GDP. The subsidies were distributed among petroleum and other oil products (US$6.24 billion), natural gas (US$3.13 billion), and electricity (US$2.17 billion). All subsidies were explicit subsidies to consumers rather than to producers. While the magnitude of fossil fuel subsidies is high, Algeria subsidizes comparatively cleaner fuels and does not subsidize coal at all.

China maintains the sixth-largest fossil fuel subsidy program in the world. In 2020 explicit subsidies amounted to US$15.73 billion (IMF 2022), which was equivalent to only about 0.1 percent of China's GDP. The subsidies were distributed among electricity (US$13.69 billion) in the form of a consumer subsidy, natural gas (US$1.34 billion) as both a consumer and a producer subsidy, and coal (US$0.37 billion) and petroleum and other oil products (US$0.33 billion) as producer subsidies. Because coal made up 64.1 percent of China's generated electricity in 2020 (IEA 2022), electricity subsidies and coal subsidies are difficult to disentangle.

Indonesia had an overall fossil fuel subsidy program amounting to US$11.96 billion in 2020 or roughly 1.1 percent of GDP (IMF 2022). The subsidies were distributed among electricity (US$5.49 billion), petroleum and other oil products (US$3.44 billion), coal (US$2.85 billion), and natural gas (US$0.17 billion). The majority of fossil fuel subsidies in Indonesia are implemented as explicit consumer subsidies, with the exception of natural gas and US$0.13 billion for petroleum, which are explicit producer subsidies. Overall, Indonesia has a relatively high level of direct subsidies for coal compared to the other three countries analyzed.

The Islamic Republic of Iran paid 22 percent of its GDP, or US$41.72 billion in 2020, to subsidize fossil fuels—more than any other country, according to IMF (2022). The subsidies are distributed among electricity (US$26.51 billion) and petroleum and other oil products (US$15.21 billion); of these, US$10.44 billion are consumer subsidies for diesel. Within this four-country sample, the Islamic Republic of Iran has the highest level of subsidies, both in absolute and in relative terms to its GDP. However, the government does not subsidize coal and hardly any coal is used to generate electricity, which means that there are no indirect coal subsidies through the electricity market.

Changing consumption

The fossil fuel subsidy reform simulations presented here translate the removal of subsidy programs into changes in fuel prices. Using estimates of fuel price elasticities from the literature and data on supply chain relationships (input-output tables), changes in fuel consumption are estimated. These changes are then translated into changes in sectoral emissions of key air pollutants. Figure 5.6 summarizes the resulting changes in air pollution emissions and concentrations for different sectors and pollutants.

Figure 5.6 summarizes the estimated change in the level of particle emissions of different pollutants following a simulated subsidy reform in 2021. The estimates suggest that particle pollution declines in all four countries as a result of the fossil fuel subsidy reforms. Which specific pollutants are reduced and by how much depend on the sectoral composition of the different economies and their main sources of pollution. For instance, in many cases, the transport sector is a key contributor to air pollution, driven by subsidized diesel and gasoline.

The strongest reductions in particle emissions are estimated to take place in Algeria and the Islamic Republic of Iran, given the large magnitude of their subsidy programs. In Iran, black carbon emissions decline by 43 percent as a result of fossil fuel reform. Similar effects can be seen for $PM_{2.5}$. In Algeria, levels of black carbon and $PM_{2.5}$ decline by about 35 percent compared to 2019. In China and Indonesia, the reduction is less drastic, but still significant. In China, a fossil fuel subsidy reform can help to reduce all air pollutants by up to 4 percent, while in Indonesia sulfur dioxide (SO_2) emissions are estimated to decrease by 14 percent and nitrogen oxide (NO_x) emissions by 6 percent. Fossil fuel reforms in Algeria and the Islamic Republic of Iran could be particularly effective in curbing air pollution and reducing the substantial health burden associated with poor air quality (WHO 2021), but less so in the other two countries. Box 5.3 discusses the effects of subsidy removal on greenhouse gases.

FIGURE 5.6 Simulated changes in SO_2, NO_x, and $PM_{2.5}$ emissions following subsidy reform across countries, 2021–35

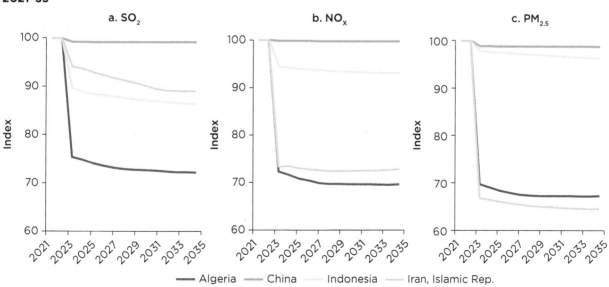

Source: Klaiber, Rentschler, and Dorband 2022, based on the World Bank CPAT (Climate Policy Assessment Tool).
Note: The vertical axis represents a standardized index of total air pollutant emissions. NO_x = nitrogen oxide; $PM_{2.5}$ = fine particulate matter; SO_2 = sulfur dioxide.

Fossil fuel subsidy reform contributes to reducing greenhouse gases

In a widely noted study, Jewell et al. (2018) estimate that removing global fuel subsidies would reduce carbon dioxide (CO_2) between 0.5 gigatons and 2 gigatons by 2030, which is equivalent to a 1–4 percent net reduction in global greenhouse gases. While this effect may appear to be limited at first, a few issues are noteworthy:

- *CO_2 reductions in perspective.* CO_2 reductions of between 0.5 gigatons to 2 gigatons amount to roughly one-quarter of the energy-related emissions reductions pledged by all countries under the Paris Agreement on Climate Change (that is, 4–8 gigatons of CO_2; Erickson et al. 2020). A single policy measure that could reduce energy-related emissions by a quarter is, in fact, remarkably effective.
- *Hidden subsidies.* Global analyses almost always underestimate the size and reach of fossil fuel subsidy schemes. While explicit consumer subsidies are often well identified, measures (both monetary and in-kind) to support fossil fuel producers and energy-intensive industries are difficult to capture.
- *Time scales.* Considering the near-term impacts of subsidy removal (for example, within 10 years, up to 2030) is useful, especially given the urgency of reducing CO_2 to stay within the warming targets of the Paris Agreement. However, removing fossil fuel subsidies changes the incentives for long-term investment and planning decisions—for example, in decarbonized power and transport systems—that are unlikely to have fully materialized within 10 years of the reform.
- *Additional steps.* The removal of explicit (that is, active) fossil fuel subsidies is just a first step toward addressing the underpricing of fossil fuels. Explicit subsidies are dwarfed by the magnitude of the social costs of fossil fuels. For this reason, removing explicit subsidies alone will not fix climate change or air pollution. In addition to reforming explicit subsidy programs, the wider societal costs of fossil fuel consumption need to be reflected in fossil fuel prices and be supported by complementary policies.

The nature of the subsidy program and the sectoral composition of pollution sources determine the air pollution benefits of subsidy reform. Different fuels are used by different users and sectors. Thus it is possible to estimate which sectors are responsible for the main reductions in air pollution resulting from subsidy removal. Figure 5.7 summarizes the sectoral contributions to estimated changes in the concentration of ambient particulate matter compared to a scenario of no reform.

The largest differences are evident in Algeria, where annual average concentrations of ambient $PM_{2.5}$ could be reduced by 3 micrograms per cubic meter ($\mu g/m^3$)—in large part from changes in the industrial sector and the road transport sector. In the Islamic Republic of Iran, the changes are slightly smaller, at roughly 2 $\mu g/m^3$, with the largest effect coming from the transport sector. In China and Indonesia, the effects of phasing out all fossil fuel subsidies translate roughly to a reduction in ambient $PM_{2.5}$ concentrations of 0.2 $\mu g/m^3$ and 0.5 $\mu g/m^3$, respectively. The biggest part of these changes results from the residential, services, and construction sector in China and the transport sector in Indonesia. These national average reductions may appear small, but they are likely to be concentrated locally (for example, in the proximity of major intersections or industrial plants) in areas where the reduction in pollution is likely to be far more substantial.

FIGURE 5.7 Reduction of PM$_{2.5}$ concentrations following subsidy reform in select countries, by source of pollution, 2020–35 (projected)

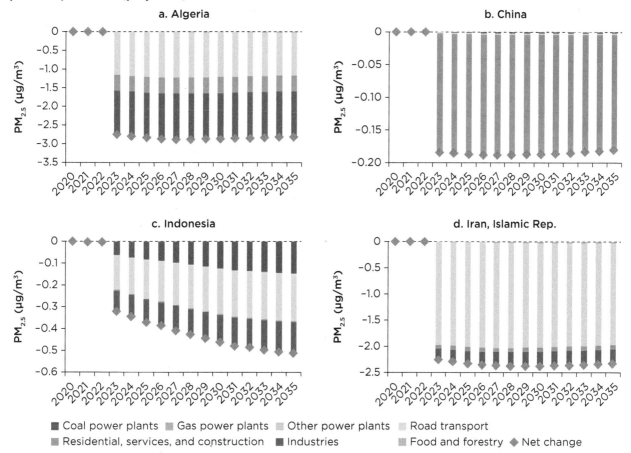

Source: Klaiber, Rentschler, and Dorband 2022, based on the World Bank CPAT (Climate Policy Assessment Tool).
Note: Changes in PM$_{2.5}$ concentrations are computed on the basis of emissions reductions (figure 5.6) using TM5-FASST, a global atmospheric source–receptor model for rapid impact analysis of emission changes on air quality and short-lived climate pollutants, as contained in CPAT (Van Dingenen et al. 2018, box 5.1). PM$_{2.5}$ = fine particulate matter; µg/m³ = micrograms per cubic meter.

Reducing adverse health effects

The relationship between particle concentrations of different pollutants and the incidence of different diseases has been documented extensively in the medical literature. Air pollution causes many severe health conditions, depending on the concentration of particulate matter and its chemical composition. For instance, heightened PM$_{2.5}$ exposure has been shown to increase the risk of chronic obstructive pulmonary disease, ischemic heart disease, lung cancer, and stroke, as well as chronic and acute respiratory diseases such as asthma; it is especially dangerous for vulnerable populations like children and the elderly (Cohen et al. 2017; WHO 2018). Besides cardiovascular and respiratory diseases, growing evidence exists on the role of PM$_{2.5}$ exposure in increasing the risk of type 2 diabetes and neurological diseases such as Alzheimer's (GBD 2019 Risk Factors Collaborators 2020; Peters et al. 2019).

By reducing fossil fuel consumption and the associated particle emissions, fossil fuel subsidies can contribute directly to reducing a country's deaths from air pollution. Yet lowering consumption should be considered as only a first step toward wider actions to reduce air pollution. Based on the estimated reductions in particle emissions and ambient concentrations, it is possible to distinguish the health impacts associated with different air pollutants and their originating sectors—and thus reconstruct which health conditions are associated with air pollution from which sector. The estimated reduction in air pollution deaths differs according to the sectoral structure of economies and the nature of the subsidy program. Figure 5.8 summarizes the estimated cumulative number of deaths that could be avoided in four countries if fossil fuel subsidies were phased out in 2021. It distinguishes deaths that are associated with different economic sectors according to their respective contribution to air pollution levels.

Figure 5.9 shows the number of deaths averted as a result of immediately phasing out fossil fuels for a larger sample of countries in 2036. The overall number of aggregate deaths averted between 2022 and 2035 in the 25 countries is 360,000 people (figure 5.9, panel a). This number represents the marginal contribution that fossil fuel subsidy reforms can make to the wider efforts to improve air quality. The analysis suggests that fossil fuel subsidies and the resulting increase in consumption of fossil fuels are major contributors to health impacts and deaths related to ambient air pollution. The estimates show that fossil fuel reforms can have a substantial and lasting positive impact on health and society. In 3 of the 25 countries, fossil fuel subsidies contribute to more than 5 percent of aggregate premature deaths caused by air pollution (figure 5.9, panel b).

FIGURE 5.8 **Cumulative deaths averted with subsidies phased out in select countries, 2020–35 (projected)**

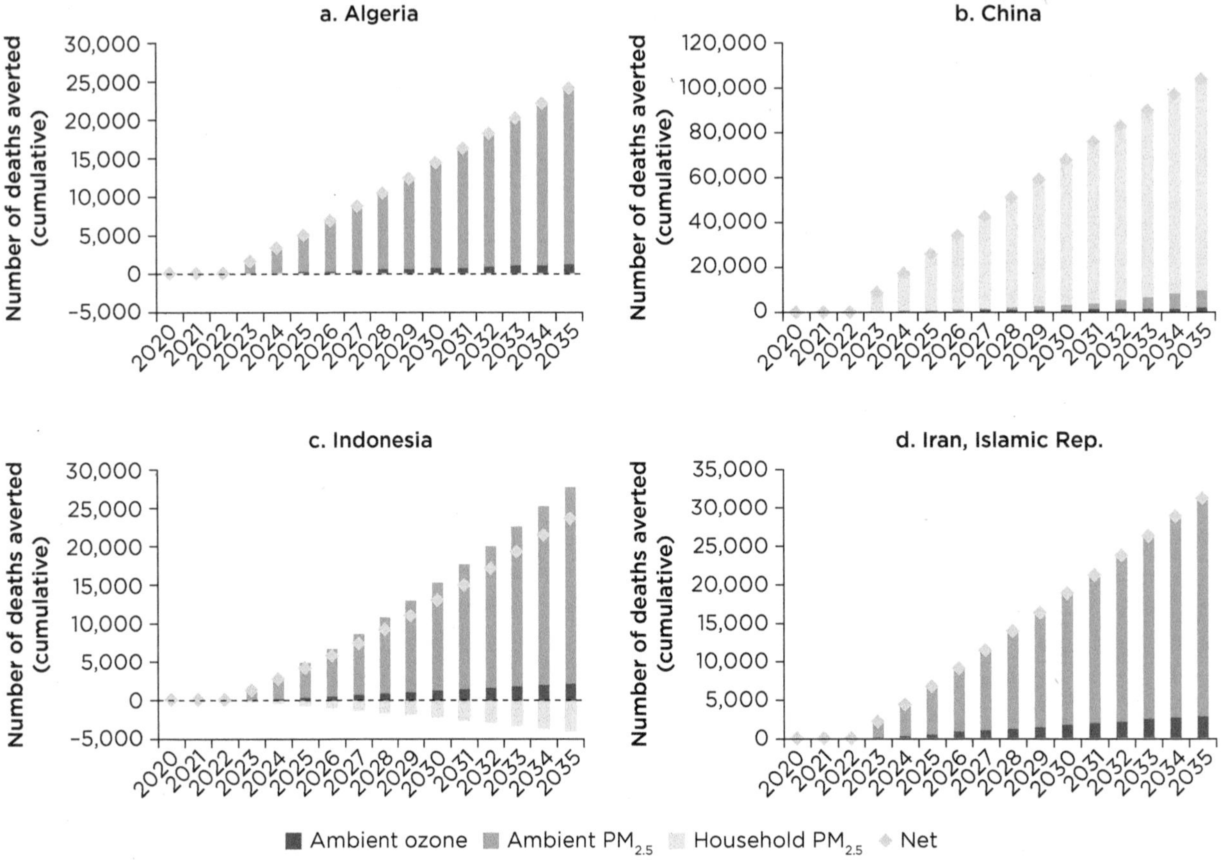

Source: Klaiber, Rentschler, and Dorband 2022, based on the World Bank CPAT (Climate Policy Assessment Tool).
Note: Pollutant concentrations are converted to mortality estimates based on concentrations-response function as part of the CPAT model (box 5.1). Air pollution deaths may increase as a result of fossil fuel subsidy reform if relative prices shift in favor of more polluting fuels. $PM_{2.5}$ = fine particulate matter.

FIGURE 5.9 Cumulative and aggregate deaths averted as a result of the removal of explicit subsidies relative to all deaths caused by air pollution in select countries, 2022–35 (projected)

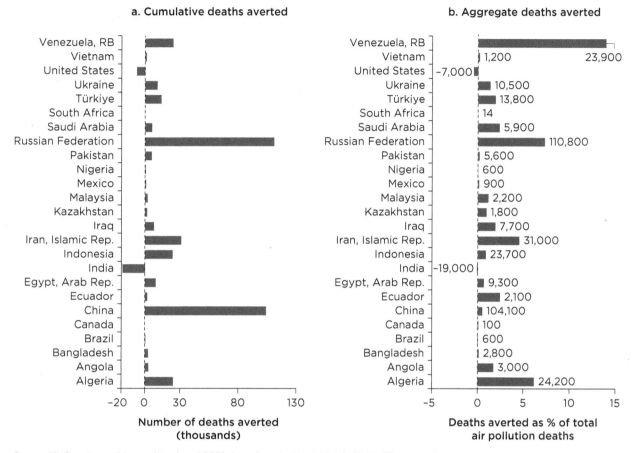

Source: Klaiber, Rentschler, and Dorband 2022, based on the World Bank CPAT (Climate Policy Assessment Tool).
Note: Data labels on panel b refer to aggregate number of deaths averted between 2022 and 2035. Air pollution deaths may increase as a result of fossil fuel subsidy reform if relative prices shift in favor of more polluting fuels.

Among the four countries analyzed more closely, the largest health benefits are estimated to be in China, where roughly 104,000 air pollution deaths could be avoided within 13 years of a fossil fuel subsidies reform. In the Islamic Republic of Iran, around 31,000 deaths could be avoided in the same time frame, compared with about 24,000 in both Algeria and Indonesia. Air pollution–related deaths may, in fact, increase from some sources. For instance, in Indonesia the removal of LPG subsidies could cause low-income households to switch to less expensive, more polluting cooking fuels, such as kerosene and charcoal. In most cases, the reduction in deaths is driven most strongly by the reduction in ambient $PM_{2.5}$ levels, which has been documented to be responsible for about 62 percent of all air pollution deaths worldwide in 2019 (Health Effects Institute 2020).

Subsidy reform simulations for a sample of 25 countries highlight the cases in which subsidy reforms can have particularly large health benefits (figure 5.9). In the Russian Federation, the estimated number of deaths avoided is roughly 111,000 within 13 years of the subsidy reform; this is followed by China, with roughly 104,000. In the case of Russia, the size can be explained by the high level of coal subsidies, which engender toxic SO_x and $PM_{2.5}$ emissions in particular. In both China and Russia, fossil fuel subsidy reforms are estimated to reduce coal consumption and thus improve air quality. Overall, the number of deaths avoided as a result of fossil fuel subsidy reforms is driven by several factors: the size of the overall population exposed to hazardous air pollution, high subsidization rates for the most polluting types of fuel (coal in particular), and avoidance of switching toward more polluting fuels.

In high-income countries, the removal of fossil fuel subsidies may not yield the same high reductions in air pollution deaths, as air pollution regulations tend to be more stringent—for example, for the industry and construction sectors. For instance, reductions in fossil fuel consumption may not reduce particle concentrations significantly if advanced air filtration systems are already reducing the pollution intensity of consumption. Moreover, high-income countries typically do not maintain large consumer fuel subsidy programs—and may be underrepresented. For instance, Canada and the United States have large implicit producer subsidy schemes, which are difficult to quantify and tend to be underreported in global databases such as the ones used here (IMF 2022).

> Air pollution deaths can rise as a consequence of fossil fuel subsidy reform—if relative prices shift in favor of the most polluting fuels.

Air pollution deaths can rise as a consequence of fossil fuel subsidy reform—if relative prices shift in favor of the most polluting fuels. In the 25 countries for which simulations of fossil fuel subsidy reform were conducted for this report, two countries are estimated to experience a net increase in air pollution deaths if no additional measures are taken—India and the United States. This stark illustration shows that the removal of explicit subsidies is not a panacea for reducing air pollution. In both countries, this situation is caused by an increase in coal consumption as a substitute for natural gas.

For India, the estimated effect could be due to a lack of detailed data on coal subsidies, which is likely to lead to an underestimation of the relative price of coal after the fossil fuel subsidy reform. The removal of natural gas subsidies could especially harm lower-income households, who may be pushed to switch to more polluting biofuels for residential uses, such as cooking and heating. In the United States, the effect is caused by increased use of the existing capacities of coal-fired power generation. In both countries, the available data suggest that subsidies are higher for natural gas than for coal; the phasing out of fossil fuel subsidies leads to the substitution away from natural gas and toward coal, which increases ambient air pollution. Accordingly, the pollution levels of $PM_{2.5}$, SO_2, black carbon, and NO_x increase, thus resulting in estimated increases in mortality rates. These hypothetical examples serve as reminders that the design of fossil fuel subsidy reforms needs to account for the possibility of fuel-switching effects—for example, by facilitating the transition to less polluting types of fuel.

Besides mortality and morbidity effects, air pollution also lowers the cognitive ability and productivity of affected populations. Empirical evidence from brain-training experiments shows that $PM_{2.5}$ exposure impairs the cognitive abilities of adults, while these effects are largest for persons in prime working age and persons with low ability (La Nauze and Severnini 2021). There is even evidence that short-term exposure to air pollution negatively affects the performance of highly skilled workers—professional baseball umpires have been documented to make more incorrect calls when exposed to high levels of carbon monoxide (Archsmith, Heyes, and Saberian 2018). Such evidence on the relationship between air quality and brain health explains the pernicious and hidden impacts on productivity that exacerbate inequalities (Peeples 2020).

Tackling air pollution, tackling vulnerabilities

The estimates provided in these chapters on energy subsidies and air pollution reinforce the case for implementing targeted measures to reduce the pollution intensity of economic growth—for instance, supporting the uptake of less polluting technologies in industry and infrastructure or facilitating the transition toward cleaner fuels (in particular, electrification). In addition, measures are warranted to address the disproportionate exposure of poor people to pollution (box 4.4). Expanding the provision of affordable and adequate health care in large urban centers in low- and middle-income countries can help to reduce mortality, bringing it closer to the levels experienced in higher-income countries. Mandating transparent accounting for environmental and health externalities in planning decisions can help to steer pollution sources (for example, industrial zones or power plants) away from low-income communities. Finally, removing incentives that perpetuate the overconsumption of fossil fuels can yield a double dividend for poor people, improving their lives and livelihoods.

BOX 5.4

Health benefits of climate change mitigation policies

Reducing carbon emissions cleans up the local air. Subsidies for polluting fossil fuels incentivize overconsumption and entrench inefficient and polluting practices. While removing subsidies can address such adverse effects to some extent, active taxation is considered to be the most efficient way of fully addressing the negative externalities of underpriced fossil fuels. With the momentum building for climate change mitigation policies, carbon emissions have become a key focus of such externality taxes. By imposing an active tax on carbon emissions, the price of fossil fuels can reflect the societal costs of climate change, and consumers can adjust their demand in line with the price signal.

However, carbon taxes have wider societal benefits that go beyond the mitigation of climate change. Many of the behavioral and technological shifts associated with higher fossil fuel prices—for example, transition to renewable energy or enhanced public transport systems—have immediate local health benefits, as they reduce not just carbon emissions, but also harmful local air pollutants that cause 4.5 million deaths worldwide (IHME 2020). Such outcomes have been documented in China (Wei et al. 2020), Eastern Europe (Taheripour et al. 2022), the European Union (Chen et al. 2020; Vandyck et al. 2018), and the United States (Saari et al. 2015) and more widely for transitions in electricity generation (for example, Tong et al. 2021). Yet what is the global scale of these health benefits, and by how much could carbon taxes reduce local air pollutants?

Models of the global economy and climate shed light on the health benefits of climate action. Computable general equilibrium (CGE) models are used widely for simulating a wide range of policy measures and scenarios. The Environmental Impact and Sustainability Applied General Equilibrium (ENVISAGE) model is a state-of-the-art multiregion, multisector CGE model that assesses the interplay between the global economy and the environment (van der Mensbrugghe 2019). Taheripour et al. (2022) use this model to

(Continued)

BOX 5.4
Health benefits of climate change mitigation policies *(continued)*

examine country-level health benefits that can be achieved by reducing greenhouse gas emissions through a carbon tax. Their analysis establishes a soft link between the ENVISAGE model and the TM5-FASST global atmospheric source-receptor model (Van Dingenen et al. 2018). It is calibrated to the Global Trade Analysis Project 10A Power Data Base, with 2014 aggregated to 53 countries and regions and 19 sectors.

To provide a sense of the health implications of meeting global climate objectives, Taheripour et al. (2022) consider two policy scenarios that correspond to the carbon prices deemed necessary to reach the 2°C and the 1.5°C global-warming targets. The former is captured through a global carbon price of US$75 per ton of CO_2-equivalent (CO_2-eq) and the latter through a US$150 per ton of CO_2-eq imposed on the CO_2 emissions from fossil-fuel combusting activities, including households.[a] By using a carbon tax to assess the health impacts of emissions reductions, the study ensures that emissions abatement is distributed efficiently globally and minimizes distortionary responses within economies that may emerge with other policy instruments.

The study estimates the impacts of climate mitigation policies on changes in the level of greenhouse gas emissions and other pollutants in 53 countries and regions. These changes in air pollutant emissions are transferred to the TM5-FASST global atmospheric source-receptor model, which links greenhouse gas emissions and other air pollutants to six types of diseases identified in the TM5-FASST model: ischemic heart disease, chronic obstructive pulmonary disease, stroke, lung cancer, lower respiratory airway infections, and type 2 diabetes. In the final step, the study estimates changes in mortality rates and the health benefits of climate mitigation policies.

A carbon tax consistent with a 1.5°C warming goal could save around 360,000 lives per year. Figure B5.4.1, panel a, offers an overview of the estimated global average reductions of the most common air pollutants—in particular, sulfur dioxide (SO_2), carbon dioxide, particulate matter (PM_{10} and $PM_{2.5}$), carbon monoxide, nonmethane volatile organic compounds, black carbon, nitrogen oxides, organic carbon, methane, nitrous oxide, and ammonia. Estimates at the country level (figure B5.4.1, panel b) show that the reduction in air pollutants could be substantial, especially in heavily polluted regions. In China, SO_2 concentrations are estimated to drop by 41 percent as a consequence of a carbon tax of US$150 per ton of CO_2-eq. In 17 countries or regions, the reduction in SO_2 would be 60 percent or higher.

Such significant reductions in air pollution also translate to a reduction in the 4.5 million annual deaths that are caused by poor outdoor air quality (IHME 2020). A carbon tax of US$75 per ton of CO_2-eq results in an estimated 229,000 fewer deaths globally each year, as a consequence of improved air quality. For a tax of US$150 per ton of CO_2-eq, annual deaths would be reduced by about 363,000 worldwide. Figure B5.4.2 provides a country and regional breakdown of the distribution of these deaths averted. Countries with large populations exposed to heavy air pollution would experience the largest reductions in deaths, especially countries in East and South Asia as well as in Central and Eastern Europe.

(Continued)

Health benefits of climate change mitigation policies *(continued)*

FIGURE B5.4.1 Reduction of common air pollutants associated with carbon taxes of US$75 and US$150 per ton of CO₂-eq

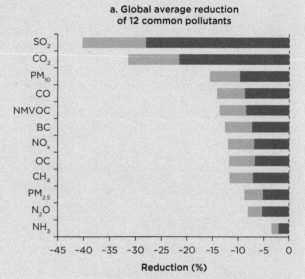

a. Global average reduction of 12 common pollutants

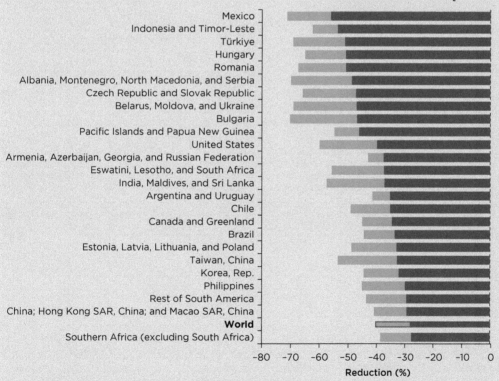

b. Country-level reduction of SO₂

■ Reduction at US$75 per ton of CO₂-eq   ■ Additional reduction at US$150 per ton of CO₂-eq

Source: Taheripour et al. 2022.
Note: BC = black carbon; CH₄ = methane; CO = carbon monoxide; CO₂ = carbon dioxide; CO₂-eq = carbon dioxide equivalent; N₂O = nitrous oxide; NH₃ = ammonia; NMVOC = nonmethane volatile organic compounds; NOₓ = nitrogen oxides; OC = organic carbon; PM₂.₅ and PM₁₀ = particulate matter; SO₂ = sulfur dioxide.

(Continued)

FIGURE B5.4.2 **Deaths averted due to air pollution associated with carbon taxes of US$75 and US$150 per ton of CO$_2$-eq**

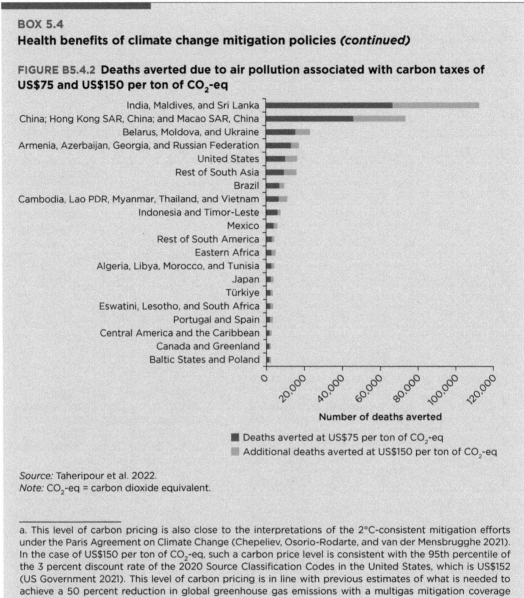

Source: Taheripour et al. 2022.
Note: CO$_2$-eq = carbon dioxide equivalent.

a. This level of carbon pricing is also close to the interpretations of the 2°C-consistent mitigation efforts under the Paris Agreement on Climate Change (Chepeliev, Osorio-Rodarte, and van der Mensbrugghe 2021). In the case of US$150 per ton of CO$_2$-eq, such a carbon price level is consistent with the 95th percentile of the 3 percent discount rate of the 2020 Source Classification Codes in the United States, which is US$152 (US Government 2021). This level of carbon pricing is in line with previous estimates of what is needed to achieve a 50 percent reduction in global greenhouse gas emissions with a multigas mitigation coverage (Peña-Lévano, Taheripour, and Tyner 2019).

References

Archsmith, J., A. Heyes, and S. Saberian. 2018. "Air Quality and Error Quantity: Pollution and Performance in a High-Skilled, Quality-Focused Occupation." *Journal of the Association of Environmental and Resource Economists* 5 (4): 827–63.

Arze del Granado, F. J., D. Coady, and R. Gillingham. 2012. "The Unequal Benefits of Fuel Subsidies: A Review of Evidence for Developing Countries." *World Development* 40 (11): 2234–48.

Chen, J., M. Chepeliev, D. Garcia-Macia, D. Iakova, J. Roaf, A. Shabunina, D. van der Mensbrugghe, and P. Wingender. 2020. "EU Climate Mitigation Policy." European Department Paper, International Monetary Fund, Washington, DC. https://www.imf.org/en/Publications/Departmental-Papers -Policy-Papers/Issues/2020/09/16/EU-Climate-Mitigation-Policy-49639.

Chepeliev, M., I. Osorio-Rodarte, and D. van der Mensbrugghe. 2021. "Distributional Impacts of Carbon Pricing Policies under the Paris Agreement: Inter- and Intra-Regional Perspectives." *Energy Economics* 102 (October): 105530.

Cohen, A. J., M. Brauer, R. Burnett, H. R. Anderson, J. Frostad, K. Estep, K. Balakrishnan, et al. 2017. "Estimates and 25-Year Trends of the Global Burden of Disease Attributable to Ambient Air Pollution: An Analysis of Data from the Global Burden of Diseases Study 2015." *The Lancet* 389 (10082): 1907–18.

Enriquez, S., B. Larsen, and E. Sanchez-Triyana. 2019. "Local Environmental Externalities due to Energy Price Subsidies: A Focus on Air Pollution and Health." ESRAF Good Practice Note 8, Energy Sector Management Assistance Program (ESMAP), World Bank, Washington, DC.

Erickson, P., H. van Asselt, D. Koplow, M. Lazarus, P. Newell, N. Oreskes, and G. Supran. 2020. "Why Fossil Fuel Producer Subsidies Matter." *Nature* 578: E1–E24.

GBD 2019 Risk Factors Collaborators. 2020. "Global Burden of 87 Risk Factors in 204 Countries and Territories, 1990–2019: A Systematic Analysis for the Global Burden of Disease Study 2019." *The Lancet* 396 (10258): 1223–49.

Health Effects Institute. 2020. "State of Global Air 2020." Special Report, Health Effects Institute, Boston, MA.

IEA (International Energy Agency). 2022. "China Country Profile." IEA, Paris. https://www.iea.org/countries/china.

IEA (International Energy Agency), OPEC (Organization of the Petroleum Exporting Countries), OECD (Organisation for Economic Co-operation and Development), and World Bank. 2010. "Analysis of the Scope of Energy Subsidies and Suggestions for the G-20 Initiative." Prepared for submission to the G-20 Summit Meeting, Toronto, Canada, June 26–27, 2010.

IHME (Institute for Health Metrics and Evaluation). 2020. *Global Burden of Disease Study 2019 (GBD 2019) Results*. Seattle: IHME. https://vizhub.healthdata.org/gbd-results/.

IMF (International Monetary Fund). 2022. Fossil Fuel Subsidies by Country and Fuel Database 2021. Washington, DC: IMF. https://www.imf.org/en/Topics/climate-change/energy-subsidies.

Jewell, J., D. McCollum, J. Emmerling, C. Bertram, D. E. Gernaat, V. Krey, L. Paroussos, et al. 2018. "Limited Emission Reductions from Fuel Subsidy Removal Except in Energy-Exporting Regions." *Nature* 554 (7691): 229–33.

Klaiber, C., J. Rentschler, and I. Dorband. 2022. "Distributional and Health Co-Benefits of Fossil Fuel Subsidy Reforms." Technical background paper prepared for this report. World Bank, Washington, DC.

La Nauze, A., and E. Severnini. 2021. "Air Pollution and Adult Cognition: Evidence from Brain Training." NBER Working Paper 28785, National Bureau of Economic Research, Cambridge, MA.

Peeples, L. 2020. "News Feature: How Air Pollution Threatens Brain Health." *Proceedings of the National Academy of Sciences (PNAS)* 117 (25): 13856–60.

Peña-Lévano, L. M., F. Taheripour, and W. E. Tyner. 2019. "Climate Change Interactions with Agriculture, Forestry Sequestration, and Food Security." *Environment and Resource Economics* 74 (2): 653–75.

Peters, R., N. Ee, J. Peters, A. Booth, I. Mudway, and K. Anstey. 2019. "Air Pollution and Dementia: A Systematic Review." *Journal of Alzheimer's Disease* 70 (S1): S145–S163.

Rentschler, J. 2016. "Incidence and Impact: The Regional Variation of Poverty Effects due to Fossil Fuel Subsidy Reform." *Energy Policy* 96 (September): 491–503.

Rentschler, J., and M. Kornejew. 2017. "Energy Price Variation and Competitiveness: Firm Level Evidence from Indonesia." *Energy Economics* 67 (September): 242–54.

Ruggeri Laderchi, C., A. Olivier, and C. Trimble. 2013. *Balancing Act: Cutting Energy Subsidies While Protecting Affordability*. Report 76820. Washington, DC: World Bank.

Saari, R. K., N. E. Selin, S. Rausch, and T. M. Thompson. 2015. "A Self-Consistent Method to Assess Air Quality Co-Benefits from U.S. Climate Policies." *Journal of the Air & Waste Management Association* 65 (1): 74–89.

Taheripour, F., M. Chepeliev, R. Damania, and J. Russ. 2022. "Employment and Emission-Reduction Priorities: Europe and Central Asia." Background paper prepared for this report, World Bank, Washington, DC.

Tong, D., G. Geng, Q. Zhang, J. Cheng, X. Qin, C. Hong, K. He, and S. J. Davis. 2021. "Health Co-Benefits of Climate Change Mitigation Depend on Strategic Power Plant Retirements and Pollution Controls." *Nature Climate Change* 11: 1077–83.

US Government. 2021. "Social Cost of Carbon, Methane, and Nitrous Oxide: Interim Estimates under Executive Order 13990." Technical Support Document, Interagency Working Group on Social Cost of Greenhouse Gases, Washington, DC, February 2021. https://www.whitehouse.gov/wp-content /uploads/2021/02/TechnicalSupportDocument_SocialCostofCarbonMethaneNitrousOxide.pdf.

van der Mensbrugghe, D. 2019. "The Environmental Impact and Sustainability Applied General Equilibrium (ENVISAGE) Model. Version 10.01." Center for Global Trade Analysis, Purdue University, West Lafayette, IN. https://mygeohub.org/groups/gtap/envisage-docs.

Van Dingenen, R., F. Dentener, M. Crippa, J. Leitao, E. Marmer, S. Rao, E. Solazzo, and L. Valentini. 2018. "TM5-FASST: A Global Atmospheric Source–Receptor Model for Rapid Impact Analysis of Emission Changes on Air Quality and Short-Lived Climate Pollutants." *Atmospheric Chemistry and Physics* 18 (21): 16173–211.

Vandyck, T., K. Keramidas, A. Kitous, J. V. Spadaro, R. Van Dingenen, and B. Saveyn. 2018. "Air Quality Co-Benefits for Human Health and Agriculture Counterbalance Costs to Meet Paris Agreement Pledges." *Nature Communications* 9: 4939.

Wei, X., Q. Tong, I. Magill, P. Vithayasrichareon, and R. Betz. 2020. "Evaluation of Potential Co-Benefits of Air Pollution Control and Climate Mitigation Policies for China's Electricity Sector." *Energy Economics* 92 (October): 104917.

WHO (World Health Organization). 2018. "Burden of Disease from Ambient Air Pollution for 2016." Version 2 (April 2018). WHO, Geneva.

WHO (World Health Organization). 2021. *WHO Global Air Quality Guidelines: Particulate Matter (PM$_{2.5}$ and PM$_{10}$), Ozone, Nitrogen Dioxide, Sulfur Dioxide, and Carbon Monoxide*. Geneva: WHO.

World Bank. 2010. "Subsidies in the Energy Sector: An Overview." Background paper for the World Bank Group Energy Sector Strategy. World Bank, Washington, DC.

PART II

LAND

CHAPTER 6

Size, Scope, and Composition of Agricultural Subsidies

> "The farmer is the only man in our economy who buys everything at retail, sells everything at wholesale, and pays the freight both ways."
>
> —*John F. Kennedy*

CHAPTER AT A GLANCE

The most common policy objectives of agricultural subsidies are to provide price stability and food security, to support farmers' incomes and livelihoods, and to improve environmental outcomes. However, these subsidies often lead to unintended consequences that are counterproductive to policy goals. How much are countries spending to achieve these goals, and what is the composition of this spending?

- Subsidies in the agriculture sector exceed US$635 billion per year, the equivalent of 0.9 percent of gross domestic product (GDP) and 18 percent of agricultural value added for the 84 countries studied.

- Most agricultural support—61 percent—is distortive and affects farmers' planting, harvesting, and input decisions. This support comes in the form of input subsidies, output payments, or market price support. As subsequent chapters show, these subsidies have the biggest spillover effects by causing harmful environmental outcomes.

- Richer countries spend more on agricultural subsidies than poorer countries, even when calculating spending as a share of total agricultural production value.

- Analysis of new data on 38 countries collected for this report finds that they spend nearly US$5 billion per year on building, operating, and maintaining irrigation infrastructure. This spending comes to approximately US$195 per year per hectare of farmland that is equipped for irrigation.

Introduction

This part of the report comprises four chapters that discuss agricultural subsidies and some of their economic, social, and environmental implications. This chapter examines the presence and levels of support to the agriculture sector around the world. It discusses different definitions of agricultural subsidies and quantifies their magnitude. Chapter 7 then examines the economic and social impacts of agricultural subsidies, exploring the impact of subsidies on the efficiency of agricultural production as well as the distributional impacts on those who receive subsidies. Chapters 8 and 9 examine the environmental externalities of subsidies, with chapter 8 focusing on water-based effects through nutrient pollution and overextraction and chapter 9 focusing on land-based effects through the lens of deforestation. Both chapters also study the spillovers of these externalities on health and human capital accumulation.

What is an agricultural subsidy?

The development, expansion, and advancement of agriculture may be the single most important driver behind all human civilizations. The switch from hunting and gathering to farming enabled humans to cultivate up to 100 times more calories per acre of land (Diamond and Ordunio 1999). As advancements like Bronze Age tools, irrigation, and animal husbandry were developed, they reduced the need for human labor and freed up workers to take on roles that advanced the economy and society. Today, just over 25 percent of the world is employed in the agriculture sector. The most advanced economies can produce enough to feed their entire population with less than 2 percent of their population employed in agriculture and yet have large surpluses to export (World Bank 2019).

Having a well-developed, robust, and efficient agriculture sector is key to achieving the kind of structural transformation (that is, a shift of economic activity from the agriculture sector to manufacturing and services) that leads to higher incomes and greater welfare. Since agriculture can play such a central and predictable role in economic development and can also be an important driver of growth through high value added production and processing, there is an argument for targeted subsidies and public investments to support this process. However, support to agriculture is not monolithic; rather it comes in a wide variety of forms. Countries provide agricultural support using different tools— including input subsidies, trade restrictions, tax breaks, publicly funded infrastructure, and research and development (R&D)—all of which can have different impacts. Some of these impacts will work toward anticipated policy priorities, and others will lead to unintended negative effects and externalities.

Agriculture is both a victim and a villain when it comes to environmental degradation and climate change. Agriculture is vulnerable to extreme weather and climatic events, and at the same time, agriculture, land use, and land use change account for about a quarter of total global emissions of greenhouse gases (Tubiello 2019). Indeed, an estimated 73 percent of all deforestation is caused by agriculture (Hosonuma et al. 2012). Thus, while much of agricultural support is intended to increase farmers' resilience to extreme weather events, such support can unintentionally expand the sector's environmental footprint, exacerbate climate impacts, and diminish its resilience to climate change. Reforms like a "climate-smart agriculture" approach—which aims to take an integrated approach to managing landscapes to increase productivity, enhance resilience to climate, and reduce emissions—can help to achieve these multiple goals simultaneously (World Bank 2018). However, before proposing reforms to achieve these goals, a better understanding of the current state of agricultural support is needed to inform and improve potential pathways of reform. To shed light on this matter, this section presents an overview of the objectives and mechanisms of agricultural support and discusses the strengths and weaknesses of these measures of support.

Policy objectives of agricultural support

Given this wide variety of objectives, there can be a gray area between what is truly a subsidy—public funding to improve private, economic production—and what is income support or a safety net program for rural residents. This report does not make an explicit distinction and uses the terms "subsidy" and "support" interchangeably. Nevertheless, the most commonly stated objectives of agricultural support programs are to (1) provide price stability and food security, (2) support farmers' incomes and livelihoods, and (3) improve environmental outcomes (OECD 2020; WTO 2006).

First, agricultural support can be used to attain price stability and food security. This goal is often targeted by guaranteeing farmers a minimum price for certain items, to incentivize them to increase production, or by providing consumers with a maximum price for food to ensure access. In 2020 alone, eight countries experienced food crises that had more than 1 million people in emergency or catastrophe and famine situations (FSIN and Global Network Against Food Crises 2021).[1] Naturally, ensuring food security is a fundamental policy goal for any government. Nevertheless, a pursuit of domestic self-sufficiency, which is often the focus of food security policies, can also be shortsighted. Blocking imports to encourage domestic production can make food supplies more costly and more vulnerable to local weather or economic shocks, reduce nutritional diversity, and limit access (FAO 2006).

Second, agricultural support can be implemented to support farmers' incomes with the ultimate goal of promoting rural development and reducing rural poverty. Rural poverty remains a challenge for many countries since about 79 percent of the world's poor live in rural areas, and the poverty rate in rural areas is 17.2 percent, triple the rate in urban areas (5.3 percent) (UN 2019). Given the importance of the agriculture sector in rural livelihoods, providing agricultural support is a natural way to support rural development while addressing poverty. Such support can include enabling farmers to increase their incomes by charging higher prices or producing greater quantities or by lowering their costs through input subsidies.

Third, agricultural support can be used to promote environmental objectives by encouraging specific farming practices. For example, subsidies to adopt climate-smart practices lower the cost of such inputs or investments. If the use of these inputs or investments requires learning or taking on additional risk, then these subsidies could encourage the adoption of good agricultural practices to establish a "virtuous cycle." Similarly, many environmentally linked agricultural subsidies involve promoting landscape management to improve land productivity and other ecosystem services like watershed management (box 6.1).

Irrespective of the stated objectives of any support policy, rent seeking plays a prominent role in shaping (1) the design and (2) the magnitude of support. Empirical work (for example, Dutt and Mitra 2010; Furtan, Jensen, and Sauer 2008; Lopez 2001) suggests that observed support often mirrors the sector's lobbying prowess. The source of influence varies across countries and subsectors within countries (for example, small versus commercial farming). In some instances, the source of influence may reflect a lack of competition and the dominance of a few oligopolistic players that often control the supply of inputs or supply chains of output. In other cases, the influence on policy may reflect the size of the labor force employed in agriculture.

BOX 6.1
Landscape restoration projects in Ethiopia

Since 2008, the World Bank has been supporting the government of Ethiopia to restore and enhance degraded landscapes. This support has been provided through a series of programs,[a] each built on and broadening experience gained over time.

Initially, these programs focused on community-based approaches to support sustainable land management practices such as constructing terraces on farmed hillsides and planting trees. Over time, these initiatives have broadened in their approach and moved from "mass mobilization" to "locally driven" participatory approaches enabling local associations and cooperatives to lead on planning and implementation. Current programs focus on delivering longer-term sustainability through major investments that strengthen land tenure, support climate-smart agricultural practices, improve livelihood security, and build value chain linkages.

The results have been impressive. More than 1.1 million hectares of land are now under sustainable land management practices, with this figure expected to exceed 3 million hectares by the end of the current round of support. More than 180,000 hectares of forests have been restored through afforestation and reforestation, with more than 123,000 hectares of forests now under participatory forest management and forest cooperatives responsible for managing them sustainably.

There is already evidence that these large-scale investments in regenerating watersheds and increasing land productivity have led to measurable changes in degraded landscapes. Preliminary analysis of the watershed areas covered by the World Bank–supported programs indicates that 62 percent now have higher scores on the normalized difference vegetation index (NDVI)[b] than at the start of interventions. NDVI is a broad measure of vegetation productivity, frequently used as a proxy for crop, grassland, and forest productivity. NDVI measurements are reliable and consistent over time and space thanks to satellite-based remote sensing.

Efforts to advance the restoration of degraded watersheds, increase land productivity, and build resilience to climate change at the required scale have required transformative, rather than incremental, changes. The World Bank's support now uses both project and results-based financing instruments. More than 3,000 cooperatives and associations have been established, and this number will grow to around 6,150 by the end of the current phase of support. With support from local extension agents, these groups are developing their own plans for watershed restoration and management and receive support for their implementation.

a. Sustainable Landscape Management Projects (SLMP I, 2009–14; SLMP II, 2014–18); Resilient Landscapes and Livelihoods Project (RLLP, 2019–24; RLLP II, 2021–25); Climate Action through Landscape Management (CALM, 2019–24); and Oromia Forested Landscape Project (OFLP, 2017–22).

b. NDVI uses remote-sensing data to identify the quality and distribution of vegetation. Changes in vegetation can be analyzed using time-series observations.

Agricultural support can come in many forms, and each form will have different benefits and disadvantages. Common support in the agriculture sector includes (1) input subsidies, (2) market price support, (3) output payments, (4) decoupled payments, (5) tax incentives, and (6) publicly funded infrastructure and R&D. These policies operate through different mechanisms, and there are potential policy trade-offs. Some of these mechanisms and trade-offs are summarized in table 6.1, with more details in box 6.2.

TABLE 6.1 Policy trade-offs of different agricultural support mechanisms

Type of support	Possible benefits	Possible disadvantages
Input subsidies Subsidies that reduce the price of inputs like fertilizers and improved seeds	Input subsidies incentivize the use of inputs that could lead to higher yields. If market failures such as lack of access to credit to purchase inputs are causing farmers to employ below-optimal levels of inputs, subsidies can correct this failure. Additionally, if land becomes more productive, this increase in productivity may reduce pressures on land expansion, an effect known as the Borlaug Hypothesis.	Input subsidies can incentivize the overuse of inputs, which can lead to increased nutrient runoff into waterways (see chapter 8) and additional pressures on land use (see chapter 9). In addition, input subsidies can lead to distorted combinations of inputs that deviate from optimally efficient levels. If input subsidy schemes are untargeted or poorly targeted, they are likely to accrue to richer households, which purchase more inputs.
Market price support Subsidies that arise from policy measures like trade restrictions, creating a gap between domestic producer prices and international prices	Import bans or tariffs reduce the quantity of a particular crop within the country, thus increasing the price. This price increase has the dual effects of incentivizing farmers to produce more of it (improving domestic self-sufficiency) and providing income support to these farmers.	Self-sufficiency policies like import bans and tariffs may hamper food security efforts by increasing the risk of a crisis in the event of a national supply shock. Market price support also distorts farmers' planting decisions, incentivizing the production of less efficient crops. In addition, it may incentivize crop extensification onto marginal lands by artificially raising the price of commodities, which can weaken overall efficiency and lead to unnecessary deforestation and habitat loss.
Output payments Direct subsidies from taxpayers to farmers for growing specific crops	As with market price support, output payments incentivize the production of specific crops, either to promote self-sufficiency or to support farmers.	Output payments are likely to have negative impacts similar to those of market price support. However, as discussed in chapter 7, the distortionary effects of output payments are likely to be larger than those of market price support. These large effects occur because the benefits of output payments to farmers are more certain and therefore more likely to influence farmers' decisions. Output payments are also likely to be highly regressive since richer farmers produce more output and therefore capture a larger share of the subsidy.
Decoupled payments Income transfers to farmers that are not linked to production or farm size	Decoupled payments do not distort farmers' cropping and input-use decisions. Such subsidies therefore do not have the harmful environmental spillover effects onto land and water use that coupled subsidies have. Decoupled payments may also help farmers to overcome credit constraints that are preventing them from optimizing production strategies. In addition, they act as safety net or welfare payments to raise economic well-being and overall incomes of farm households.	If subsidies are intended to incentivize the production of certain crops, decoupled payments are not an effective instrument, since they do not have a direct effect on production. However, since they are funded through taxes, these income transfers still have an overall impact on allocative efficiency in the economy by shifting resources from taxpayers to payment recipients.
Public goods provision Expenditures on public goods like infrastructure, research and development, and so forth	Public goods provision can generate several benefits, including research and development of new farming technologies and techniques, provision of infrastructure like irrigation to increase yields and strengthen resilience to weather shocks, and improved access to information and technologies through agricultural extension facilities. Due to the public nature of the benefits provided by these investments, as well as the large fixed costs associated with them, spending on these items is unlikely to be done by the private sector.	Some care must be taken when providing public goods to ensure that they do not lead to unintended impacts. For instance, while the provision of irrigation infrastructure is generally beneficial to farmers, it can lead to overconsumption of water and less resilience in the long run if complementary policies like water markets or pricing are not implemented.

Source: World Bank.

BOX 6.2

A simple profit-maximization model to illustrate the policy impacts of support mechanisms

The following profit-maximization model illustrates the potential mechanisms of agricultural support programs:

$$\max y \times p - (p_L L + p_F F + p_K K + p_T T), \tag{B6.2.1}$$

where $y = f(L,K,T,F)$, L is labor, K is capital, T is land, F is other inputs, and p denotes price.

The first term of this equation, $y \times p$, or production times price, represents the total revenue that farmers receive for their products. The remaining terms ($p_L L + p_F F + p_K K + p_T T$) are the costs of production, with the quantity of each input (L, F, K, and T) multiplied by its price per unit. Farmers therefore seek to maximize the value of their revenue minus their costs—that is, their profits.

An input subsidy—for example, through a fertilizer subsidy—would lower p_F. The farmer would be expected to increase F, changing the input mix for production, which would increase production or decrease production costs. The increased use of inputs like fertilizer can lead to intensification (that is, higher yields), which can lead to a variety of trade-offs. Fertilizer overuse can also lead to increased nutrient runoff into water bodies, adversely affecting water quality and health (Brainerd and Menon 2014; Zaveri et al. 2019). This issue is discussed in depth in chapter 8. Likewise, if intensification affects T, the amount of land that is employed, such intensification can put pressures on the environment and on forests in particular. Chapter 9 discusses this issue in more depth. Another example of an input subsidy is unpriced, or underpriced, water provided through irrigation. The benefits of irrigation are significant and, in the right circumstances, can greatly increase yields and strengthen climate resilience. However, when complementary policies fail to send the right water scarcity signals, this failure can lead to the cultivation of water-intensive crops in inappropriate areas, depleting water and reducing resilience during times of drought (Damania et al. 2017).

Market price support, sometimes implemented through trade measures and policies like an import ban, can affect the prices that farmers receive for their crops. In the setting of an import ban, the policy would reduce the quantity of a particular crop within the country, thus increasing the price, p, that domestic farmers can receive. Therefore, farmers would be incentivized to increase production, y, and sell in the domestic market. This mechanism has been used to promote domestic self-sufficiency, which may be inconsistent with food security. Increased production may lead to increased land use for agriculture, T, which is often associated with deforestation. In addition, it may incentivize farmers to produce on more marginal lands, lowering the overall efficiency of production. See chapter 7 for a deeper discussion of this effect.

Output payments linked to the quantity of the final product produced could have a similar effect as market price support, increasing the price of the final product, p, and thereby increasing production, y. However, decoupled payments—that is, support to farmers that is not linked to overall production—would not alter the farmer's production decision since input and output prices remain unchanged, making this policy less distortionary. For a given increase in production, input subsidies are likely to be more distortionary than output-price payments since input subsidies cause distortions in the input-mix decision while also encouraging overproduction.

Tax incentives may alter farmers' income (total revenue) or total costs, depending on their implementation. For example, income tax reductions could lead to increased investments in the future, and some of these investments may benefit productivity, the environment, or both. Tax incentives that promote certain inputs would lower input prices,

(Continued)

A simple profit-maximization model to illustrate the policy impacts of support mechanisms *(continued)*

and some of these incentives may harm or benefit the environment, depending on the specific inputs they target. However, tax incentives, even those that promote good agricultural practices, may increase future production, which may increase extensification and deforestation.

Finally, publicly funded infrastructure and research and development can influence several factors that will determine production and spillover effects. Irrigation infrastructure, for instance, will affect the input mix that farmers employ, by increasing the inputs that are complementary to water and decreasing those that are substitutes. While the specificities are complex, in some contexts, water can be a substitute for other inputs for high-value crops, but a complement to inputs for low-value crops (Cai, Ringler, and You 2008). In other words, when irrigation becomes available, higher-value crops like fruits and vegetables become less input intensive, while lower-value crops, like wheat and maize, become more input intensive. Irrigation infrastructure can also affect land use decisions, T, and other decisions related to adaptation to weather shocks (Zaveri, Russ, and Damania 2020).

What is the magnitude of subsidies in the agriculture sector?

The remainder of this chapter explores the magnitude of agricultural support and subsidies around the world. However, doing so is not a straightforward exercise, since definitions vary, as can the availability of data. The chapter first describes the various definitions used by organizations to quantify agricultural support, how they relate to each other, and the benefits and drawbacks to using them to classify subsidies. It then explores data on the magnitude of subsidies across several of these definitions. While some of these definitions may include partial estimates of public expenditures, most do not include the irrigation sector. Therefore, the chapter concludes with new results on the magnitude of irrigation subsidies collected for this report.

Measures of agricultural support

Producer support in agriculture is usually defined based on the measures used either to provide producer protection or to seek contributions to revenue through taxation. Generally, there are three types of measures: (1) those that account for agricultural tariffs for imports and exports of agricultural commodities; (2) those that account for the full set of distortions to agricultural prices, both domestic and due to trade; and (3) those that account for distortions in both inputs (seeds, fertilizers, and so forth) as well as outputs (price and quantity of agricultural products).

The most basic definition of producer support is related to tariff measures—that is, support to domestic farmers that either (1) raises the price of imports to make it easier for domestic farmers to compete in the domestic market or (2) lowers the costs of exports to make it easier for domestic farmers to compete in the international market. The World Trade Organization's (WTO) definition of support includes applied tariff rates and schedules of commitments made by countries in WTO negotiations (box 6.3). This measure is used largely to estimate how agricultural support can distort trade relationships.

While tariff-based measures are relatively easy to interpret and analyze, nontariff barriers, such as licenses and sanitary measures, are also commonly used in agriculture and

BOX 6.3

Domestic support in agriculture: The WTO "boxes"

For categorizing subsidies both in agriculture as well as in other sectors, the World Trade Organization (WTO) uses categories that follow a "traffic light system": green for support permitted under WTO regulations, amber for support that should be reduced, and red for support that is forbidden. In agriculture, the red box has been replaced with a blue box, indicating that, for agricultural support, no subsidies are expressly forbidden.

- *Green box.* Green box subsidies do not distort trade or at most cause minimal distortion. They must be government funded and must not involve price support. Green box subsidies are allowed without limits. Examples include decoupled support and environmental protection programs.

- *Amber box.* Nearly all domestic support measures, except those in the green and blue boxes, considered to distort production and trade fall into the amber box. These measures include measures to support prices or subsidies directly related to production quantities.
- *Red box.* None in agriculture.
- *Blue box.* Blue box measures are "amber box with conditions" designed to reduce distortion. Any support that would normally be in the amber box is placed in the blue box if the support also requires farmers to reduce production. There are no limits on spending for blue box subsidies.

Source: World Trade Organization (https://www.wto.org/english/tratop_e/agric_e/agboxes_e.htm).

can be more complex. A second definition of support, known as the nominal rate of protection (NRP), takes these various measures of support into account. NRP estimates the tariff equivalents of policies that change the domestic price of agricultural products relative to their reference price, which is defined as the border price evaluated at the official nominal exchange rate, adjusted for distribution, storage, transport, other marketing costs, and quality differences (Krueger, Schiff, and Valdés 1988).[2]

The Organisation for Economic Co-operation and Development's (OECD) producer support estimate (PSE) is more comprehensive, since the PSE includes support for both outputs (final crops produced) and inputs (seeds, fertilizer, and so forth). Therefore, it includes market price support, similar to what is included in the NRP, as well as payments based on outputs, inputs, planted area, and so forth. Table 6.2 describes the advantages and disadvantages of the NRP and PSE approaches.

Related to the PSE measure is the total support estimate (TSE), which includes the PSE plus estimates of other general services support. These estimates can include development and maintenance of agricultural infrastructure, marketing and promotion, public stockholding, and investments in R&D for agriculture. Therefore, TSE is the most comprehensive measure of agricultural support. Box 6.4 provides a more detailed description of how the NRP, PSE, and TSE are calculated.

Agricultural support can also be defined in several ways based on its potential market distortion: border measures that provide market price support, coupled subsidies, and decoupled subsidies (Mamun, Martin, and Tokgoz 2019). Border measures create price gaps, which would be the equivalent of tariffs and taxation. These measures, such as market price support, create incentives for producers to increase output, while encouraging consumers in protected markets to reduce consumption of the protected product. Coupled subsidies provide direct subsidies on output or inputs that create incentives to

TABLE 6.2 Comparison of the nominal rate of protection and producer support estimate

Characteristic	Nominal rate of protection	Producer support estimate
Availability of data estimates	Large number of countries; all countries in the OECD database	OECD countries and a few non-OECD countries; covers about 60% of agricultural products
Scope of estimates	Focus on outputs only; input subsidies are missing, although they are a relatively small fraction of total agricultural support (Anderson et al. 2008)	Takes both output and input support into account; input support may be important for externalities
Granularity of estimates	Limited information on commodity-specific support for countries not on the OECD list	Commodity-specific support, which varies by country

Source: World Bank, based on OECD 2000, 2020.
Note: OECD = Organisation for Economic Co-operation and Development.

BOX 6.4

The nominal rate of protection and producer support estimate

The *nominal rate of protection (NRP)* estimates the tariff equivalents of policies that change the domestic price of agricultural products relative to their reference price, which is defined as the border price evaluated at the official nominal exchange rate, adjusted for distribution, storage, transport, and other marketing costs. The reference price can also be adjusted for quality differences based on the availability of input data. The NRP is measured as:

$$NRP = \frac{PP}{RP} - 1,$$

(B6.4.1)

where PP is the producer price at farm gate and RP is the reference price. The NRP is essentially a tariff equivalent, t, of the measures used for domestic prices.

The NRP can be defined for each commodity c in country i,

$$NRP_{commodity} = \frac{\sum_i (PP_i \times Q_i)}{\sum_i (RP_i \times Q_i)} - 1.$$

(B6.4.2)

The NRP for a commodity is the ratio between the price gap and the observed reference price measured at the same point in the value chain.

Similarly, for each country i, the total NRP in agriculture, NRP_{i_agri}, is the sum of all the commodity support:

$$NRP_{i\_agri} = \frac{\sum_c (PP_c \times Q_c)}{\sum_c (RP_c \times Q_c)} - 1.$$

(B6.4.3)

A related definition, the effective rate of protection (ERP), takes into account the value added from the sale of the commodity. In this case, the value added, VA, is given by the price of the output, P, minus the cost of materials, C (that is, $VA = P - C$). The ERP is calculated based on the value added with the tariff (value added at domestic prices), VA_t, and the value added at world prices in a common currency, VA_f:

$$ERP = \frac{VA_t - VA_f}{VA_f} \times 100.$$

(B6.4.4)

(Continued)

BOX 6.4
The nominal rate of protection and producer support estimate *(continued)*

The ERP is a better measure of the true impact of distortions on incentives, since it accounts for the effect of subsidies and taxes on inputs. However, it only considers policies that affect output and input through trade barriers.

The *producer support estimate (PSE)* is measured by summing up the following:

- Market price support; payments based on output
- Payments based on input
- Payments based on area planted or number of animals; production is required
- Payments based on historical area planted or number of animals; production is required
- Payments based on historical area planted or number of animals; production is not required
- Payments based on noncommodity criteria
- Miscellaneous payments.

The *total support estimate* (TSE) measures the overall transfers associated with agricultural support that is financed by consumers and taxpayers. It is the sum of the explicit and implicit gross transfers from consumers of agricultural commodities to agricultural producers. Formally, it is given by:

$$TSE = PSE + TCT + GSSE, \qquad (B6.4.5)$$

where *TCT* is total taxpayer-to-consumer transfers, and *GSSE* is the general services support estimate, which mainly tracks public goods provision. These measures include transfers to:

- Improve agricultural production (research and development)
- Provide agricultural training and education (agricultural schools)
- Control the quality and safety of food, agricultural inputs, and the environment (inspection services)
- Improve off-farm collective infrastructure, including downstream and upstream industry (infrastructure)
- Assist marketing and promotion (marketing and promotion)
- Meet the costs of depreciation and disposal of public storage of agricultural products (public stockholding)
- Provide other general services that cannot be disaggregated and allocated to the above categories due, for example, to a lack of information (miscellaneous).

Source: OECD 2000.

increase output. Decoupled subsidies avoid altering incentives to change output or input levels, but provide direct income support to producers, thus acting as lump-sum subsidies. Coupled subsidies are more distorting than decoupled subsidies because they affect supply and demand. Chapters 8 and 9 examine how these subsidies affect environmental outcomes.

Estimating the magnitude of agricultural support

This report uses data compiled by Gautam et al. (2022) from three sources to analyze the magnitude of agricultural support. The OECD started the effort of measuring agricultural support in the 1980s for OECD countries. This database now includes 54 countries and regions[3] that are both OECD and non-OECD members, including large producers such as China and India (OECD 2020). To expand the presence of low- and middle-income countries in the analysis, this database was then merged with data from two other sources: the Inter-American Development Bank's Agrimonitor program (Anríquez et al. 2016), which contains data from 17 countries in Latin America, and the Food and Agriculture Organization's Monitoring and Analysing Food and Agricultural Policies (MAFAP) program (Angelucci et al. 2013), which contains data from 13 countries in Sub-Saharan Africa. In total, 84 countries are included in the analysis. In 2016, these countries accounted for 67 percent of the global value of agricultural production (FAO 2022).[4]

On average between 2016 and 2018, the annual TSE for these 84 countries amounted to US$635 billion per year, which equals approximately 0.9 percent of GDP and 18 percent of agricultural value added for these countries. The share of this TSE that was transferred to individual producers—that is, the PSE—was about 71 percent, with the remaining share split between the general services support estimate (GSSE) (18 percent) and taxpayer-to-consumer transfers (TCT) (11 percent) (see box 6.4 for definitions).

US$635 billion: Each year, 84 countries that account for two-thirds of global agricultural production spend US$635 billion on agricultural subsidies.

Thus the bulk of support goes to producers. Around 61 percent of this support is in the form of coupled support, such as market price support or payments for input use, which distort producers' decisions. As discussed in chapters 8 and 9, this type of support is responsible for much of the harmful environmental impacts. However, recent trends suggest that some countries have increased funding for *decoupled* support, through direct payments and agricultural investments in public goods such as extension services and infrastructure.

TSE is positive for 78 of the 84 countries in the analysis. In the remaining countries—Argentina, Ethiopia, Ghana, Malawi, Mali, and Vietnam—net support to agricultural production is negative, implying that policies that reduce farmers' net revenues from agricultural products exceed the value of any support provided.[5] By a wide margin, China provides the most support, with a TSE of US$237 billion in 2016, which was 37 percent of the global total and more than the EU-28 countries and the United States combined. Figure 6.1 shows the TSE for the largest 20 countries and regions, broken down by PSE, GSSE, and TCT. PSE is the dominant form of support in most countries, while for most countries, public goods provision tracked by GSSE is quite small. The exceptions to this trend are countries like (1) India, where producer support is actually negative[6] and public goods provision and transfers to consumers are the major forms of support to agriculture, and (2) the United States, where most support comes in the form of TCT, which is generally intended to reduce food prices for consumers.

While figure 6.1 shows the distribution of total TSE, the picture changes quite a bit when it is put in terms of the ratio of support to agricultural production value—the level of support relative to the size of the agriculture sector. Figure 6.2 compares this ratio to GDP per capita of the country. This measure is perhaps more accurate for comparing the magnitude of support across countries since it standardizes for the size of the sector in each country. It shows that there is generally a weakly positive relationship between a country's level of development and the relative level of support to agriculture. Given the often-stated purposes of agricultural support—to support food security and promote rural development—the opposite relationship might be expected, since those policy goals are more critical in low- and middle-income countries. Nevertheless, the relationship may also be driven by an affordability constraint in low- and lower-middle-income countries.

FIGURE 6.1 Total support estimate, by country, 2016-18

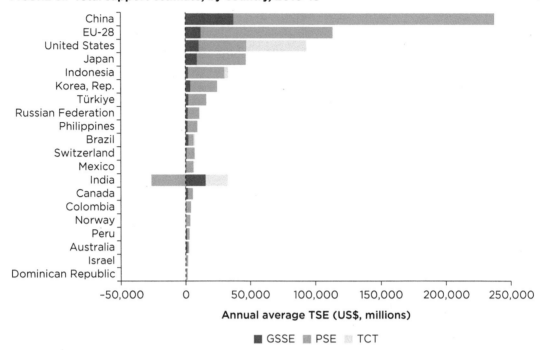

Annual average TSE (US$, millions)

■ GSSE ■ PSE ■ TCT

Sources: Data are from OECD 2020, the Inter-American Development Bank's Agrimonitor program (Anríquez et al. 2016), and the Food and Agriculture Organization's Monitoring and Analysing Food and Agricultural Policies (MAFAP) program (Angelucci et al. 2013).

Note: The figure shows TSEs for the 20 countries and region with the largest values, for the average of the years 2016-18. GSSE = general services support estimate; PSE = producer support estimate; TCT = taxpayer-to-consumer transfers; TSE = total support estimate.

FIGURE 6.2 Total support estimate as a share of agricultural production value compared to GDP per capita, 2016-18

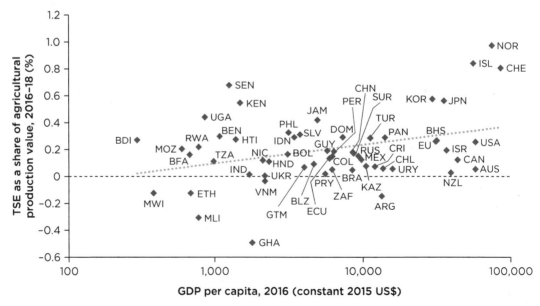

GDP per capita, 2016 (constant 2015 US$)

Sources: TSE data are from OECD 2020, the Inter-American Development Bank's Agrimonitor program (Anríquez et al. 2016), and the Food and Agriculture Organization's Monitoring and Analysing Food and Agricultural Policies (MAFAP) program (Angelucci et al. 2013). Agricultural value added and GDP per capita are from the World Bank's World Development Indicators database.

Note: The figure shows total support estimate (TSE) as a share of agricultural value added (2016-18 mean) versus GDP per capita (2016).

Additionally, once the size of the agriculture sector is accounted for, regions like China, the EU-28, and the United States are no longer outliers in terms of the magnitude of agricultural support. Indeed, the countries that become outliers are generally wealthy countries with difficult climates for agriculture, like Iceland, Norway, and Switzerland. Among low- and middle-income countries, Senegal is an outlier, with a TSE value that is 67 percent as large as the total value of production in the country. Nearly half of that support is in GSSE and is for the support of public goods, rather than for payments to farmers or consumers.

61%: The share of agricultural subsidies that are coupled to production and thus distort the economy and have detrimental spillovers into the environment

Another important way to measure agricultural support is by whether it is coupled or decoupled from production decisions. Coupled support (for example, market price support or input subsidies) has the effect of distorting the incentives for farmers to (1) increase output, (2) change the area under cultivation of a specific crop, or (3) change the amount or ratio of inputs used. This distortion can have significant unintended effects on the environment and other externalities. However, decoupled support (for example, unconditional cash transfers) is not linked to production and is therefore less distortionary, since it only expands the budget of producers, and the farmer can choose how to spend the additional money. Across the 84 countries included in the analysis, 61 percent of all subsidies are coupled, 28 percent are decoupled, and 11 percent are uncategorized.

Despite most global subsidies being coupled support—and therefore leading to distortions in agricultural production—significant heterogeneities are evident across countries and regions. Figure 6.3 shows the TSE by region and income group, broken down by coupled, decoupled, and uncategorized support. For most regions, coupled support makes up the majority of TSE, in line with the global figures. However, in South Asia and Sub-Saharan Africa, aggregate coupled support is actually negative, meaning that there is a net tax on producers. The result in Sub-Saharan Africa is driven largely by Ethiopia, Ghana, and Mali, which have large net coupled taxes. In South Asia, the result is driven by India. Similarly, when it comes to income groups, coupled support tends to be large in wealthier countries.

Estimating the magnitude of irrigation support

Support to irrigation facilities is a major component of agricultural support that receives comparatively little attention in discussions of agricultural subsidies. The establishment, operation, and maintenance of many irrigation schemes have been financed by-and-large by governments, development banks, and donor agencies. Irrigation is a critical component of agricultural inputs, particularly in areas that are prone to fluctuations in annual rainfall or that grow very water-intensive crops like rice, sugarcane, and cotton. Globally, 40 percent of total food is produced through irrigated agriculture on a footprint of only 20 percent of agricultural land. Thus, through intensification of land use, irrigated agriculture has the potential to limit the land requirements of agriculture and could potentially lead to improved sustainability.[7]

Nevertheless, subsidies to the irrigation sector can also be harmful. Irrigation diverts water toward agricultural production and away from the environment, causing damage, much of which is not accounted for in a market system (resulting in a negative externality). Subsidies that encourage irrigation lower the price of water used, which encourages

FIGURE 6.3 **Agricultural support, by type of support, region, and country income group, 2016–18**

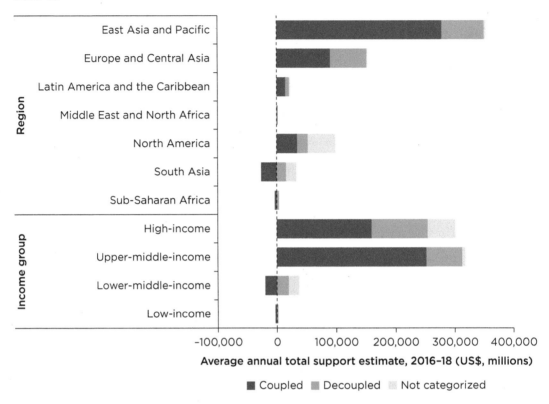

Sources: OECD 2020, the Inter-American Development Bank's Agrimonitor Program (Anríquez et al. 2016), and the Food and Agriculture Organization's Monitoring and Analysing Food and Agricultural Policies (MAFAP) program (Angelucci et al. 2013).
Note: The figure shows the total support estimate (TSE), broken down by coupled, decoupled, and not categorized support, by region and country income group.

overuse and profligacy. Furthermore, subsidies change the distribution of value between producers and consumers, distort the prices and markets that surround the irrigation sector, and result in deadweight social losses to society (see Hellegers et al. 2022 for a deeper discussion).

Knowing the level of the subsidy is important if one is to understand its impact on the marginal costs of water and consequently the variable costs of production. Nevertheless, there has been no systematic attempt to collect information on the magnitude of spending and support in the irrigation sector for the past several decades. The most recent attempt at estimating such magnitudes was done in the 1990s, when the magnitude of global subsidies was estimated at US$33 billion per year (Brown 2001).

A new survey was conducted to understand expenditures and subsidies in the irrigation sector.[8] Data were gathered from the relevant ministries on expenditures in the irrigation sector, payments collected from farmers and other irrigation water users, and other key information about the sector. For most countries, the financial information collected is from the national or federal level. Thus, if additional expenditures occurred at provincial, state, or local levels, these expenditures are not included. Therefore, the figures reported here should be considered a conservative, lower bound of total public expenditures in the sector. Data were obtained from 20 countries on capital expenditures (that is, investments in building irrigation infrastructure itself) and from 13 countries on operational expenditures (that is, expenditures on operations and maintenance). This information was then complemented by a similar effort to collect public expenditures in the irrigation sector, which relied on BOOST data (Joseph et al., forthcoming). Between these two sources, data were available for 38 countries that account for approximately 145.5 million hectares of land equipped for irrigation, or 43 percent of the global total.

In total, these 38 countries spent US$4.95 billion on capital expenditures for irrigation annually during 2015–19. The 30 countries with available data on operational expenditures spent a recorded US$867.3 million per year over the same period. This figure equates to approximately 0.31 percent of total GDP being spent annually on irrigation and 6.6 percent of value added coming from agriculture. Put another way, the average country is spending US$195 per hectare of farmland equipped for irrigation per year.

> On average, countries annually spend US$195 per hectare of farmland equipped for irrigation on both capital expenditures and operations and maintenance.

For the 20 countries from the survey,[9] additional data were collected not just on government spending on irrigation, but also on total spending by irrigation users. This information enables calculation of the share of spending on irrigation that is subsidized versus paid for by farmers. In all 20 countries, farmers do not pay out of pocket at all for capital expenditures, whereas in 4 of the 13 countries where data on operational expenditures are available, farmers do not pay for them at all. In total, 94 percent of total irrigation spending is not recouped through charges to farmers.

Even in this small sample, there is quite a lot of variation in spending across countries. Sri Lanka, spending US$460.41 per hectare per year, spends the largest amount on irrigation per hectare of farmland equipped for irrigation (including both farmer payments and subsidies). In contrast, Indonesia spends the least, at US$3.82 per hectare per year. In terms of expenditures per farmer, Kazakhstan spends the most, at US$2,335, and Indonesia spends the least, at US$1.26. Finally, in terms of average spending per cubic meter of water allocated for irrigation, Georgia spends the most, at US$0.40, while Indonesia spends the least, at US$0.001. Such variation reinforces the need for regular, comprehensive surveys of irrigation expenditure to track levels of spending and investigate the effectiveness of these often-significant investments in countries with mounting debt levels.

Notes

1. Those countries are the Democratic Republic of Congo (5.6 million people), Afghanistan (4.3 million), the Republic of Yemen (3.6 million), Sudan (2.2 million), South Sudan (1.7 million), Ethiopia (1.4 million), Haiti (1.2 million), and Zimbabwe (1.0 million). "Emergency" implies households that either (1) have large food consumption gaps reflected in very high levels of acute malnutrition and excess mortality or (2) are able to mitigate large food consumption gaps but only by employing emergency livelihood strategies and liquidating assets. "Catastrophe or famine" implies that households have an extreme lack of food or other basic needs even after fully employing coping strategies and that starvation, death, and extremely critical levels of acute malnutrition are evident.

2. A similar definition of support, the nominal rate of assistance (NRA), includes the ad valorem tax on competing imports (that is, tariffs) and the direct production subsidy (or tax) for farmers (Anderson et al. 2008). However, the NRA database has been discontinued.

3. Data on the EU-28 are aggregated into a single region. Thus, OECD includes stand-alone data for 26 countries and the 28 European Union countries (including the United Kingdom, prior to Brexit) combined as a 27th data point.

4. The results reported in this section are, with a few exceptions, averaged for the years 2016–18, unless otherwise indicated. The exceptions are six countries from the MAFAP database for which data end before 2016. For these countries, the latest year of data was used: The Bahamas (2014), Belize (2014), Guyana (2014), Haiti (2012), Jamaica (2014), and Panama (2015).

5. Argentina's export taxes and unstable macroeconomic conditions (that is, depreciation of the currency in 2018) have depressed producers' domestic prices, causing a substantial net negative TSE (−2.3 percent of GDP in 2018). In Ghana, although import tariffs are imposed on certain items to protect domestic producers, most producers are still affected by a negative PSE. In Ethiopia and Vietnam, producers who grow products that are ineligible for protective tariffs are implicitly taxed, leading to average producer prices that are below world prices (negative PSE) and a subsequent negative net TSE. In Malawi and Mali, agricultural support such as market price support or input subsidies is excessively concentrated on certain items, limiting support for other items. Producers of those unsupported items are disadvantaged in price (Angelucci et al. 2013; OECD 2020).

6. The latest data for India, covering the years 2018–20, show that support to producers includes budgetary spending of 8.6 percent of gross farm receipts, positive MPS of +2.0 percent of gross farm receipts for supported commodities, and negative MPS of −17 percent for those that are taxed. Overall, net support was negative, at −6.4 percent of gross farm receipts. Support to producers was negative throughout the last two decades but fluctuated markedly. The negative producer support estimate shows that domestic producers, on average, were implicitly taxed, since budgetary payments to farmers did not offset the price-depressing effect of complex domestic marketing regulations and trade policy measures (OECD 2020).

7. Whether or not irrigated agriculture leads to a smaller agricultural land footprint is a subject of much debate. Key to this debate is whether the Borlaug Hypothesis holds—that improvements in agricultural technology will enable farmers to produce more food from a given piece of land, thereby enabling growth in food supply without leading to increased deforestation—or whether "Jevons' effects" will dominate. Jevons' effects imply that, as intensification and irrigation increase yields, those circumstances increase the profitability of land use and thus induce the use of additional land for agriculture.

8. This survey is unscientific, and countries were not randomly sampled. Countries were included where World Bank staff had access to key figures within agriculture, water, or other ministries that could provide financial data on irrigation expenditures. Thus, the report makes no claims about the representativeness of this survey's results.

9. These countries are Afghanistan, Albania, Armenia, Cambodia, China, Ethiopia, Georgia, Ghana, Indonesia, Kazakhstan, the Lao People's Democratic Republic, Mali, Myanmar, Nigeria, Pakistan, the Philippines, Somalia, Sri Lanka, Tajikistan, and Uzbekistan.

References

Anderson, K., M. Kurzweil, W. Martin, D. Sandri, and E. Valenzuela. 2008. "Measuring Distortions to Agricultural Incentives, Revisited." *World Trade Review* 7 (4): 675–704.

Angelucci, F., H. Gourichon, A. Mas Aparisi, and A. Witwer. 2013. "Monitoring and Analysing Food and Agricultural Policies in Africa." MAFAP Synthesis Report, Food and Agriculture Organization, Rome.

Anríquez, G., W. Foster, J. Ortega, C. Falconi, and C. P. De Salvo. 2016. "Public Expenditures and the Performance of Latin American and Caribbean Agriculture." IDB Working Paper IDB-WP-722, Inter-American Development Bank, Washington, DC.

Brainerd, E., and N. Menon. 2014. "Seasonal Effects of Water Quality: The Hidden Costs of the Green Revolution to Infant and Child Health in India." *Journal of Development Economics* 107 (March): 49–64.

Brown, L. R. 2001. *State of the World 2001: A Worldwatch Institute Report on Progress toward a Sustainable Society*. New York: W. W. Norton.

Cai, X., C. Ringler, and J. Y. You. 2008. "Substitution between Water and Other Agricultural Inputs: Implications for Water Conservation in a River Basin Context." *Ecological Economics* 66 (1): 38–50.

Damania, R., S. Desbureaux, M. Hyland, A. Islam, S. Moore, A.-S. Rodella, J. Russ, and E. Zaveri. 2017. *Uncharted Waters: The New Economics of Water Scarcity and Variability*. Washington, DC: World Bank.

Diamond, J. M., and D. Ordunio. 1999. *Guns, Germs, and Steel: The Fates of Human Society*. Vol. 521. New York: W. W. Norton.

Dutt, P., and D. Mitra. 2010. "Impacts of Ideology, Inequality, Lobbying, and Public Finance." In *The Political Economy of Agricultural Price Distortions*, edited by K. Andersen, 278–303. Cambridge, UK: Cambridge University Press.

FAO (Food and Agriculture Organization). 2006. "Food Security." Policy Brief 2, June 2006. FAO, Rome. https://www.fao.org/fileadmin/templates/faoitaly/documents/pdf/pdf_Food_Security _Cocept_Note.pdf.

FAO (Food and Agriculture Organization). 2022. FAOSTAT Statistical Database. Rome: FAO.

FSIN (Food Security Information Network) and Global Network Against Food Crises. 2021. *Global Report on Food Crises 2021*. Rome: World Food Programme. https://www.wfp.org/publications /global-report-food-crises-2021.

Furtan, W. H., M. S. Jensen, and J. Sauer. 2008. "Rent Seeking and the Common Agricultural Policy: Do Member Countries Free Ride on Lobbying?" Paper prepared for the 107th Seminar of the European Association of Agricultural Economists, January 30–February 1, 2008, Sevilla, Spain.

Gautam, M., D. Laborde, A. Mamun, W. Martin, V. Piñeiro, and R. Vos. 2022. "Repurposing Agricultural Policies and Support: Options to Transform Agriculture and Food Systems to Better Serve the Health of People, Economies, and the Planet." World Bank and International Food Policy Research Institute, Washington, DC.

Hellegers, P., B. Davidson, J. Russ, and P. Waalewijn. 2022. "Irrigation Subsidies and Their Externalities." *Agricultural Water Management* 260 (February 1): 107284.

Hosonuma, N., M. Herold, V. De Sy, R. S. De Fries, M. Brockhaus, L. Verchot, A. Angelsen, and E. Romijn. 2012. "An Assessment of Deforestation and Forest Degradation Drivers in Developing Countries." *Environmental Research Letters* 7 (4): 044009.

Joseph, G., L. Andres, A. Bahuguna, Q. Wang, and Y. Rong Hoo. Forthcoming. *Public Spending in the Water Sector*. Washington, DC: World Bank.

Krueger, A. O., M. Schiff, and A. Valdés. 1988. "Agricultural Incentives in Developing Countries: Measuring the Effect of Sectoral and Economywide Policies." *World Bank Economic Review* 2 (3): 255–71.

Lopez, R. A. 2001. "Campaign Contributions and Agricultural Subsidies." *Economics & Politics* 13 (3): 257–79.

Mamun, A., W. Martin, and S. Tokgoz. 2019. "Reforming Agricultural Subsidies for Improved Environmental Outcomes." IFPRI Report, International Food Policy Research Institute, Washington, DC.

OECD (Organisation for Economic Co-operation and Development). 2000. *Agricultural Policies in OECD Countries: Monitoring and Evaluation*. Paris: OECD.

OECD (Organisation for Economic Co-operation and Development). 2020. *Agricultural Policy Monitoring and Evaluation 2020*. Paris: OECD. https://doi.org/10.1787/928181a8-en.

Tubiello, F. N. 2019. "Greenhouse Gas Emissions due to Agriculture." In *Elsevier Encyclopedia of Food Systems*. Amsterdam: Elsevier.

UN (United Nations). 2019. *The Sustainable Development Goals Report 2019*. New York: United Nations.

World Bank. 2018. "Realigning Agricultural Support to Promote Climate-Smart Agriculture." World Bank, Washington, DC. http://documents1.worldbank.org/curated/en/734701543906677048/pdf/132660-REPLACEMENT-PUBLIC-Realigning-Agricultural-Support-CSA-v2.pdf.

World Bank. 2019. "Employment in Agriculture (% of Total Employment) (Modeled ILO Estimate)." World Development Indicators (database). Washington, DC: World Bank. https://data.worldbank.org/indicator/SL.AGR.EMPL.ZS.

WTO (World Trade Organization). 2006. *World Trade Report, 2006: Exploring the Links between Subsidies, Trade, and the WTO*. Geneva: WTO.

Zaveri, E., J. Russ, and R. Damania. 2020. "Rainfall Anomalies Are a Significant Driver of Cropland Expansion." *Proceedings of the National Academy of Sciences* 117 (19): 10225–33.

Zaveri, E., J. Russ, S. Desbureaux, R. Damania, A.-S. Rodella, and G. Ribeiro. 2019. *The Nitrogen Legacy: The Long-Term Effects of Water Pollution on Human Capital*. Washington, DC: World Bank.

CHAPTER 7

Inefficient, Unequal, and Unwise

The Economic and Distributional Impacts of Agricultural Subsidies

> "Farming isn't a battle against nature, but a partnership with it.
> It is respecting the basics of nature in action and ensuring that they continue."
> —*Jeff Koehler*

CHAPTER AT A GLANCE

Improving agricultural productivity and providing support to poor farmers are two common goals of agricultural subsidy programs. Nevertheless, the evidence shows that effectiveness in achieving either of these goals is limited.

Agricultural subsidies reduce technical efficiency:

- *New evidence shows that agricultural subsidies reduce farm-level efficiency.* Several new analyses described in this chapter—including analysis at the global level, meta-analyses of recent academic literature, and country deep dives—show that, while agricultural subsidies may lead to higher output and yields overall, they tend to reduce technical efficiency.

- *Subsidies cause farmers to use more inputs in less efficient ways.* The evidence strongly points to subsidies lowering total factor productivity, which is a concern in a resource-constrained world where the demand for food is projected to increase by more than 50 percent by 2050.

Agricultural subsidies are rarely pro-poor, with important nuances:

- *Evidence on the distributional impacts of agricultural subsidies is mixed.* Both input and output subsidies are tied to the level of production and therefore almost always benefit richer households—which tend to be larger farms with higher levels of production—more than poorer households. New analyses on the distributional impacts of these subsidies confirm these results, with some important nuances.

- *Analyses of output subsidies across 16 countries find that in 10 countries, such subsidies tend to accrue to poorer regions, which receive a higher subsidy per unit of agricultural production.* Although richer areas receive more of the subsidy—because they produce more agricultural output—they receive a lower subsidy than poorer regions for each unit of output produced. Therefore, these subsidies exhibit some measure of spatial progressivity in some countries.

• *Input subsidies tend to accrue to wealthier households at much higher rates than to poorer households, even when they are intended to target poorer households.* Participation rates and aggregate benefits of input subsidy programs are analyzed in six low- and middle-income countries for which data are available at the household level. In general, poorer households are found to participate at lower rates and to receive a smaller share of the total subsidy. Nevertheless, the subsidy makes up a significant share of household income for the poorer quartile of households, implying that removing these subsidies without compensation would be very harmful.

Introduction

As discussed in chapter 6, the three most commonly stated objectives of agricultural subsidies are to (1) provide price stability and food security, (2) support farmers' incomes and livelihoods, and (3) improve environmental outcomes. This chapter examines the effects of subsidies on the first two of these objectives to see if, and to what extent, agricultural subsidies meet those standards. The first half of the chapter examines the impact of subsidies on agricultural productivity, a critical determinant of price stability and food security. It investigates several definitions of the word "productivity," which are used in different contexts. It combines new analyses with surveys of the literature to draw overall conclusions.

The second half of the chapter then examines the distributional impact of subsidies. Indeed, if an important objective of agricultural subsidies is to support rural development and the incomes and livelihoods of farmers, then one would expect these subsidies to be progressive in nature—that is, to support the lower ends of the income distribution more than the upper ends. To examine this question, the results of two new analyses are presented. The first examines the spatial distribution of output subsidies in 16 countries to draw conclusions about whether wealthier or poorer areas of countries tend to benefit from them. The second analysis looks at country case studies using microdata for several countries for which information is available to examine who receives agricultural input subsidies and how these subsidies compare across income levels.

Agricultural subsidies and productivity

In general, economists frown on subsidies for the following reasons. Subsidies distort the incentives of farms, firms, and families and cause them to make consumption or production decisions that are different from what they otherwise would. When compared to what economists consider a "first best" world—that is, one with a perfectly competitive market and no market imperfections or distortions—taxes and subsidies will always lead to a less efficient outcome, where less output is produced for a given level of input and aggregate welfare is lower.

Nevertheless, reality is more complex, and examples of market distortions in the agriculture sector abound, complicating the picture and making the impact of subsidies on efficiency somewhat ambiguous. Uncertainty from extreme weather events and climate change, information constraints around markets and prices, credit constraints, large fixed costs of technology and infrastructure adoption (box 7.1), and environmental externalities can all distort the free-market outcome and lead to a world in which well-designed agricultural subsidies can generate more efficient outcomes. These market failures and non-convexities can distort the free market in such a way that subsidies can have the potential to boost efficiency relative to the status quo. In economics, this result derives from the "theory of second best," which states that correcting one policy distortion (such as removing a subsidy) in the presence of several market failures may worsen outcomes and magnify inefficiencies. Given the presence of myriad distortions, theory can say very little about the impact of subsidies on efficiency, making this a question that must be answered empirically, with real-world data and estimation techniques that allow for plausibly causal inference (see Adetutu and Weyman-Jones 2019 for an example of the theory of second best on fuel subsidies).

BOX 7.1

Agricultural subsidies and technology adoption

The low adoption rates of modern inputs by farmers in large parts of the developing world are a challenge that has long puzzled economists and policy makers alike. While the green revolution took off in the Americas, Asia, and Europe, it has failed in large parts of Sub-Saharan Africa, where the adoption of inputs like fertilizers and modern seed varieties has lagged adoption rates in the rest of the world. To address this failure, many Sub-Saharan African countries have implemented input support programs, which subsidize improved seeds (Mason and Ricker-Gilbert 2013; Mason and Smale 2013) and inorganic fertilizers (Carter, Laajaj, and Yang 2021; Jayne et al. 2013; Ricker-Gilbert, Jayne, and Chirwa 2011; Xu et al. 2009) in an attempt to accelerate the diffusion of technologies and adoption of inputs.

Nevertheless, the success of these programs in facilitating long-term technology adoption is very mixed. While questions remain, some insights can be drawn from the literature on what is needed to accelerate adoption. In general, farmers' perceptions of new technology are a crucial determinant of acceptance. The belief that appropriate technologies are available and accessible (Kernecker et al. 2020); the lack of adaptability of new technology to local conditions (Mottaleb 2018); and the compatibility of a new technology with farmers' needs, environment, and climate (Mignouna et al. 2011; Nyang'au et al. 2021; Singh 2020) are among the factors influencing farmers' perceptions. Economic factors, human-specific characteristics, and technological and institutional factors have all been found to influence the adoption of agricultural technology (Damania et al. 2017; Mwangi and Kariuki 2015).

The design and duration of subsidy programs are also important in determining the success of technology adoption. Input subsidies may have a negative impact on the adoption of new technology because subsidy recipients may anchor on subsidized prices and be unwilling to buy products at market prices (Simonsohn and Loewenstein 2006), or they may anticipate future subsidies and refuse to buy commercially (Omotilewa, Ricker-Gilbert, and Ainembabazi 2019). However, a one-time or short-term subsidy can make it easier for farmers to adopt new technologies by lowering uncertainty and disseminating information about the benefits of the technology (Carter, Laajaj, and Yang 2021; Omotilewa, Ricker-Gilbert, and Ainembabazi 2019).

Another important feature of subsidy programs is how subsidies are provided—in-kind (that is, the government provides the subsidized products directly to the farmer) or through some kind of voucher system—and, if the latter, how extensive the menu of options is from which farmers can choose to spend the voucher. In-kind programs crowd out private sector activity in input distribution, depriving farmers of the benefits of efficiency as well as opportunities to get information from input suppliers that is tailored to their needs. Voucher systems allow for private sector development in the distribution of inputs, but if the vouchers can be redeemed only for specific products, farmers are still unable to make their own choices regarding input technologies. Goyal and Nash (2017, ch. 3) include a more comprehensive discussion of the pros and cons of different designs of subsidy schemes as well as a survey of evidence on the overall benefits and costs of these schemes.

A sole focus on providing agricultural subsidies as incentives to accelerate technology adoption and agricultural output has proven to be ineffective in many contexts. For example, Jayne et al. (2018) point out that the second wave of African input subsidy programs focused too much on providing fertilizers to farmers and not enough on teaching them how to use fertilizers effectively. In general, there is no single reason for the slow adoption of new technology and low agricultural output; thus, a combination of measures may be the most effective (Suri and Udry 2022).

One challenge with examining the impact of agricultural subsidies on productivity is that definitions of the term "productivity" vary. In many cases, when policy makers refer to the goals of improving productivity, they are simply referring to increasing the amount of total production of agricultural products. This goal might be appropriate if the objective is to increase self-sufficiency, which itself is an obsolete indicator of food security. However, if producing a small amount of additional agricultural products requires a significant and greater increase in inputs—such as land, labor, or machinery that might be allocated more efficiently to other sectors or in different combinations—then this is a poor indicator of "productivity" in a meaningful sense.

A better, but still not sufficient, measure would be based on the growth of yields—that is, total agricultural production per unit of land. Here, production per a single input, land, is considered in the equation. Given that land is a finite resource and that changes in land use contribute to many of the environmental externalities caused by agriculture, measuring the productivity of land may be a useful way of tracking agricultural productivity. However, since yields depend on critical inputs like labor, seeds, water, fertilizers, pesticides, and machinery, all of which have a cost both to the farmer and to society at large, this measure is still incomplete.

More comprehensive measures exist for tracking productivity, such as total factor productivity (TFP). These measures compare the ratio of total output in agriculture to the total amount of inputs. This metric is the one most closely related to efficiency of resource use and (as long as market prices are not distorted by policy interventions) to farmer profits and agricultural gross domestic product (GDP). Hence, these measures are more comprehensive and meaningful than just yields, since they account not just for land inputs, but also for fertilizers, seeds, machinery, and so forth. Indeed, productivity growth depends not just on the increase in the use of inputs, but also on how inputs are combined. Changes in agricultural technical efficiency and TFP account for this growth and indicate overall improvements in farming technologies or techniques that allow a smaller amount of inputs to produce the same or more output. The difference between these two measures is that the technical efficiency can be compared between farmers or countries, whereas TFP is often estimated as a growth index, which can be used to compare the efficiency of a farmer or a country from one time period to another.

This report adopts multiple approaches to study the impact of agricultural subsidies on efficiency. The analysis begins with a high-level, cross-country study of the impact of subsidies on TFP. Next, it turns to meta-analyses in the literature that look at the impacts of subsidies on both yields and productivity. Finally, two countries for which adequate household-level data are available, Malawi and Nigeria, provide an opportunity to examine the impacts of reforms of input subsidy programs. The results of all of these analyses are broadly consistent. Subsidies generally lead to increases in both the value of agricultural production as well as yields. However, subsidies also significantly decrease the technical efficiency of farming practices and TFP due to their distortive nature, causing farmers either to use inputs in combinations that are suboptimal or to expand cropping onto marginal lands. Stated differently, although yields may increase with subsidies, TFP typically decreases.

Agricultural subsidies slow global TFP growth

To motivate the broader analysis, this subsection begins with a high-level examination of the impact of subsidies on agricultural productivity. To do so, it uses a global, country-level database on agricultural inputs and production value provided by the United States

Department of Agriculture Economic Research Service.[1] The data set provides estimates for 179 countries and regions from 1961 to 2019.[2] The data are then matched with the country-level information on subsidies discussed in chapter 6 from the Organisation for Economic Co-operation and Development (OECD 2020), the Inter-American Development Bank's Agrimonitor program, and the Food and Agriculture Organization's Monitoring and Analysing Food and Agricultural Policies (MAFAP) program. Regression analysis is then conducted using the *stochastic frontier analysis* technique to isolate the impact of these subsidies (box 7.2).

BOX 7.2
Technical spotlight: Estimating the impact of agricultural subsidies on total factor productivity at the country level

Stochastic frontier analysis is an econometric technique used to estimate a production function as well as an inefficiency parameter simultaneously. This inefficiency parameter accounts for the fact that producers, or in this case countries, might behave suboptimally and fail to maximize production with given inputs.

To estimate the impact of agricultural subsidies on countries' technical efficiency, a quadratic production function is used:

$$\ln Y_{it} = \beta_0 + \beta_1 \ln\left(land_{it}\right) + \beta_3 \ln\left(labor_{it}\right) + \beta_3 \ln\left(Machinery_{it}\right) + \beta_4 \ln\left(Livestock_{it}\right)$$
$$+ \beta_5 \ln\left(Fertilizer_{it}\right) + \beta_{11}\left(\ln land_{it}\right)^2 + \beta_{22}\left(\ln labor_{it}\right)^2 + \beta_{33}\left(\ln Machinery_{it}\right)^2$$
$$+ \beta_{44}\left(\ln Livestock_{it}\right)^2 + \beta_{55}\left(\ln Fertilizer_{it}\right)^2 + \beta_t\left(Year\right) + \beta_R\left(RainDev\right)$$
$$+ \beta_T\left(TempDev\right) + RegionFE + V_{it} - U_{it}. \tag{B7.2.1}$$

The dependent variable is the value of total agricultural output measured in constant 2015 prices from FAOSTAT (FAO 2022). Input variables are also from FAOSTAT, where *land* is total agricultural land in rainfed-cropland-equivalent, *labor* is the total number of persons economically active in agriculture, *Machinery* is the total stock in horsepower, *Livestock* is the total number of standard livestock units, and *Fertilizer* is total inorganic nitrogen, phosphorous, and potash nutrients and organic nitrogen used. *Year* indicates the year of the observation involved, and *RegionFE* are regional fixed effects. *RainDev* is the absolute deviation in rainfall from a long-run mean, and *TempDev* is the absolute deviation in temperature from a long-run mean (Ortiz-Bobea et al. 2021). V_{it} is an error term that captures symmetric random noise.

The final term, U_{it}, represents the technical inefficiency variable. It is used to estimate an inefficiency-effects model to infer the determinants of inefficiency across countries. Since the focus is on the impacts of subsidies, that model is described by the following equation:

$$U_{it} = \delta_0 + \delta_1 MPS_{it} + \delta_2 MPS_{it-1} + \delta_3 MPS_{it-2} + \delta_4 PS_{it} + \delta_5 PS_{it-1} + \delta_6 PS_{it-2}$$
$$+ \delta_7 DC_{it} + \delta_8 DC_{it-1} + \delta_9 DC_{it-2}. \tag{B7.2.2}$$

The inefficiency-effects model includes key policy variables of interest that capture the impact of subsidies on technical efficiency. As is common in the research literature, the subsidy variables

(Continued)

BOX 7.2

Technical spotlight: Estimating the impact of agricultural subsidies on total factor productivity at the country level (*continued*)

are defined in terms of the share of the value of production. The focus is on three broad categories of subsidies: market price support (MPS); producer support (PS) through input or output subsidies; and decoupled (DC) or income support, which is not tied to production. Results from estimating this equation are presented in table B7.2.1, where *L*. indicates that the variable is lagged one year, and *L*2. indicates a two-year lag.

TABLE B7.2.1 Coefficients of technical efficiency drivers

Variable	PS	L. PS	L2. PS	MPS	L. MPS	L2. MPS	DC	L. DC	L2. DC
Coefficient	0.957***	−0.066	0.322	0.255***	0.226***	0.209	−0.106	−0.075	0.878
Standard error	(0.306)	(0.312)	(0.237)	(0.092)	(0.083)	(0.163)	(0.558)	(0.519)	(0.614)

Source: World Bank.
Note: DC = decoupled support; L = one-year lag; L2 = two-year lag; MPS = market price support; PS = producer support.
***$p < 0.01$

Coefficients estimated by the inefficiency-effects model indicate the direction of the effect of each variable on technical inefficiency (that is, a negative coefficient for a given variable means a positive effect on technical efficiency and vice versa). Table B7.2.1 shows that a larger share of producer support in the total value of production is correlated with higher technical inefficiency for the year in which this subsidy is received. Similarly, a larger share of MPS in the value of production is also correlated with higher technical inefficiency in both the current and the one-year-lagged periods. Finally, decoupled support does not have a statistically significant correlation with technical efficiency.

While these results are indicative, they are not definitive evidence of the impact of agricultural subsidies on technical efficiency. Indeed, this analysis has a few important limitations. First, the stochastic frontier approach assumes that the units of production—countries in the analysis—have a common technology, which is not always true. Second, due to data limitations, it is not possible to use less restrictive production functions, such as the translog and control for other exogenous environmental factors, that affect technical efficiency and production. Finally, analysis at the country level can mask significant heterogeneities that underlie the data. For instance, two countries may have similar-looking subsidy schemes at the macro level—that is, they may have similar levels of market price support or investments in public infrastructure, but their subsidy schemes may be implemented in meaningfully different ways. Cross-country regressions like the ones presented here can observe overall trends in these relationships but cannot account for specific nuances. The next two subsections dig deeper into these questions using country-level analyses to explore this issue further.

The results of this exercise align with what would be expected from theory: producer support increases technical inefficiency by distorting incentives and affecting farmers' decisions. Output subsidies artificially increase returns, encouraging expansion of production by bringing suboptimal land into production, shifting crop choices toward those that are subsidized, or encouraging intensification of inputs used beyond their optimal level. Similarly, input subsidies lower the cost of the inputs and encourage overuse, resulting in a suboptimal combination in which factors of production reduce efficiency.

The results show that market price support (MPS) has a similar impact on decision-making since it increases the prices received for production and reduces overall efficiency. However, there are two important distinctions. First, the impact of MPS on inefficiency is about one-fourth as large as the impact of producer support. And second, unlike output and input subsidies, MPS also has a lagged impact, where the share of support received in the previous year is correlated with choices in the current year. These distinctions might be due to the differences in the mechanisms employed. Input and output subsidies are budgetary support measures aimed at lowering the cost of production or supplementing the revenues that the farmer would receive from the market. At the same time, MPS alters the price that the farmer receives in the market; hence the choices made today incorporate the market price received in the previous year. In addition, from a farmer's point of view, producer support might be more certain than MPS. If MPS support comes through trade restrictions, farmers may know that the price will be artificially higher or lower than the market price, but not necessarily by how much. In contrast, with producer support, farmers typically know exactly how much subsidy they will receive. This difference in the certainty of benefits leads farmers to be more likely to make changes in their production decisions based on producer support than on MPS, thus making producer support more distortionary. However, providing MPS only for some crops distorts production in favor of that commodity. One consequence of this distortion that is observed in many regions around the world is the fact that so much rice production occurs in areas where growing conditions are very unsuitable.

Unlike the results of these coupled support measures, there is no statistically significant relationship between technical inefficiency and decoupled support. The lack of a significant relationship does not imply that decoupled support has no impact. Indeed, decoupled subsidies may cover fixed costs and keep farmers in business who otherwise would not be competitive. In addition, decoupled subsidies may lower technical inefficiency by indirectly increasing investments due to the relaxation of financial constraints and by reducing risk aversion due to higher income. However, since decoupled support programs can vary so significantly from one country to another (or from one year to another within the same country), it is likely that this effect is too noisy to be picked up in the data.

In addition to reducing efficiency in agriculture, there is also evidence that agricultural subsidies can reduce efficiency in other sectors as well. Subsidies incentivize the reallocation of resources toward agricultural production. In doing so, they reduce the availability of resources in other sectors. For instance, Krishnaswamy (2018) finds that rice and wheat subsidies in India increase the intensity of inputs dedicated to those crops. In addition to land and other inputs, they also increase demand for agricultural labor, raising wages in the agriculture sector relative to the nonagriculture sector and reducing output in nonagricultural firms. Such unintended consequences are particularly worrisome in low- and lower-middle-income countries, which are struggling to develop productive and competitive manufacturing and service sectors. The same is true for other

Market price support and direct, coupled subsidies reduce farmers' efficiency. Market price support may be less disruptive than direct subsidies because its impact on prices is less certain, and it does not distort the relative prices of inputs.

critical inputs like water and land. Subsidizing agriculture leads to more water being extracted and additional land area being put under cultivation, with enormous consequences for sustainability. These two important issues are discussed in chapters 8 and 9, respectively.

Meta-analyses of the literature find input subsidies increase production but at the cost of efficiency

Agricultural input subsidies have been widely implemented in low- and middle-income countries, starting with fertilizer subsidies in the 1970s in countries like India and Indonesia. The adoption of fertilizers and improved seeds has been credited with the quadrupling of yield in Asia and South America, while fertilizer use and yields have remained low in Sub-Saharan Africa (Carter, Laajaj, and Yang 2021). In response, many Sub-Saharan African countries began implementing input subsidies in the 1990s with the goal of increasing agricultural productivity. One of the primary objectives of input subsidy programs is to encourage technology adoption and increase agricultural productivity (discussed in box 7.1). The argument is that doing so has the potential to begin a "virtuous cycle" where farmers—who are too poor to afford modern inputs and are unaware of their benefits—learn the benefits of adopting new technology or better farming practices and, with their higher profits, can afford to buy the inputs. However, input subsidy programs can also lead farmers to use a suboptimal mix of inputs, which may lead to no gain, or even a reduction, in productivity. It is also an empirical question whether farmers actually realize benefits large enough to persuade them to incur the true cost of inputs once the subsidy ends, and there is evidence that they often do not.

Understanding the impact of input subsidies in agriculture on outcomes such as efficiency, yield, and income remains an important policy question for many countries that continue to allocate resources to such programs or attempt to reform existing programs. An extensive literature has shown the effects of input subsidy programs on various outcomes of interest, ranging from input use (Duflo, Kremer, and Robinson 2011; Kim et al. 2021); technology adoption (Koppmair, Kassie, and Qaim 2017; Mason and Smale 2013); yield (Carter, Laajaj, and Yang 2021; Wossen et al. 2017); income (Mason et al. 2017; Yi, Lu, and Zhou 2016); welfare (Gemessa 2022); and labor markets (Ricker-Gilbert 2014). However, a synthesis of the impacts of input subsidies is limited (Jayne et al. 2018; Ricker-Gilbert, Jayne, and Shively 2013).

Meta-analysis is a technique used to summarize the results and conclusions from an overall body of evidence from multiple studies. The method is a statistical technique that combines results from existing studies to create a "pooled" common estimate and identify sources of disagreement (see online appendix C).[3] In a recent meta-analysis, Minviel and Latruffe (2017) study the impact of public subsidies on farm technical efficiency using stochastic frontier analysis (described in box 7.2) as well as an approach called data envelopment analysis (DEA). Both of these techniques estimate efficiency frontiers against which they rank farmers. It is possible to draw implications about the efficiency impact of the subsidy by comparing performance against the efficiency frontier for farmers who received the subsidy and for those who did not. The meta-analysis includes 195 distinct results from 68 different studies conducted between 1986 and 2014. Overall, the analysis finds that agricultural subsidies reduce farm technical efficiency, consistent with the cross-country analysis presented above. There is also evidence that this negative effect becomes smaller over time, implying that recent subsidy programs are less likely to reduce agricultural efficiency than older programs. This finding may be due to methodological differences in the studies themselves, or it may be due to the fact that more recent subsidy programs are more likely to include decoupled subsidies, which are less likely to cause the distortions in production decisions that lead to efficiency losses.

To complement this work, a more focused meta-analysis was conducted on the effects of agricultural input subsidies on agricultural productivity as well as on living standards, as measured by income or expenditures. The methodology and results are described in box 7.3, and more details are given in Triyana and Nguyen (2022). The meta-analysis finds that, overall, input subsidy programs can increase yields and farmers' incomes, despite having a deleterious effect on technical efficiency. Farmers are incentivized to use larger amounts, but also inefficient mixes, of inputs because they do not pay the full costs of agricultural production. This situation results in higher yields but lower technical efficiency. In the next subsection, these issues are studied at a more granular level, by examining several recent subsidy reforms and implementation efforts and estimating the impacts of these reforms on agricultural efficiency.

BOX 7.3

Technical spotlight: Meta-analysis of the effects of agricultural input subsidies on agricultural production and farmers' incomes

Triyana and Nguyen (2022), in a background paper for this report, begin their meta-analysis with a database search using three databases—Web of Science, Scopus, and JSTOR—followed by title and abstract screening, full-text screening, and finally the meta-analysis regression. The database search involves identifying keywords based on index articles that were previously identified. The search is restricted to articles in economics journals published between 2000 and 2021 in low- and middle-income countries. The keyword search generally aims to find input subsidy programs in agriculture and their impacts on productivity or income or expenditure. Figure B7.3.1 describes the number of articles reviewed at each screening stage.

Contacting authors yielded one study, which brings the total to 12 studies for the meta-analysis. The top three reasons for exclusion in the abstract-screening stage are "not an input subsidy or price intervention" (199 articles), "not developing countries" (82 articles), and "not related to agriculture" (39 articles). The top three reasons for exclusion in the full-text screening stage are "not causal method" (11 articles), "no impact on output, yield, or income" (8 articles), and "nonoriginal research" (5 articles).

Initially, 655 articles were identified for title and abstract screening. After the full-text screening, 12 studies were used in the meta-analysis regression. Based on these articles, input subsidy programs are found, on average, to be associated with an 18 percent increase in yield. For the secondary outcome of interest, input subsidy programs are found to be associated with a 16 percent increase in income or expenditure, which suggests that the increase in yield translates to improvements in living standards. The analysis includes four articles from Malawi, so a back-of-the-envelope calculation is performed based on these estimates and the costs of Malawi's input subsidy program. In this example, a benefit-to-cost ratio of 1.2 is calculated, which implies that a US$1 subsidy is associated with US$1.20 of benefits.

(Continued)

Technical spotlight: Meta-analysis of the effects of agricultural input subsidies on agricultural production and farmers' incomes (*continued*)

Although this calculation suggests that the benefits through higher yield and income outweigh the program costs, the calculation does not consider the longer-term issue of whether farmers continue to use fertilizer when they no longer receive the input subsidies. There is considerable evidence that the rate of attrition is high (Duflo, Kremer, and Robinson 2011, for example), raising questions about whether farmers really find the fertilizer to be worth the cost. In addition, and as discussed above, improving yields and income, while laudable policy goals, say nothing about overall efficiency. Indeed, converting the fertilizer subsidy to cash transfer programs, investments in public goods, agricultural extension facilities, or other types of rural development programs may be more beneficial from an efficiency standpoint.

FIGURE B7.3.1 Number of articles at each screening stage of the meta-analysis

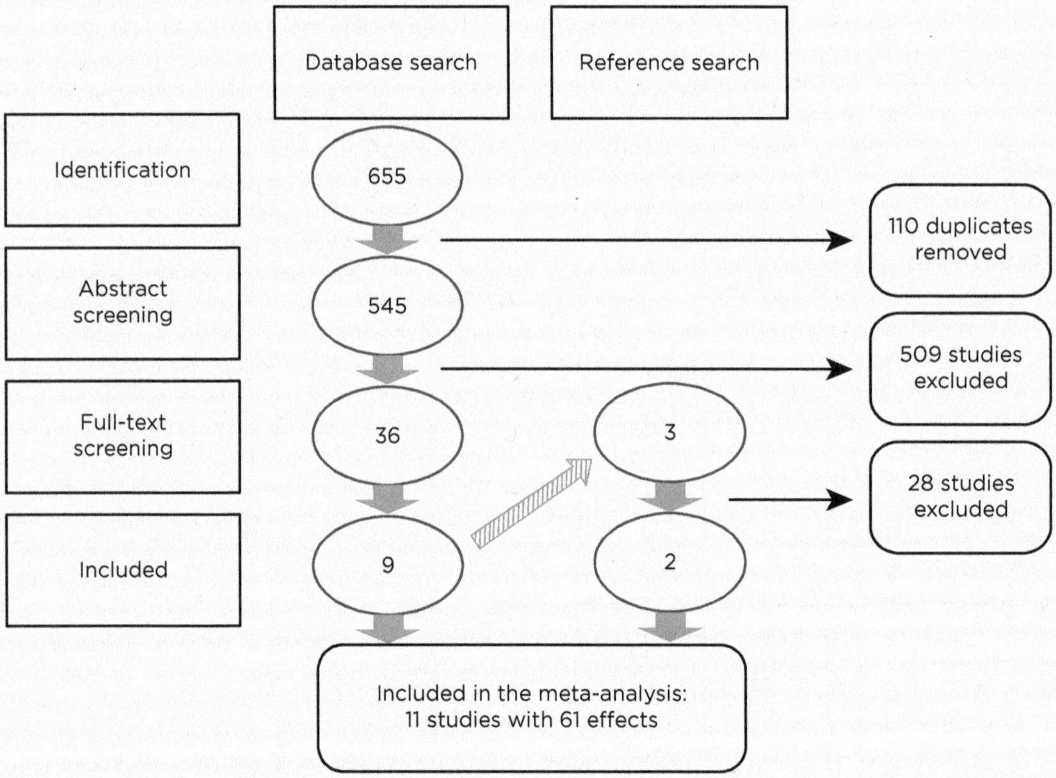

Source: Triyana and Nguyen 2022.
Note: The initial database search yielded 213 articles from Web of Science, 149 articles from Scopus, and 293 articles from JSTOR. Searching the references of the 9 articles yielded 3 additional potentially relevant articles, 2 of which are included in the meta-analysis.

Source: Triyana and Nguyen 2022.

Agricultural input subsidy programs are one of the most pervasive tools that governments use to try to strengthen the self-sufficiency of smallholder farmers and reduce rural poverty in low- and middle-income countries (Duflo, Kremer, and Robinson 2011; Kim et al. 2021). For farmers with limited access to agricultural inputs, input subsidy policies are an enticing way to attempt to increase production in agriculture (Jayne et al. 2018). Most studies that have examined this relationship focus on the effect of programs on yields driven by the increased use of fertilizers or hybrid seeds.[4] As demonstrated in the previous subsection, even though investigations often find that these policies lead to an increase in yields, it is important to examine not only the short-term changes in yields, but also the long-term changes in productivity and efficiency that promote sustainability and enable smallholder farmers to be self-sufficient. Because these policies often constitute a large share of national budgets, they cannot be sustained indefinitely, and thus a gradual phaseout that achieves gains in sustainability after the subsidy ends must be a clear goal (Jayne and Rashid 2013). Furthermore, agricultural input-driven yield might lead to unintended consequences, such as environmental pollution due to excessive input use and excessive land use or deforestation. These issues are investigated in greater detail in subsequent chapters.

This subsection summarizes new research by Park and Kim (2022), which investigates the effects of two recent policy changes in Malawi and Nigeria. The results provide further evidence on the effects of recent input subsidy reforms on farm-level TFP and technical efficiency and the secondary effects on extensification—that is, the expansion of arable land into forested areas. The Malawian and Nigerian governments implemented policy reforms in 2015 and 2012, respectively, which moved in opposite directions: Malawi reduced the magnitude and scope of their input subsidy, while Nigeria increased the magnitude and scope of its subsidy. Several different methodologies—including propensity score matching (PSM), difference-in-differences (DID), and two-stage DEA analysis—are used to explore the plausibly causal impacts of the two reforms and provide consistent results on the impact of the policy changes. The reforms and study methodologies are discussed in more detail in box 7.4 (see also Park and Kim 2022 for additional details).

The estimation using PSM-DID finds that Malawi's Farm Input Subsidy Program (FISP) reform in 2015, which increased subsidized fertilizer prices sevenfold (from MK 500 in 2013 to MK 3,500 in 2015), reduced households' total fertilizer use by 34 percent (see table D.1 in online appendix D for full results). Correspondingly, total production (kilograms) and yields (kilograms per hectare) of participating households also fell by 29 percent and 28 percent, respectively. Nevertheless, no negative impact is found for productivity measured by farm-level TFP. Furthermore, the decline in total production and yields due to the reform did not cause any extensification, as measured by both land expansion and deforestation, implying that farmers did not employ more arable land to compensate for the adverse yield shock and, conversely, did not reduce the amount of arable land as a result of decreased fertilizer use.

DEA analysis provides complementary evidence explaining the mechanism behind the results. The first-stage results with DEA show that the overall efficiency of Malawi's farm households is very low.

Less is more?
Reforms in Malawi that significantly reduced input subsidies resulted in farmers using less fertilizer but had no measurable impact on their technical efficiency.

BOX 7.4
Technical spotlight: Farm input subsidy reforms in Malawi and Nigeria

In 1998 the Malawian government launched an agricultural input subsidy program to improve farm productivity and enhance food security by distributing small packs of improved maize or legume seeds and fertilizers to smallholder farmers (Harrigan 2008). This scheme was criticized for hindering the development of private input suppliers, and so in 2005, a new subsidy policy—the Farm Input Subsidy Program (FISP)—was introduced to limit beneficiaries to smallholders. The most significant change was that farmers received government-provided vouchers that they could redeem for fertilizer at agricultural outlets.

In 2015, another major policy reform was implemented to address concerns about the efficiency and effectiveness of the program. The beneficiary-selection method, previously entrusted to community or religious leaders, was altered to become a centralized, less biased system of targeting beneficiaries (Chirwa et al. 2016). The reform significantly increased farmers' contributions for the purchase of 50-kilogram bags of subsidized fertilizer from MK 500 prior to the reform to MK 3,500 after it.

In addition, the new policy required farmers to pay MK 1,000 and MK 500 for maize and legume seeds, which previously had been free of charge. Lastly, the government implemented this policy through private retailers, not through direct government intervention, to improve the efficiency of logistics (for more detailed information, see Park and Kim 2022, table 1).

In Nigeria, in contrast, various policies have been implemented to improve agricultural productivity. Among them was the fertilizer subsidy policy, similar to the subsidy policies in Malawi. The nontargeted fertilizer subsidy policy has been in effect in Nigeria since the 1970s. Both federal and local governments play an essential role in procuring and distributing fertilizer. However, this scheme was considered inefficient due to its irregular and unpredictable implementation and the limited role of the private sector. In addition, after introducing paper vouchers, forgery and coupon loss became obstacles to effective policy implementation.

In 2012, the Nigerian government launched the Growth Enhancement Support Scheme (GESS) with Electronic Wallet, a so-called "smart" subsidy, to address these problems. Under this innovative mobile phone–based agricultural input subsidy program, farmers could obtain a 50 percent subsidy on each 50-kilogram bag of nitrogen, phosphorus, and potassium and urea fertilizer and a 90 percent subsidy on each 50-kilogram bag of improved seeds (Alabi and Oshobugie 2020).

Data and methodology

To analyze the impact of the reforms implemented in 2015 and 2012 in Malawi and Nigeria, the study uses four waves of Malawi's Integrated Household Panel Survey (2010, 2013, 2016, and 2019) and three waves of Nigeria's General Household Survey (2010, 2012, and 2015), which provide information on the characteristics of agricultural farm households before and after the reforms.

The study employs two methodologies to estimate the impact of the policy reforms on farm-level productivity: (1) propensity score matching (PSM) with difference-in-differences (DID) estimation to find the causal relationship between the reform and outcome variables of interest—yields, total factor productivity (TFP), and cropland—and (2) two-stage data envelopment analysis (DEA) to investigate reform-driven farm-level changes in efficiency.

(Continued)

Technical spotlight: Farm input subsidy reforms in Malawi and Nigeria (*continued*)

DID is a widely used econometric model that captures the statistical difference between changes in outcome variables of interest of the beneficiary group and the unaffected group before and after the event. However, if the treatment is not randomly assigned or is distributed more toward a specific class—as is the case in the subsidy programs in Malawi and Nigeria—the two groups cannot be inherently comparable. Therefore, the study first calculates the farmer's "propensity score," which is the probability of being exposed to the policy based on the observed characteristics of the farmer. Then the DID model is applied by assigning a comparable control group with the same propensity score but no treatment. This approach allows for an interpretation of the results that is plausibly causal—that is, the changes in output and efficiency that are observed were plausibly caused by the change in policy.

The study also uses two-stage analysis with DEA to explore how the reform changes the technical efficiency of individual farms. Using DEA as a first step, the technical efficiency of the individual farmers of each country is derived. This approach measures how efficiently households are able to turn a set of agricultural inputs into output, relative to other households. A level of inefficiency is defined as the distance between efficient producers and inefficient producers. Similar to the stochastic frontier analysis approach discussed in box 7.2, the determinants of efficiency are explored in the second stage, focusing on the influences of the input subsidy program.

Source: Park and Kim 2022.

Around 70 percent of farming households produce at less than 20 percent of their achievable efficient production (table D.3 in online appendix D). This finding may indicate that, more than subsidized inputs, there are other binding constraints on productivity, such as lack of information on cropping techniques and technology or the ineffectiveness of some inputs (see box 7.8 later in this chapter). The second-stage regression shows how the FISP reform changed farmers' efficiency.[5] With higher subsidies, more fertilizer was used, but in ways that were inefficient and did not have the intended impact of increasing yields. When FISP was reformed in 2015, FISP participants reduced their fertilizer use, but improved their farm efficiency (table D.4 in online appendix D). Thus, by raising the subsidized price of fertilizer, farmers used less of it, which increased the technical efficiency of FISP beneficiaries. The result highlights the need for identifying the constraints on productivity and delivering policy packages that target these constraints.

In Nigeria, the introduction in 2012 of the e-voucher Growth Enhancement Support Scheme (GESS) expanded access to fertilizer subsidies, resulting in a tenfold increase in the total amount of fertilizer used by participating farmers and a 43 percent increase in total output. However, no evidence is found that increased fertilizer use led to higher productivity, measured by TFP. Instead, participating farmers expanded their arable land by 10 percent and applied the additional fertilizer to the new farmland. This finding highlights the need for complementary measures—in this case, higher yields led to undesirable rebound (Jevons) effects into natural habitats that needed to be controlled through a different package of enforceable policies.

More is less?

A subsidy reform
in Nigeria that
increased the size
of, and access
to, a fertilizer
subsidy resulted
in more fertilizer
being used and
an expansion of
cropland, but
a significant
reduction
in technical
efficiency.

The DEA analysis for Nigeria provides consistent evidence regarding farm efficiency. Around one-third of households have an efficiency level below 20 percent of their achievable production (table D.3 in online appendix D). The input subsidy is found to have further detrimental effects on farmers' technical efficiency. Specifically, farmers participating in the e-voucher program exhibited a 4.6 percent reduction in farm efficiency. In contrast to the results from Malawi, changes in fertilizer use were not the main driver of efficiency loss. Instead, land use expansion was the major contributor to farm-level inefficiency, reducing efficiency by 27 percent (table D.4 in online appendix D).

The findings from the study (using two reforms that work in opposite directions) provide a consistent lesson. First, subsidy-driven fertilizer use does not necessarily lead to improvements in farm-level productivity or technical efficiency. In Malawi, the increase in farmers' out-of-pocket cost for subsidized fertilizer and the subsequent significant reduction in the fertilizer application rate led to an increase in farm-level technical efficiency. In this regard, the Malawi reform achieved a certain level of success by reducing the fiscal outlays of the subsidy, while increasing farm-level efficiency. In Nigeria, increasing the size of, and access to, the fertilizer subsidy led to an increase in yields, but the effect was much smaller than the increase in fertilizer application, demonstrating how the subsidy led to the inefficient use of fertilizers.

These results offer several important takeaways for policy. Providing input support without providing comprehensive education and consultations or ensuring that appropriate technology and complementary inputs are available can result in the inputs being used improperly, leading to wasted fertilizers and government spending. In addition, providing excessive inputs can lead to unintended environmental consequences—an issue addressed in subsequent chapters. While increased fertilizer use in Nigeria led to an expansion of arable land, potentially causing loss of forested areas, decreased fertilizer use in Malawi did not reduce the use of arable land that was already cleared. This finding likely follows from the fact that expanding farmland requires a large fixed cost, which, once paid, cannot be recovered by reducing land use. More attention needs to be paid to the extensification promoted by agricultural support programs (Houssou and Chapoto 2015).

Recent years have seen the feminization of agriculture, especially in low- and middle-income countries. Women comprise up to 48 percent of the agricultural workforce in these regions, reaching close to 60 percent in several Sub-Saharan African countries (FAO 2020; Giroud and Huaman 2019); yet together with other marginalized groups, women have considerably less access to landownership and a weaker voice in resource management decisions. Furthermore, agriculture for these groups is characterized by smaller farms, which are devoted to household consumption and have less access to input markets, compared to male head-of-household farms (Phiri et al. 2022). Given the lack of data, taking a closer look at the agricultural gender gap in terms of subsidies was not possible. Nevertheless, box 7.5 describes the results of an analysis examining the impact of an irrigation scheme in the Peruvian Andes and likewise finds that TFP and yields were unaffected, but access to irrigation did lead to an expansion of cropland. These case studies draw attention to the need for complementing input interventions with policies that address incentives to expand land use.

BOX 7.5
Technical spotlight: The Mi Riego highland irrigation program in Peru

Irrigation subsidies have many benefits, but also unintended costs. Irrigation provides farmers with greater stability and control over their harvests, reducing the risk of weather shocks and allowing for the expansion and diversification of cultivated crops. Stability and growth enhance food security and increase labor opportunities, which often benefit the rural poor. Furthermore, positive externalities arise when markets for inputs, labor, and financial services develop around the agriculture industry. However, not all irrigation externalities are positive, not all benefits are distributed efficiently or equitably, and not all markets are able to adjust optimally. As a result, hydrological investments can lead to the overuse of water, environmental damage, net economic losses, and even increased poverty (Duflo and Pande 2007).

While only a broad analysis of the full benefits and costs of irrigation projects—which accounts for the internalization of externalities as well as estimates of direct, indirect, and implicit subsidies—can reveal the desirability of such public expenditures, a more focused examination of farm-level agricultural production is an essential starting point. Do irrigation projects lead to higher yields, productivity, and, more important for the farmer, profits? Undoubtedly, the answer to this question depends on the scale, location, design, and implementation of each specific project. Yet valuable and generalizable lessons can be taken from each study in the context of economic development. This box seeks to draw lessons from an irrigation project carried out in the highlands of Peru.

The Mi Riego fund in Peru (later known as Sierra Azul) is an ongoing initiative funded by the government of Peru with the purpose of reducing the gap in irrigation infrastructure for agriculture in areas with the most potential to decrease poverty. The program began operations in 2013. At that time, benefits were limited to farmers with plots of land located 1,500 or more meters above sea level. The project was also initially limited to marginal high-altitude agricultural areas in the Peruvian Sierra, where small-scale agriculture is disconnected from industry-level national and international markets. Indeed, these farms are owned mostly by rural, poor indigenous Aymara and Quechua communities, producing for own consumption or for localized trade. Limiting the original fund to high-altitude areas, relevant to the data used here, was meant to target regions with limited agricultural potential because of the land's characteristics (the altitude requirement decreased and eventually disappeared through time).

To evaluate the impact of this subsidized irrigation, the study looks at information on parcels of farmland from the National Agricultural Survey for the years 2016 through 2019. By georeferencing the exact location of more than 500 projects carried out by Mi Riego up to the year 2017, the distance of each parcel to the closest irrigation project is used as part of a difference-in-differences (DID) strategy to compare the before-and-after effects of irrigation according to distance. The crop-level data considered are only for conglomerates that were sampled continuously throughout each survey year, thus creating a pseudo-panel at the conglomerate level.

A regression specification that includes department-specific time trends, conglomerate, year and crop fixed effects, and controls for farmer demographic and labor market characteristics (education, experience, age, training, native language, gender, access to credit) as well as temperature and precipitation suggests that Mi Riego irrigation projects may have increased output, but likely through the extensification of croplands, since yields were not affected (table B7.5.1).

(Continued)

BOX 7.5
Technical spotlight: The Peruvian highland irrigation program Mi Riego (*continued*)

Moreover, based on information regarding the revenues and costs incurred by each farmer given in the survey, the regression analysis finds that farmer revenues and profits were not affected by the irrigation projects. Extensification and total production increased only marginally—3.6 percent and 2.8 percent, respectively. This increase was not enough to push up revenues or profits for small farmers.

TABLE B7.5.1 Regression results of the effect of the Mi Riego program on several outcomes, Peru

Variable	Yield	Output	Area	Revenue	Profits	TFP (IV)
DID	−0.00774	0.0284*	0.0361**	0.000133	0.000318	−0.0132*
	(0.00493)	(0.01580)	(0.01610)	(0.01630)	(0.03850)	(0.00795)
DDD	0.0307**	0.00667	−0.0240	0.0320	0.0589	0.00156
	(0.01320)	(0.02310)	(0.02250)	(0.02970)	(0.07880)	(0.02190)
Labor (IV)	No	No	No	No	No	Yes
Harvest area (IV)	No	No	No	No	No	Yes
Observations	5,669	5,669	5,669	2,399	1,196	3,816
R^2	0.843	0.673	0.627	0.574	0.394	0.782

Source: World Bank.
Note: Standard errors are in parentheses. Robust standard errors are clustered at the conglomerate level. Conglomerate, year, and crop fixed effects are included in all regressions. Weather around the time of planting and farmer characteristics are included in all regressions. The sample is limited to Mi Riego projects completed by the end of 2017 and only for transitory crops. DDD = difference-in-difference-in-difference (Mi Riego × Post × low rainfall); DID = difference-in-difference (Mi Riego × Post); IV = instrumental variable; TFP = total factor productivity.
*$p < 0.1$ **$p < 0.05$

To be sure, the focus can be shifted from yields to total factor productivity (TFP), which will measure increases in output when controlling for inputs. Following Aragón, Oteiza, and Rud (2021), an instrumental variable strategy is used to control for labor and land, since omitted variables could affect both input use and TFP. The results confirm the suggestion that, despite greater production, farmer productivity has not improved. Indeed, table B7.5.1 reports a possible negative effect on TFP, which could be a sign of imperfect input markets that lead to input substitution as agriculture expands, resulting in inefficient mixes of irrigation with other inputs (Jones et al. 2020; also Dillon and Fishman 2019). Moreover, it is likely that the expansion was to less productive land, also explaining a drop in TFP.

Finally, the identification includes a DID setup in order to account for unexpected dry spells. Here, yields are found to increase for farms closer to irrigation projects during dry months, although TFP remains unaffected in general.

For the years covered in this analysis, the overall suggestion is that improved irrigation for poor farmers in Peru may have led to extensification of agricultural land, without improving farming efficiency. At the same time, irrigation exhibited the desirable effect of protecting yields during times of low rainfall. With these findings in mind, the results of the analysis should be interpreted with caution, as they likely point to the difficulty of increasing farmer productivity for marginalized groups living in difficult terrain, more than being an externally valid characterization of all irrigation schemes.

Agricultural subsidies and distributional goals

Efficiency and productivity gains are not the only justification for implementing agricultural subsidies. Indeed, 80 percent of the world's extreme poor and 75 percent of the moderate poor live in rural areas (Castaneda et al. 2016), engaging largely in agriculture. In many cases, agricultural subsidy programs double as rural development, livelihood, and safety net programs. Therefore, another important indicator of the usefulness of agricultural subsidies is how progressive they are—that is, what share of the subsidy is captured by the lowest income quantiles. Nevertheless, global estimates on the equity impacts of agricultural subsidies are lacking, particularly in low- and middle-income countries.

In richer countries like Canada, the United States, and European Union members, evidence suggests that support to agriculture is distributed disproportionately to larger farms (Moreddu 2011). This outcome is natural when agricultural support is tied to production or factors of production like land (FAO, UNDP, and UNEP 2021). If inputs and outputs are subsidized in proportion to their quantity, larger farms and richer farmers will always receive a disproportionate share of the benefits. Female-headed farms are also less likely than male-headed farms to benefit from fiscal subsidies, because female-headed farms tend to be smaller (FAO 2011).

To contribute to this literature, the following discussion presents new analyses examining the distributional impacts of agricultural subsidies in multiple countries. First, the spatial incidence of output subsidies in 16 countries is examined to determine whether these subsidies are more likely to accrue to poorer regions within countries. Then, input subsidy programs in six countries for which sufficient data are available at the household level—Bangladesh, Malawi, Nigeria, Tanzania, Vietnam, and Zambia—are examined to estimate the distributional impacts of these subsidies.

The spatial incidence of subsidies

This subsection presents a new global analysis examining the distributional impact of output subsidies. Despite the dearth of microdata on output subsidies, the creative use of crop production data and crop-level national statistics on output subsidies makes it possible to assess the broad contours of the distributional impact of these subsidies, even if these estimates are not precise. As with any subsidy linked to production amounts, output subsidies will naturally be regressive, since richer farmers will always produce more output and therefore receive a larger share of the subsidy than poorer farmers. Nevertheless, this analysis uses a lower threshold for what is considered progressive. It examines whether poorer regions in each country receive a relatively larger share of output subsidies than their share of agricultural production value. For instance, if the poorer half of the country produces 20 percent of total agricultural production, but receives 40 percent of the subsidy, the subsidy can be said to be spatially progressive. This information is especially useful for addressing development challenges in lagging regions, where poverty tends to be spatially concentrated. In addition, because poorer farmers are more likely to live in poorer regions, assessing the broad contours of subsidies' distributional impact can give some indication of the overall progressiveness or regressiveness of the subsidy scheme. The analysis finds that output subsidies are relatively progressive in 10 countries and regressive in 6, based on this low bar for what is considered progressive. The analysis also finds that if countries were to shift their output subsidies on rice, a very water-intensive crop responsible for exacerbating harmful externalities like water scarcity, salinization, and carbon dioxide emissions, to maize, then their subsidies would become significantly more spatially progressive.

The analysis assumes that a spatially distributionally neutral subsidy of output subsidies for a particular crop throughout a country is equal to the spatial distribution of that crop. Put another way, if a particular region in a country produces 5 percent of the country's maize, then that region likely receives 5 percent of the output subsidy for maize. This assumption will be generally true enough to make inferences about the distribution of subsidies, but clearly may not always hold. Reasons why it might not hold and their implications are discussed further below.

The analysis uses three data sets: the Spatial Production Allocation Model from the International Food Policy Research Institute (Wood-Sichra, Joglekar, and You 2016), which provides gridded data on agricultural production by crop; OECD's data set on output subsidies by country and crop (OECD 2022); and gridded data on GDP per capita (Kummu, Taka, and Guillaume 2018). Box 7.6 describes these data sets and details about the methodology.

BOX 7.6

Technical spotlight: Estimating the distributional impact of output subsidies

Three data sets are used to examine the distributional impact of output subsidies. First, disaggregated crop production data are taken from the Spatial Production Allocation Model (SPAM) (Wood-Sichra, Joglekar, and You 2016). SPAM estimates the spatial distribution of crop production for 42 crops by combining crop production statistics from national or subnational administrative regions with crop-specific suitability information based on the local landscape, climate, and soil conditions. In doing so, coarse data from countries or subnational units are disaggregated into grid cells with a 10 × 10-kilometer resolution. The most recent SPAM statistics for the year 2010 are used.

The Organisation for Economic Co-operation and Development (OECD) database, discussed in chapter 6, provides crop-level data on output subsidies (OECD 2022). In particular, the producer single commodity transfer (PSCT) indicator is used, which measures the annual monetary value of gross transfers from consumers and taxpayers to agricultural producers. PSCT includes both market price support (MPS)—which results largely from trade policies like import or export bans—as well as output subsidies that are fiscal in nature, such as direct transfers to farmers per unit of crop produced. PSCT can be positive or negative (while fiscal subsidies can only be positive, MPS can be positive or negative, depending on whether trade policies increase or decrease market prices). Positive values represent benefits to farmers, whereas negative values represent costs to farmers.

OECD reports PSCT for 65 commodities in 20 countries. After matching SPAM data with the OECD data, 23 commodities in 16 countries remain in the analysis. These crops are wheat, rice, maize, barley, potatoes, sorghum, cassava, lentils, other pulses, soybeans, groundnuts, coconut, palm oil, sunflower, rapeseed, sugar, cotton, coffee, cocoa beans, tea, tobacco, bananas, and plantains. These 23 commodities represent 91 percent of PSCT paid to all commodities in 2010 in countries in the OECD data set, excluding livestock, meat, and dairy products. They also account for 34 percent of total agricultural production values worldwide.

(Continued)

BOX 7.6

Technical spotlight: Estimating the distributional impact of output subsidies (*continued*)

As noted, PSCT transfers to each crop are assumed to be distributed spatially in the same way as crop production. Data are aggregated into 0.5 × 0.5 degree grid cells (approximately 56 × 56 kilometers at the equator) to map the distribution of PSCT globally and measure PSCT transfers to each region. Grid cells with less than 50 percent of cropland according to the European Space Agency's land cover–land use database[a] are excluded from the analysis. Total PSCT transfers are measured for each grid cell by summing all PSCT paid to each crop in that grid cell (see figure E.1 in online appendix E).

Finally, data on income per capita for each grid cell are from Kummu, Taka, and Guillaume (2018). Using these data, grid cells are ranked within each country from poorest to richest (see map E.2 in online appendix E), making it possible to examine whether relatively poorer grid cells within a country receive a disproportionate share of PSCT.

Following the literature on estimating income inequality, Lorenz curves are calculated to plot the cumulative PSCT received for a given percentile of GDP per capita. These curves are illustrated in figure B7.6.1. In the example, the cumulative amount of total PSCT (plotted in blue) measures the share of PSCT at each point received by grid cells in the GDP per capita percentile on the x-axis or lower. For instance, by tracing upward from the point 0.2 on the x-axis, it is evident that the bottom 20 percent of grid cells by income receive about 90 percent of total PSCT. The total value of all agricultural production (the red line) is plotted in the same way. The red line shows that these 20 percent of grid cells produce about 10 percent of agricultural production.

Borrowing from the risk-aversion literature, PSCT is said to *first-order stochastically dominate* production value if the blue line sits above the red line for the full distribution. This term implies that the country's poorer regions receive a larger share of PSCT than their share of agricultural production. In this case, it is said that the subsidy is *progressive*. If the red line stochastically dominates the blue line (that is, if the red line sits above the blue line for the full distribution), then PSCT is considered to be *regressive*. If the lines intersect at least once, then neither stochastically dominates the other of the first order. In this case, it is possible to turn to second-order stochastic dominance, where the area under each of the curves is compared, and whichever curve is larger second-order stochastically dominates the other. If PSCT is negative (as is the case for Argentina, India, and Ukraine), then the analysis is flipped—if the blue line stochastically dominates the red line (first or second order), then PSCT is regressive and vice versa. This situation occurs because negative PSCT acts more like a tax than a subsidy.

(Continued)

Technical spotlight: Estimating the distributional impact of output subsidies (*continued*)

FIGURE B7.6.1 **Example of Lorenz curve for PSCT**

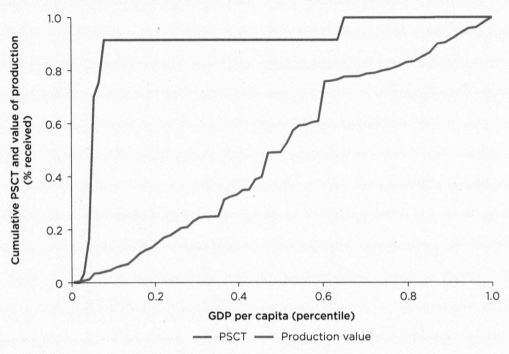

Source: World Bank.
Note: The figure shows an example of the analysis for a single country. The blue line plots the cumulative amount of total PSCT received by each income percentile along the x-axis. The red line plots the same information for total agricultural production value. PSCT = producer single commodity transfer.

———————
a. See https://www.esa-landcover-cci.org.

For each country, the distribution of the producer single commodity transfer (PSCT) subsidy and the distribution of output values are plotted following the description in box 7.6. The results of the analysis are shown in table 7.1. The first column shows the area under the curve (AUC) for production value (the red line in figure B7.6.1). The second column shows the AUC for PSCT (the blue line in figure B7.6.1). The third column is the difference between the two. If the difference is negative (positive)—that is, the AUC for PSCT is smaller (greater) than the AUC for production value—then the subsidy is regressive (relatively progressive).[6] The results show that PSCT is relatively progressive in 10 countries and regressive in 6, according to second-order stochastic dominance. For the full distribution of commodities in all 16 countries, see online appendix F.

The countries in table 7.1 are ranked according to how regressive their subsidies are (column 3), where larger positive numbers are more progressive, and larger negative numbers are more regressive. The Philippines has the most regressive subsidies, whereas Brazil has the most progressive. Figure E.1 in online appendix E plots the differences between AUCs and GDP per capita. While there is a slight inverted-U trend, the relationship is not very strong. While the differences in the areas under the curve may seem small, even a value of 0.01 is meaningfully different from zero. A score of

TABLE 7.1 Distribution of producer single commodity transfer relative to agricultural production value

Country	Current analysis			Policy experiment (subsidy allocation switched from rice to maize)		
	(1) AUC production value	(2) AUC PSCT	(3) Difference between AUCs	(4) AUC PSCT with policy	(5) Difference between AUCs with policy	(6) Policy impact
Philippines	0.64	0.51	−0.13	0.62	−0.02	0.11
Mexico	0.58	0.52	−0.05	0.52	−0.05	0.00
Canada	0.59	0.54	−0.05	0.51	−0.08	−0.03
Japan	0.36	0.33	−0.02	0.69	0.33	0.36
Vietnam	0.47	0.46	−0.01	0.55	0.08	0.09
Türkiye	0.55	0.54	−0.01	0.54	−0.01	0.00
Argentina[a]	0.50	0.49	0.01	0.49	0.01	0.00
Korea, Rep.	0.51	0.51	0.01	0.55	0.05	0.04
China	0.58	0.60	0.02	0.59	0.02	0.00
Colombia	0.52	0.55	0.02	0.61	0.09	0.06
United States	0.50	0.55	0.05	0.55	0.05	0.00
Ukraine[a]	0.61	0.56	0.06	0.55	0.07	0.01
India[a]	0.57	0.51	0.06	0.49	0.08	0.02
Indonesia	0.63	0.71	0.08	0.72	0.08	0.01
Kazakhstan	0.67	0.77	0.10	0.74	0.08	−0.02
Brazil	0.48	0.62	0.15	0.64	0.16	0.02

Source: World Bank.
Note: The countries in the table are ranked and color-coded according to how progressive or regressive their subsidies are, as per the values in column 3. Larger positive numbers are more progressive, and larger negative numbers are more regressive. AUC = area under the curve; PSCT = producer single commodity transfer.
a. Identifies countries with negative PSCT values. The analysis is flipped for these countries to be consistent with measured values for other countries. In the case of negative PSCT, differences between AUCs are multiplied by the negative one.

zero implies that the subsidy is received in equal proportion to production value. Consequently, if a region produces 10 percent of total agricultural production, it also receives 10 percent of the output subsidy.[7] A score of 0.01 is the equivalent of starting with this equal proportion distribution and shifting 50 percent of the subsidy received by one quintile of the distribution to the quintile below (for example, from the 40th income percentile to the 20th income percentile). Likewise, a score of 0.02 is the equivalent of shifting 50 percent of the subsidy two quintiles below or 100 percent of the subsidy one quintile below.

Rice receives the most subsidies among all commodities across countries (see table E.1 in online appendix E). For example, in Japan, the Republic of Korea, and Vietnam, more than 80 percent of all subsidies are paid to the production of rice. However, in China and Kazakhstan, PSCT is negative for the production of rice (implying that trade restrictions artificially lower the price of rice below the market price), while the PSCT is positive overall. In India, overall PSCT is negative, with 33 percent going toward the production of rice. Following rice, the crops that receive the largest share of PSCT are maize, wheat, sugar, and soybeans. Although the distributions of PSCT transfers on crops across countries vary considerably, the distribution of production tends to be more progressive for maize than for other crops—that is, maize tends to be produced in poorer areas of the country.

One possible policy experiment is to estimate the impact of a shift in subsidy from rice to maize on the distribution of the subsidy. Columns 4, 5, and 6 of table 7.1 show the results of such an experiment. Column 4 recalculates the AUC of PSCT if each country were to switch its allocation of PSCT from rice to maize. Column 5 then recalculates the difference between the production value AUC and the PSCT AUC (column 4 minus column 1). Column 6 shows the overall impact of this policy change (column 5 minus column 3). The results are quite stark. Nine countries would achieve more progressive output subsidies from this policy change, while only Canada and Kazakhstan would have more regressive subsidies. Japan and Vietnam would move from overall regressive subsidies to relatively progressive subsidies. Japan, the Philippines, and Vietnam would benefit the most from shifting their PSCT from rice to maize.

With a few additional assumptions, this analysis can be repeated for input subsidies. Box 7.7 describes how this analysis is done and presents the results, which are mixed. About half of countries have spatially progressive input subsidies, and half have spatially regressive subsidies.

The findings here are subject to several important caveats. First, the bar for a subsidy to be spatially progressive in this exercise is quite low and is focused more on regional equity than on individual impacts. The analysis examines whether poorer areas receive a larger proportion of subsidy relative to their proportion of agricultural production. Nevertheless, in absolute terms, poorer farmers tend to receive a smaller share of output

---

BOX 7.7

Technical spotlight: Estimating the distributional impact of input subsidies

With a few additional assumptions, the analysis can be extended to examine the spatial incidence of input subsidies. While data on crop-specific input subsidies are not available, the data sets described in chapter 6 provide country-level data on input subsidies. In addition, the same data sets used to examine the distributional impact of output subsidies are used here.

Since input subsidies are reported at the country level and not the crop level, two additional assumptions are made to estimate their spatial incidence. First, it is assumed that input subsidies are used mainly for the production of five crops—potatoes, sugarcane, maize, rice, and wheat. This assumption is based on literature surveys (Dorward 2009; Hemming et al. 2018), which found that these crops are the most likely to be targeted by such programs as well as the fact that they tend to be among the highest-yielding crops. Second, it is assumed that the share of input subsidies that these crops receive is the same as the share of their values. For instance, if maize accounted for 10 percent of the overall value of production among these crops in a country, then maize would receive 10 percent of the country's total input subsidies. If a grid cell produces 20 percent of total maize in that country, that grid cell would be estimated to receive 10 percent × 20 percent = 2 percent of the country's total input subsidy. In sum, these

(Continued)

BOX 7.7

Technical spotlight: Estimating the distributional impact of input subsidies (*continued*)

MAP B7.7.1 **Estimated distribution of input subsidies across countries**

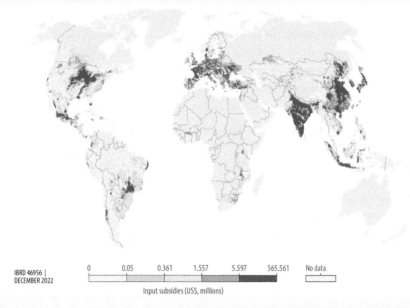

IBRD 46956 |
DECEMBER 2022

| 0 | 0.05 | 0.361 | 1.557 | 5.597 | 565.561 | No data |

Input subsidies (US$, millions)

Source: World Bank.
Note: Data are aggregated into 0.5 × 0.5 degree grid cells (approximately 56 × 56 kilometers at the equator) to map the distribution of input subsidies globally and measure the subsidies at each region. Grid cells with less than 50 percent cropland are excluded from the analysis.

assumptions assign input subsidies based on the value produced by selected crops in each region. Based on these assumptions, map B7.7.1 shows the distribution of input subsidies across the world.

Using these data, the methodology employed for output subsidies is repeated, as described in box 7.6. Table B7.7.1 shows that the spatial incidence of input subsidies is spatially regressive in 9 countries (which includes the EU-28 as a single unit), spatially progressive in 11 countries, and flat in the 3 remaining countries. The countries in table B7.7.1 are ranked according to how progressive their subsidies are. As with output subsidies, the results are mixed and can vary significantly from one country to the next. However, some patterns do emerge. Countries with progressive input subsidy schemes tend to be wealthier—that is, upper-middle income and high income. Countries in Africa tend to have spatial incidences that are more regressive. The clear exceptions to these patterns are Kenya, which has the most progressive spatial incidence of all countries, and the EU-28, which is among the most regressive. The European Union result is not surprising given that the input subsidy spatial distribution correlates with output and its value. The European Union Common Agricultural Policy has also received significant criticism for how regressive it is, particularly in its early stages (Kaditi and Nitsi 2011; OECD 2003).

(Continued)

BOX 7.7

Technical spotlight: Estimating the distributional impact of input subsidies (*continued*)

TABLE B7.7.1 Distribution of input subsidies relative to agricultural production value

Country	AUC production value	AUC input subsidy	Difference between AUCs
Philippines	0.48	0.63	−0.15
Brazil	0.34	0.48	−0.14
EU-28	0.54	0.61	−0.07
South Africa	0.52	0.57	−0.05
Burkina Faso	0.48	0.51	−0.03
Kazakhstan	0.64	0.67	−0.03
Malawi	0.54	0.56	−0.02
Bolivia	0.48	0.49	−0.01
Ukraine	0.60	0.61	−0.01
Vietnam	0.47	0.47	0.00
Korea, Rep.	0.52	0.52	0.00
Mozambique	0.73	0.73	0.00
India	0.58	0.57	0.01
Türkiye	0.56	0.54	0.01
United States	0.52	0.50	0.02
Argentina	0.52	0.50	0.03
Canada	0.62	0.59	0.03
Colombia	0.56	0.52	0.03
Indonesia	0.69	0.64	0.06
China	0.64	0.58	0.06
Mexico	0.67	0.58	0.09
Japan	0.45	0.35	0.09
Kenya	0.82	0.68	0.15

Source: World Bank.
Note: The countries in the table are ranked and color-coded according to how spatially progressive or regressive their subsidies are, as per the values in the last column. Larger positive numbers are more progressive, and larger negative numbers are more regressive. AUC = area under the curve.

subsidies overall because they produce less output than richer farmers. Thus, if a subsidy is marked as being progressive, this designation is only true in a regional sense.

The analysis likely overestimates the relative progressivity of a subsidy because it assumes that the distribution of output subsidies for a single crop is the same as the distribution of production of that crop. Since poorer farmers are less connected to markets and more likely to produce for household consumption, they are less likely to benefit from output subsidies.

The spatial nature of the analysis provides an additional reason why the progressiveness of PSCT subsidies is likely to be overestimated. The analysis examines variation across grid cells of differing GDP per capita. However, within grid cells, there is also heterogeneity of GDP per capita as well as of crop production and PSCT received. Furthermore, it is likely that richer farmers within grid cells are going to receive a disproportionate amount of PSCT subsidy. By aggregating to the grid cell as the unit of analysis, that variation within grid cells is lost.

Finally, the analysis is based on equity objectives. Subsidies are considered progressive if they are paid in greater proportions to poorer areas and smaller proportions to wealthy areas. However, governments may consider other objectives such as food security and environmental preservation when allocating subsidies across regions. In this case, a government may be willing to accept a trade-off where the subsidy is regressive but fulfills other policy goals (for instance, governments may try to safeguard nationwide food security by supporting large agricultural producers in certain regions).

The regressive nature of input subsidy schemes

Countries provide agricultural input subsidies for a variety of reasons. These subsidies tend to focus on two key inputs: inorganic fertilizer and hybrid or improved seeds. While the main objective of such subsidies is to increase long-term national productivity by encouraging the application of agricultural inputs, in many cases governments specifically target poor households, particularly those lacking access to these inputs due to affordability constraints or lack of access to markets (Jayne et al. 2018). These subsidies have the potential to be important sources of welfare and safety nets by simultaneously reducing production costs and increasing the revenues that farmers generate (Hemming et al. 2018; Morris 2007).

Nevertheless, contrary to their stated intention, input subsidies tend to benefit richer farmers more than poorer farmers (Ricker-Gilbert and Jayne 2012). This situation is perhaps inevitable when subsidies are not targeted, but even occurs with subsidies targeted at the poor. Access to subsidies is not always uniform. Those with better political connections (Dionne and Horowitz 2016) or better physical connections to markets (Mason and Ricker-Gilbert 2013) are more likely to access the subsidy. Recent reform efforts like those in Malawi and Nigeria have attempted to reduce these biases. Even when access is prima facie equal, co-spending requirements can make accessing the subsidy unaffordable to the poor. For example, if subsidies lower the price of fertilizers or seeds by 50 percent, farmers' 50 percent share of the cost may still be too much of a burden for farmers with lower incomes. Lack of experience with certain inputs or past experiences perceived as negative may discourage use among the poor (box 7.8).

Distributional impact of subsidy schemes in six countries

To shed more light on these issues, this subsection examines the distributional impact of various forms of agricultural input subsidy schemes implemented in four African countries (Malawi, Nigeria, Tanzania, and Zambia) and two Asian countries (Bangladesh and Vietnam). Although there are differences in time and scale, each country included in this analysis has provided a farm-input subsidy using a similar framework. The specifics of each of the subsidy schemes are discussed in online appendix G. While there is heterogeneity across different countries, several important takeaways emerge from the analysis. First, participation rates among the poor tend to be lower, even in countries where subsidies are designed specifically to target the poor. Second, the value of subsidies received tends to be

The varying effectiveness of inorganic fertilizer application on smallholder-managed fields

Fertilizer application is a straightforward remedy to increase agricultural productivity and subsequently reduce food insecurity among smallholder farmers in low- and middle-income countries. Some governments have attempted to promote fertilizer adoption through subsidies to smallholder farmers (Holden 2019). However, productivity gaps persist in smallholder-managed fields, leaving the effectiveness of these policies unclear (Jayne et al. 2018).

In some regions, particularly in East and South Asia, overuse of fertilizer is the primary factor hindering the effectiveness of fertilizer use. Smallholder farmers often do not have information about fertilizer recommendations such as the appropriate amount of fertilizer and how densely to plant crops after fertilizer application. For example, in some Asian countries, the application of heavily subsidized nitrogen fertilizers far exceeds the recommended dosage, while application rates of unsubsidized fertilizers like potassium and phosphorus are far below recommended levels (Basak 2010; Islam et al. 2022). Evidence also shows that farmers may knowingly overuse fertilizer to compensate for lack of other inputs in order to achieve their yield goals, which increases inefficiency since the simple addition of fertilizer does not necessarily lead to higher yields (Ren et al. 2021). When use is excessive, not all fertilizer will be absorbed by crops, leading to nutrient leakage that can inhibit the growth of crops (Izaurralde, McGill, and Rosenberg 2000; Wang et al. 2019). In India, for example, only 32 percent of nitrogen is absorbed by plants, compared with 52 percent in Europe and 68 percent in Canada and the United States (Zhang et al. 2015). In Shandong Province, China's program showed that fertilizer use could be reduced by 22 percent without any loss of yield (Huang et al. 2012). Moreover, excessive use of inorganic fertilizer causes environmental damage by increasing greenhouse gas emissions and ground and water pollution (Izaurralde, McGill, and Rosenberg 2000; Mujeri et al. 2012).

The untimely use of fertilizer is another problem that limits its effectiveness and needs to be an essential component of farmer education and consultation processes. Crops benefit the most from fertilizer applied at specific times of the planting cycle. However, the slow distribution of fertilizers that are part of input subsidy programs causes farmers to miss the appropriate time to apply the fertilizer to crops (Namonje-Kapembwa, Black, and Jayne 2015). Accordingly, governments often try to improve the distributional process by including the private sector as part of their reform (Jayne et al. 2018).

Applying fertilizer without understanding local soil or agronomic conditions may also influence the fertilizer's effect on yields. Soil characteristics—biological, chemical, and physical—are crucial factors that drive the effectiveness of yield-increasing inputs such as improved seed varieties and inorganic fertilizer (Burke, Jayne, and Black 2017; Kim and Bevis 2019). Fertilizers are generally less effective on soils low in organic matter or minerals, such as the soils in many parts of Africa (Marenya and Barrett 2009a, 2009b; Zingore et al. 2008). Additionally, acidic soils impede the retention of potassium and limit the phosphorus and nitrogen that crops can take up (Jones et al. 2013; Kasim, Ahmed, and Majid 2011). More important, the excessive and repeated use of inorganic fertilizer without proper farming techniques fosters soil degradation by reducing the soil's organic matter, carbon, and biodiversity—a situation that increases concerns about the sustainability of agriculture in Africa (Lehmann et al. 2020). Inappropriate fertilizer application without proper treatment (such as liming and the addition of manure and organic matter) will be ineffective and may cause adverse environmental consequences, which threaten the sustainability of agriculture (Bationo et al. 2007).

Finally, limited access to modern agricultural technology may reduce the potential benefits of applying fertilizers. For instance, when smallholder farmers face water shortages due to lack of irrigation or drought, applying fertilizer may not improve productivity (Jayne and Rashid 2013; Wiedenfeld 1995). In addition, since fertilizer application requires additional labor, lack of access to machinery negatively affects the effectiveness of fertilizer use (Mgbenka, Mbah, and Ezeano 2016). Therefore, particularly for rain-fed crops, if the proper amount of water is not supplied at the right time or other yield-augmenting inputs are not applied, then fertilizer application will inevitably be less effective (Mujeri et al. 2012).

skewed toward higher-income households. Nevertheless, the subsidy tends to make up a larger share of the poor's household income and is therefore important for income support. Consequently, addressing barriers to adoption by the poor and improving the targeting of these programs are critical for achieving the goals of input subsidy programs.

A consistent result of the assessment is that participation in subsidy schemes is not greater among poorer households. Figure 7.1 displays the ratio of beneficiaries in each income quartile to the total participants within each country. In most countries, middle-income farmers are most likely to participate in the program. The exceptions are Tanzania, where the highest income quartile has the highest participation rate, and Vietnam, where the lowest quartile has the highest participation rate. Notably, the lowest income quartile has the lowest participation rate in Malawi, Nigeria, and Tanzania. In Malawi and Tanzania, participation in this group is particularly low, at nearly less than half the rate of Q2 or Q3. These patterns suggest that input subsidies are often poorly designed to assist the poorest farmers.

Participation rates are not the only thing that matters when determining policy progressiveness; distribution of the subsidy also matters. If lower-income households receive a disproportionately large amount of the subsidy, this might compensate for low participation rates. Figure 7.2 describes the proportion of the total subsidy amount that is distributed to households in each of the income quartiles. The average amount of subsidy varies across countries, ranging from US$14.70 (Tanzania) to US$73.90 (Zambia) per year (see table G.1 in online appendix G).[8] Consistent with the trends in participation rates, more subsidies are distributed toward middle-income households.

In Malawi and Tanzania, for every US$1 in input subsidy given to support a household in the bottom 20 percent of the country's income distribution, at least US$5 is spent supporting a household in the top 20 percent of the country's income distribution.

FIGURE 7.1 **Share of participation in agricultural input subsidy programs in select countries, by income quartile**

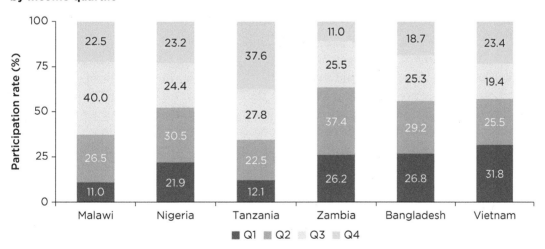

Source: World Bank, using data from 2019 Malawi Integrated Household Panel Survey, 2015 Nigeria General Household Survey, 2012 Tanzania National Panel Survey, 2015 Zambia Living Conditions Monitoring Survey, 2015 Bangladesh Household Income and Expenditure Survey, and 2014 Vietnam Living Standards Survey.
Note: The figure shows the share of input subsidy participation by household income quartile, where Q1 is the lowest income quartile and Q4 is the highest.

FIGURE 7.2 **Proportion of agricultural input subsidy distributed in select countries, by income quartile**

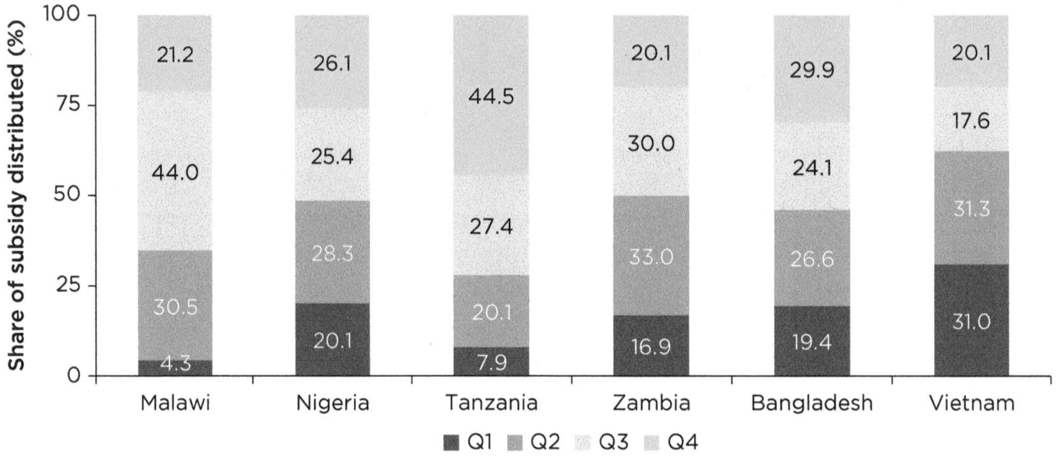

Source: World Bank, using data from 2019 Malawi Integrated Household Panel Survey, 2015 Nigeria General Household Survey, 2012 Tanzania National Panel Survey, 2015 Zambia Living Conditions Monitoring Survey, 2015 Bangladesh Household Income and Expenditure Survey, and 2014 Vietnam Living Standards Survey.
Note: The figure shows the proportion of agricultural input subsidy by household-income quartile, where Q1 is the lowest income quartile and Q4 is the highest.

In all countries, the amount of subsidy received by the lowest income-quartile group is disproportionately lower than the participation rate. For instance, in Malawi, the poorest households receive less than 4.3 percent of the total subsidy amount, while their proportion of participation is 11 percent. This pattern can be explained by the need to purchase at least two bags of fertilizer and make a co-payment to benefit from the subsidy. Collectively, these payments are too expensive for the poorest farmers. This regressive pattern is most evident in countries with relatively large average subsidy payments (Malawi, Nigeria, Tanzania, and Zambia).

Finally, the absolute value of a subsidy may not tell the whole story. A US$5 subsidy constitutes a larger share of the income of a poorer farmer than a richer farmer and therefore benefits the poorer farmer more. Figure 7.3 shows the amount of subsidy received as a share of total household income by income quartile for program participants. It indicates that the subsidy amount consists of a significant share of total income for those who participate, particularly for the lowest income-quartile group, despite lower participation rates and smaller subsidies in absolute value. This picture shows how important those benefits are for poor farmers. However, even if the amount of benefit is substantial relative to their total income, participation rates are low due to the burden of having to pay the same amount to get the subsidy. In addition, the fixed costs of collecting the subsidized fertilizer—filling out paperwork, driving to the collection spot from remote locations, and so forth—may simply not be worthwhile for the small amount of subsidy provided to small producers. Therefore, the large share of benefits relative to total income seems not to be very attractive to poor farmers, and those in the greatest need may choose to opt out. In addition, other barriers may be playing a role, such as inaccessibility or lack of information, which implies that communication and targeting campaigns are insufficient (Dionne and Horowitz 2016; Jayne et al. 2018; Mason and Ricker-Gilbert 2013).

FIGURE 7.3 Subsidy as a share of household income in select countries, by income quartile

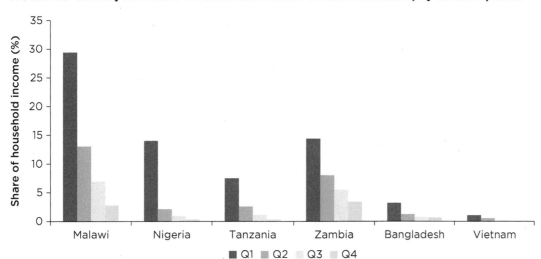

Source: World Bank, using data from 2019 Malawi Integrated Household Panel Survey, 2015 Nigeria General Household Survey, 2012 Tanzania National Panel Survey, 2015 Zambia Living Conditions Monitoring Survey, 2015 Bangladesh Household Income and Expenditure Survey, and 2014 Vietnam Living Standards Survey.
Note: The figure shows the value of the subsidy that accrues to each income quartile as a share of total household income of households that participate in the subsidy program, where Q1 is the lowest income quartile and Q4 is the highest.

Two important takeaways arise from these results. First, if these subsidies are intended to target lower-income farmers, they are not performing well in the countries considered. Even in Vietnam, the country with the most progressive input subsidy of the six programs, households in the bottom 50 percent by income only capture 62 percent of the subsidy. For all other countries, these households capture less than half of the subsidy—and often significantly less than half. There may be several reasons for this finding, but one important reason is that the amount of co-payment required may make it unaffordable or undesirable for lower-income households to participate. Second, the results in figure 7.3 make it clear that removing these subsidies would be quite painful for the bottom income quartile. In Malawi, the subsidy's share of household income is nearly 30 percent of Q1 household income, and in Nigeria and Zambia, the share is nearly 15 percent. Given these two observations, a cash-transfer program might be more effective at providing targeted benefits to lower-income households in ways that benefit them the most.

Finally, while this analysis has been done for only six countries, the results are likely generalizable, for several reasons. First, subsidies that scale with production are always going to go disproportionately to large producers. Large producers use disproportionately more inputs, so even a "flat" subsidy will benefit richer households more than poorer households. Second, subsidies that require significant out-of-pocket expenditures will always discourage lower-income households from participating. Most input subsidy programs do not give away inputs completely for free. By requiring farmers to pay a proportion of the cost, such programs ensure that farmers who receive the subsidy actually want to use the inputs, reducing waste. Nevertheless, if the co-payment is too high, poor farmers may be priced out of the market. Finally, subsidies in Malawi and Nigeria have been reformed recently to try to improve access to poorer households, and yet they still fail to do so. It is difficult to target these households through mechanisms like input subsidies, which may be the wrong instrument for the job. Recent experiences from an input subsidy program supported by the World Bank may shed light on how to improve the efficacy and targeting of such programs (box 7.9).

BOX 7.9

Lessons from e-voucher programs in Guinea, Mali, and Niger

Following the 2008 global food crisis, three West African countries—Guinea, Mali, and Niger—launched programs aimed at providing agricultural assistance to smallholder farmers. Specifically, these programs attempted to address inefficiencies in agricultural productivity stemming from inadequate use of yield-boosting inputs such as fertilizers and improved seed varieties. Despite their best intentions and very generous subsidies, the programs saw modest, if any, impact on growth of agricultural productivity. The programs were also criticized for lack of efficiency, traceability, and transparency.

Accordingly, around the 2015-16 agricultural season, the three governments introduced pilot e-voucher programs aimed at improving the traditional approach. The pilot programs in all three countries were supported by the World Bank–funded West Africa Agricultural Productivity Program. Several key characteristics were included in the programs to try to overcome the criticisms of the prior subsidy programs. First, to improve transparency, the programs introduced a text messaging e-voucher to communicate directly with farmers and input dealers, avoiding the distributional problems of the previous approach. Second, to improve efficiency, the programs targeted vulnerable populations and encouraged the participation of private dealers. Third, to improve the identification of beneficiaries and their needs, a pre-implementation survey was conducted, enabling the targeting of vulnerable farmers. Fourth, to improve the quality of inputs and efficiency of the distribution process, the private sector was engaged through a database of agricultural dealers who met stringent technical and commercial standards. Finally, to achieve traceability, monitoring and evaluation were enhanced.

Current experience emerging from these e-voucher pilot programs in West Africa draws some important lessons for the efficiency and distributional impact of similar agricultural input programs.

First, and most important, targeting is a key factor in successful input subsidy programs. If the policy aims to reduce poverty, governments can intensively support poor households to help them to improve their agricultural yields and thus act as a safety net. If the goal is to enhance national productivity or support a particular crop, governments can target beneficiaries accordingly, potentially leading to more support for middle-income households than for lower-income households. The targeting systems of the three countries varied as a result of differences in their beneficiary selection criteria. In Guinea, 3,500 beneficiaries were selected out of 96,000 registered farmers based on the size of their farmland, with a high proportion of female beneficiaries (36 percent) and smallholders (39 percent). In Mali, 24,583 farmers growing targeted crops in each region were selected as beneficiaries from a total of 93,000 farming households that applied, resulting in a high participation rate of 74.3 percent (18,273 farmers). In Niger, the program was restricted to beneficiaries of the existing social safety net program who were residents of municipalities with vulnerable populations exceeding 50 percent, resulting in 26 percent of all 30,838 beneficiary households being headed by women.

Second, although the e-voucher program compensates for the limitations of the traditional system of distributing physical coupons, older generations, particularly in rural areas and areas with high illiteracy rates, still have limited access to the subsidy. Solutions such as voice-messaging services or near-field communication technology used in Ethiopia can help to overcome barriers due to illiteracy and limited phone ownership.

Finally, input procurement and timely provision of inputs are essential factors in the effectiveness of the policy. However, late delivery was often observed due to delays in the procurement of inputs from the public sector or organizational problems at some dealers who were not yet familiar with the new system.

Source: World Bank 2019.

The way forward

The results presented in this chapter demonstrate that the core motivations behind most agricultural subsidy programs—to increase efficiency through technology adoption and to provide critical income support to lower-income households—are rarely met. Furthermore, when these programs meet these goals, they do so in suboptimal ways.

On the production side, subsidies often tend to increase total production or yields. Naturally, when the price of inputs decreases or the price of outputs increases, farmers are incentivized to produce more. The cross-country regressions, as well as the meta-analysis and country case studies, show that production is often increased by distorting the ratio of inputs beyond what is most efficient, leading to a decline in technical efficiency. In addition, this distortion also leads to a misallocation of resources across sectors, where labor, land, and water are all allocated in larger shares to agriculture where they may not be most efficiently put to use.

Nevertheless, the findings that subsidies do increase overall production must not be dismissed. If removal of the subsidy were to improve technical efficiency but lower production, then food supply might fall nonetheless. Depending on the scale of the decline in food supplies and links to international markets, local food prices may rise even if technical efficiency improves. Thus, policy makers must keep this possibility in mind when reforming subsidy programs and have plans in place to counteract a potential rise in food prices or to ensure that vulnerable populations do not become food insecure.

Experience shows that input subsidy programs are most likely to be effective at increasing technology and input adoption if they follow several important principles. They should be temporary, with the size of the subsidy and its duration communicated clearly to prevent farmers from anchoring on prices and holding out on buying commercially in anticipation of future subsidies. They should also be implemented alongside education and information programs that inform farmers about the benefits of the technology and how best to implement it given local conditions. Recent research demonstrates that social networks may also be critical for disseminating information about fertilizer use and crowding in new farmers (Carter, Laajaj, and Yang 2021). Finally, they must ensure that they are subsidizing the appropriate inputs and behavior. Soil characteristics—biological, chemical, and physical—are all important determinants of how much nutrients a crop can absorb from the soil. Subsidizing the wrong type of fertilizer or subsidizing fertilizers in countries where fertilizer use is already well above levels that maximize its value will achieve a limited impact on both farmers' technical efficiency as well as yields.

Most subsidies are poorly targeted toward poorer households. On the output subsidy side, while a majority of countries see levels of PSCT going to poorer regions of countries relative to their production value, the progressiveness of these policies tends to be low. Since output subsidies are typically crop-specific, policy makers can improve on this situation by reevaluating which crops receive subsidies and prioritizing those that tend to be produced by poorer households. In many countries, this approach implies a shift away from subsidizing rice, which would entail other environmental co-benefits like reducing agriculture's water and carbon footprints.

The situation is even starker on the input subsidy side. Even programs that intend to target poorer households often fail to do so. In Malawi's and Tanzania's subsidy programs, for instance, for every US$1 given to support a household in the bottom 20 percent (the bottom quintile) of the country's income distribution, the government spends around

US$5 subsidizing households in the top 20 percent (the top quintile) of the country's income distribution. In an era when public spending is in great need and government budgets are shrinking, the opportunity cost of poorly targeted programs like these is extremely high.

This chapter has focused only on the direct economic impacts of agricultural subsidies. Nevertheless, subsidies also have significant environmental externalities that can be as meaningful as—or more meaningful than—their fiscal impacts. The next two chapters explore these impacts, with chapter 8 examining the impact of subsidies on water quality and withdrawals, and chapter 9 examining the impact of agricultural subsidies on land use and deforestation.

Notes

1. Data are described in Fuglie, Jelliffe, and Morgan (2021), which is available at https://www.ers .usda.gov/data-products/international-agricultural-productivity/.
2. The data series for each country or region is an index with the base year of 2015, and the value in any given year is the level of TFP in that country relative to 2015. Therefore, comparing TFP indexes among countries or regions shows where TFP is growing faster or slower but does not indicate where productivity is higher or lower.
3. The online appendixes for this chapter can be found at https://openknowledge.worldbank.org /handle/10986/39423.
4. In fact, the effectiveness of the policy is questionable. Many international organizations are conducting trials providing agricultural education along with inputs. More recently, researchers have begun to deal with unintended consequences such as constrained crop diversity (Thériault and Smale 2021).
5. Notably, both fertilizer use and FISP participation are negatively correlated with farm efficiency.
6. The analysis is reversed for Argentina, India, and Ukraine because they have negative PSCTs.
7. It could also imply offsetting impacts. That is, if the subsidy is regressive in the lower half of the income distribution, but progressive in the upper half of the income distribution, these deviations can offset one another, resulting in an AUC difference of zero.
8. Given that all agricultural subsidy policies covered in this chapter provide uniform amounts of subsidies to all beneficiaries, the amount of subsidy distributed to each income-quartile group is expected to follow the same pattern as the participation rate by income-quartile groups.

References

Adetutu, M. O., and T. G. Weyman-Jones. 2019. "Fuel Subsidies Versus Market Power: Is There a Countervailing Second-Best Optimum?" *Environmental and Resource Economics* 74 (4): 1619–46.

Alabi, R. A., and O. A. Oshobugie. 2020. "The Impact of E-wallet Fertilizer Subsidy Scheme and Its Implication on Food Security in Nigeria." African Economic Research Consortium, Nairobi.

Aragón, F. M., F. Oteiza, and J. P. Rud. 2021. "Climate Change and Agriculture: Subsistence Farmers' Response to Extreme Heat." *American Economic Journal: Economic Policy* 13 (1): 1–35.

Basak, J. K. 2010. "Future Fertiliser Demand for Sustaining Rice Production in Bangladesh: A Quantitative Analysis." Unnayan Onneshan–The Innovators, Dhaka.

Bationo, A., B. Waswa, J. Kihara, and J. Kimetu. 2007. "Advances in Integrated Soil Fertility Management in Sub Saharan Africa: Challenges and Opportunities." *Nutrient Cycling in Agroecosystems,* February 1, 2007. https://doi.org/10.1007/s10705-007-9096-4.

Burke, W. J., T. S. Jayne, and J. R. Black. 2017. "Factors Explaining the Low and Variable Profitability of Fertilizer Application to Maize in Zambia." *Agricultural Economics* 48 (1): 115–26.

Carter, M., R. Laajaj, and D. A. Yang. 2021. "Subsidies and the African Green Revolution: Direct Effects and Social Network Spillovers of Randomized Input Subsidies in Mozambique." *American Economic Journal: Applied Economics* 13 (2): 206–29. https://doi.org/10.1257/app.20190396.

Castaneda, R., D. Doan, D. L. Newhouse, M. Nguyen, H. Uematsu, and J. P. Azevedo. 2016. "Who Are the Poor in the Developing World?" Policy Research Working Paper 7844, World Bank, Washington, DC.

Chirwa, E., M. Matita, P. Mvula, and W. Mhango. 2016. *Evaluation of the 2015/16 Farm Input Subsidy Program in Malawi: 2015/16 Reforms and Their Implications.* Final Report for the Ministry of Agriculture, Irrigation and Water Development, Lilongwe.

Damania, R., C. Berg, J. Russ, A. F. Barra, J. Nash, and R. Ali. 2017. "Agricultural Technology Choice and Transport." *American Journal of Agricultural Economics* 99 (1): 265–84.

Dillon, A., and R. Fishman. 2019. "Dams: Effects of Hydrological Infrastructure on Development." *Annual Review of Resource Economics* 11 (1): 125–48.

Dionne, K. Y., and J. Horowitz. 2016. "The Political Effects of Agricultural Subsidies in Africa: Evidence from Malawi." *World Development* 87 (November): 215–26.

Dorward, A. 2009. "Rethinking Agricultural Input Subsidy Programmes in a Changing World." Report prepared for the Food and Agriculture Organization of the United Nations, Rome.

Duflo, E., M. Kremer, and J. Robinson. 2011. "Nudging Farmers to Use Fertilizer: Theory and Experimental Evidence from Kenya." *American Economic Review* 101 (6): 2350–90. https://doi .org/10.1257/aer.101.6.2350.

Duflo, E., and R. Pande. 2007. "Dams." *Quarterly Journal of Economics* 122 (2): 601–46.

FAO (Food and Agriculture Organization of the United Nations). 2011. "The State of Food and Agriculture 2010–11: Women in Agriculture." FAO, Rome.

FAO (Food and Agriculture Organization of the United Nations). 2020. "FAO Policy on Gender Equality 2020–2030." FAO, Rome.

FAO (Food and Agriculture Organization of the United Nations). 2022. FAOSTAT Statistical Database. Rome: FAO.

FAO (Food and Agriculture Organization of the United Nations), UNDP (United Nations Development Programme), and UNEP (United Nations Environment Programme). 2021. *A Multi-Billion-Dollar Opportunity—Repurposing Agricultural Support to Transform Food Systems.* Rome: FAO.

Fuglie, K., J. Jelliffe, and S. Morgan. 2021. "Slowing Productivity Reduces Growth in Global Agricultural Output." In *Amber Waves: The Economics of Food, Farming, Natural Resources, and Rural America, 2021.* Washington, DC: US Department of Agriculture.

Gemessa, S. A. 2022. "An Alternative Approach to Measuring the Welfare Implications of Input Subsidies: Evidence from Malawi." *Journal of Agricultural Economics* 73 (1): 112–38.

Giroud, A., and J. S. Huaman. 2019. "Investment in Agriculture and Gender Equality in Developing Countries." *Transnational Corporations* 26 (3): 89–113.

Goyal, A., and J. Nash. 2017. *Reaping Richer Returns: Public Spending Priorities for African Agriculture Productivity Growth.* Washington, DC: World Bank.

Harrigan, J. 2008. "Food Insecurity, Poverty, and the Malawian Starter Pack: Fresh Start or False Start?" *Food Policy* 33 (3): 237–49.

Hemming, D. J., E. W. Chirwa, A. Dorward, H. J. Ruffhead, R. Hill, J. Osborn, L. Langer, et al. 2018. "Agricultural Input Subsidies for Improving Productivity, Farm Income, Consumer Welfare and Wider Growth in Low- and Lower-Middle-Income Countries: A Systematic Review." *Campbell Systematic Reviews* 14 (1): 1–153.

Holden, S. T. 2019. "Economics of Farm Input Subsidies in Africa." *Annual Review of Resource Economics* 11 (October): 501–22.

Houssou, N., and A. Chapoto. 2015. "Adoption of Farm Mechanization, Cropland Expansion, and Intensification in Ghana." Paper for the International Association of Agricultural Economists 2015 Conference, August 9–14, 2015, Milan, Italy. http://purl.umn.edu/211744.

Huang, J., C. Xiang, X. Jia, and R. Hu. 2012. "Impacts of Training on Farmers' Nitrogen Use in Maize Production in Shandong, China." *Journal of Soil and Water Conservation* 67 (4): 321–27.

Islam, M. S., R. W. Bell, M. M. Miah, and M. J. Alam. 2022. "Unbalanced Fertilizer Use in the Eastern Gangetic Plain: The Influence of Government Recommendations, Fertilizer Type, Farm Size, and Cropping Patterns." *PloS One* 17 (7): e0272146.

Izaurralde, R. C., W. B. McGill, and N. J. Rosenberg. 2000. "Carbon Cost of Applying Nitrogen Fertilizer." *Science* 288 (5467): 809.

Jayne, T. S., N. M. Mason, W. J. Burke, and J. Ariga. 2018. "Taking Stock of Africa's Second-Generation Agricultural Input Subsidy Programs." *Food Policy* 75 (February): 1–14.

Jayne, T., D. Mather, N. Mason, and J. Ricker-Gilbert. 2013. "How Do Fertilizer Subsidy Programs Affect Total Fertilizer Use in Sub-Saharan Africa? Crowding Out, Diversion, and Benefit/Cost Assessments." *Agricultural Economics* 44 (6): 687–703.

Jayne, T. S., and S. Rashid. 2013. "Input Subsidy Programs in Sub-Saharan Africa: A Synthesis of Recent Evidence." *Agricultural Economics* 44 (6): 547–62.

Jones, A., H. Breuning-Madsen, M. Brossard, A. Dampha, J. Deckers, O. Dewitte, T. Gallali, et al. 2013. *Soil Atlas of Africa*. Brussels: European Commission.

Jones, M., F. Kondylis, J. Loeser, and J. Magruder. 2020. "Factor Market Failures and the Adoption of Irrigation in Rwanda." NBER Working Paper w26698, National Bureau of Economic Research, Cambridge, MA.

Kaditi, E. A., and E. I. Nitsi. 2011. "Vertical and Horizontal Decomposition of Farm Income in Equality in Greece." *Agricultural Economics Review* 12 (1): 69–80.

Kasim, S., O. H. Ahmed, and N. M. A. Majid. 2011. "Effectiveness of Liquid Organic-Nitrogen Fertilizer in Enhancing Nutrients Uptake and Use Efficiency in Corn (Zea Mays)." *African Journal of Biotechnology* 10 (12): 2274–81.

Kernecker, M., A. Knierim, A. Wurbs, T. Kraus, and F. Borges. 2020. "Experience Versus Expectation: Farmers' Perceptions of Smart Farming Technologies for Cropping Systems across Europe." *Precision Agriculture* 21 (1): 34–50.

Kim, J., N. M. Mason, D. Mather, and F. Wu. 2021. "The Effects of the National Agricultural Input Voucher Scheme (NAIVS) on Sustainable Intensification of Maize Production in Tanzania." *Journal of Agricultural Economics* 72 (3): 857–77.

Kim, K., and L. Bevis. 2019. "Soil Fertility and Poverty in Developing Countries." *Choices* 34 (2): 1–8.

Koppmair, S., M. Kassie, and M. Qaim. 2017. "The Influence of Farm Input Subsidies on the Adoption of Natural Resource Management Technologies." *Australian Journal of Agricultural and Resource Economics* 61 (4): 539–56.

Krishnaswamy, N. 2018. "At What Price? Price Supports, Agricultural Productivity, and Misallocation." Working paper, Columbia University, New York.

Kummu, M., M. Taka, and J. H. Guillaume. 2018. "Gridded Global Datasets for Gross Domestic Product and Human Development Index over 1990–2015." *Scientific Data* 5 (1): 1–15.

Lehmann, J., D. A. Bossio, I. Kögel-Knabner, and M. C. Rillig. 2020. "The Concept and Future Prospects of Soil Health." *Nature Reviews Earth & Environment* 1 (10): 544–53.

Marenya, P. P., and C. B. Barrett. 2009a. "Soil Quality and Fertilizer Use Rates among Smallholder Farmers in Western Kenya." *Agricultural Economics* 40 (5): 561–72.

Marenya, P. P., and C. B. Barrett. 2009b. "State-Conditional Fertilizer Yield Response on Western Kenyan Farms." *American Journal of Agricultural Economics* 91 (4): 991–1006.

Mason, N. M., and J. Ricker-Gilbert. 2013. "Disrupting Demand for Commercial Seed: Input Subsidies in Malawi and Zambia." *World Development* 45 (May): 75–91.

Mason, N. M., and M. Smale. 2013. "Impacts of Subsidized Hybrid Seed on Indicators of Economic Well-Being among Smallholder Maize Growers in Zambia." *Agricultural Economics* 44 (6): 659–70.

Mason, N. M., A. Wineman, L. Kirimi, and D. Mather. 2017. "The Effects of Kenya's 'Smarter' Input Subsidy Programme on Smallholder Behaviour and Incomes: Do Different Quasi-Experimental Approaches Lead to the Same Conclusions?" *Journal of Agricultural Economics* 68 (1): 45–69. https://doi.org/10.1111/1477-9552.12159.

Mgbenka, R. N., E. N. Mbah, and C. I. Ezeano. 2016. "A Review of Smallholder Farming in Nigeria: Need for Transformation." *International Journal of Agricultural Extension and Rural Development Studies* 3 (2): 43–54.

Mignouna, D. B., V. M. Manyong, J. Rusike, K. Mutabazi, and E. M. Senkondo. 2011. "Determinants of Adopting Imazapyr-Resistant Maize Technologies and Its Impact on Household Income in Western Kenya." *AgBioForum* 14 (3): 158–63.

Minviel, J. J., and L. Latruffe. 2017. "Effect of Public Subsidies on Farm Technical Efficiency: A Meta-Analysis of Empirical Results." *Applied Economics* 49 (2): 213–26.

Moreddu, C. 2011. "Distribution of Support and Income in Agriculture." Food, Agriculture, and Fisheries Paper 46, Organisation for Economic Co-operation and Development, Paris.

Morris, M. L. 2007. *Fertilizer Use in African Agriculture: Lessons Learned and Good Practice Guidelines.* Washington, DC: World Bank.

Mottaleb, K. A. 2018. "Perception and Adoption of a New Agricultural Technology: Evidence from a Developing Country." *Technology in Society* 55 (November): 126–35.

Mujeri, M. K., S. Shahana, T. T. Chowdhury, and K. T. Haider. 2012. "Improving the Effectiveness, Efficiency and Sustainability of Fertilizer Use in South Asia." South Asia: Global Development Network, New Delhi. http://www.gdn.int/admin/uploads/editor/files/SA_3_ResearchPaper _Fertilizer_Efficiency.pdf.

Mwangi, M., and S. Kariuki. 2015. "Factors Determining Adoption of New Agricultural Technology by Smallholder Farmers in Developing Countries." *Journal of Economics and Sustainable Development* 6 (5): 208–16.

Namonje-Kapembwa, T., R. Black, and T. S. Jayne. 2015. "Does Late Delivery of Subsidized Fertilizer Affect Smallholder Maize Productivity and Production?" Paper prepared for the joint annual meeting of the Agricultural and Applied Economics Association (AAEA) and the Western Agricultural Economics Association (WAEA), July 26–28, 2015, San Francisco, CA.

Nyang'au, J. O., J. H. Mohamed, N. Mango, C. Makate, and A. N. Wangeci. 2021. "Smallholder Farmers' Perception of Climate Change and Adoption of Climate Smart Agriculture Practices in Masaba South Sub-County, Kisii, Kenya." *Heliyon* 7 (4): e06789.

OECD (Organisation for Economic Co-operation and Development). 2003. *Farm Household Income: Issues and Policy Responses.* Paris: OECD.

OECD (Organisation for Economic Co-operation and Development). 2020. *Agricultural Policy Monitoring and Evaluation 2020.* Paris: OECD. https://doi.org/10.1787/928181a8-en.

OECD (Organisation for Economic Co-operation and Development). 2022. "Monitoring and Evaluation: Single Commodity Indicators." OECD, Paris.

Omotilewa, O. J., J. Ricker-Gilbert, and J. H. Ainembabazi. 2019. "Subsidies for Agricultural Technology Adoption: Evidence from a Randomized Experiment with Improved Grain Storage Bags in Uganda." *American Journal of Agricultural Economics* 101 (3): 753–72.

Ortiz-Bobea, A., T. R. Ault, C. M. Carrillo, R. G. Chambers, and D. B. Lobell. 2021. "The Historical Impact of Anthropogenic Climate Change on Global Agricultural Productivity." arXiv preprint arXiv:2007.10415.

Park, Y., and K. Kim. 2022. "Agricultural Input Subsidies and Efficiency: Evidence from Reforms in Malawi and Nigeria." Background paper prepared for this report, World Bank, Washington, DC.

Phiri, A. T., H. M. A. C. Toure, O. Kipkogei, R. Traore, P. M. K. Afokpe, and A. A. Lamore. 2022. "A Review of Gender Inclusivity in Agriculture and Natural Resources Management under the Changing Climate in Sub-Saharan Africa." *Cogent Social Sciences* 8 (1): 2024674.

Ren, C., S. Jin, Y. Wu, B. Zhang, D. Kanter, B. Wu, X. Xi, et al. 2021. "Fertilizer Overuse in Chinese Smallholders due to Lack of Fixed Inputs." *Journal of Environmental Management* 293 (September 1): 112913.

Ricker-Gilbert, J. 2014. "Wage and Employment Effects of Malawi's Fertilizer Subsidy Program." *Agricultural Economics* 45 (3): 337–53. https://doi.org/10.1111/agec.12069.

Ricker-Gilbert, J., and T. S. Jayne. 2012. "Do Fertilizer Subsidies Boost Staple Crop Production and Reduce Poverty across the Distribution of Smallholders in Africa? Quantile Regression Results from Malawi." Paper prepared for presentation at the Triennial Conference of the International Association of Agricultural Economists, Foz do Iguaçu, Brazil, August 18–24, 2012.

Ricker-Gilbert, J., T. S. Jayne, and E. Chirwa. 2011. "Subsidies and Crowding Out: A Double-Hurdle Model of Fertilizer Demand in Malawi." *American Journal of Agricultural Economics* 93 (1): 26–42.

Ricker-Gilbert, J., T. S. Jayne, and G. Shively. 2013. "Addressing the 'Wicked Problem' of Input Subsidy Programs in Africa." *Applied Economic Perspectives and Policy* 35 (2): 322–40.

Simonsohn, U., and G. Loewenstein. 2006. "Mistake: The Effect of Previously Encountered Prices on Current Housing Demand." *Economic Journal* 116 (508): 175–99.

Singh, S. 2020. "Farmers' Perception of Climate Change and Adaptation Decisions: A Micro-Level Evidence from Bundelkhand Region, India." *Ecological Indicators* 116 (September): 106475.

Suri, T., and C. Udry. 2022. "Agricultural Technology in Africa." *Journal of Economic Perspectives* 36 (1): 33–56.

Thériault, V., and M. Smale. 2021. "The Unintended Consequences of the Fertilizer Subsidy Program on Crop Species Diversity in Mali." *Food Policy* 102 (July): 102121.

Triyana, M., and L. Nguyen. 2022. "The Effect of Agricultural Input Subsidies on Productivity: A Meta-Analysis." Background paper prepared for this report, World Bank, Washington, DC.

Wang, P., W. Zhang, M. Li, and Y. Han. 2019. "Does Fertilizer Education Program Increase the Technical Efficiency of Chemical Fertilizer Use? Evidence from Wheat Production in China." *Sustainability* 11 (2): 543.

Wiedenfeld, R. P. 1995. "Effects of Irrigation and N Fertilizer Application on Sugarcane Yield and Quality." *Field Crops Research* 43 (2–3): 101–08.

Wood-Sichra, U., A. B. Joglekar, and L. You. 2016. "Spatial Production Allocation Model (SPAM) 2005: Technical Documentation." HarvestChoice Working Paper, HarvestChoice, International Food Policy Research Institute (IFPRI), Washington, DC.

World Bank. 2019. "AFCW3 Economic Update, Spring 2019: Digitizing Agriculture; Evidence from E-Voucher Programs in Mali, Chad, Niger, and Guinea." World Bank, Washington, DC. https://openknowledge.worldbank.org/handle/10986/31576.

Wossen, T., T. Abdoulaye, A. Alene, S. Feleke, J. Ricker-Gilbert, V. Manyong, and B. A. Awotide. 2017. "Productivity and Welfare Effects of Nigeria's e-Voucher-Based Input Subsidy Program." *World Development* 97 (September): 251–65.

Xu, Z., W. J. Burke, T. S. Jayne, and J. Govereh. 2009. "Do Input Subsidy Programs 'Crowd In' or 'Crowd Out' Commercial Market Development? Modeling Fertilizer Demand in a Two-Channel Marketing System." *Agricultural Economics* 40 (1): 79–94.

Yi, F. J., W. Y. Lu, and Y. H. Zhou. 2016. "Cash Transfers and Multiplier Effect: Lessons from the Grain Subsidy Program in China." *China Agricultural Economic Review* 8 (1): 81–99. https://doi.org/10.1108/CAER-07-2015-0078.

Zhang, X., E. A. Davidson, D. L. Mauzerall, T. D. Searchinger, P. Dumas, and Y. Shen. 2015. "Managing Nitrogen for Sustainable Development." *Nature* 528 (7580): 51–59.

Zingore, S., R. J. Delve, J. Nyamangara, and K. E. Giller. 2008. "Multiple Benefits of Manure: The Key to Maintenance of Soil Fertility and Restoration of Depleted Sandy Soils on African Smallholder Farms." *Nutrient Cycling in Agroecosystems* 80 (3): 267–82.

CHAPTER 8

Reap What You Sow

The Water Footprint of Agricultural Subsidies

> "The farthest star and the mud at our feet are a family. ...
> We are at risk together, or we are on our way to a sustainable world together.
> We are each other's destiny."
> —*Mary Oliver*

CHAPTER AT A GLANCE

Subsidies are a potent policy lever in agriculture and are employed all over the world. Poorly designed subsidies, however, are enmeshed in trade-offs.

Input and output subsidies distort farmers' incentives and have the potential to scar the quality and quantity of freshwater resources:

- By substantially lowering the price of nitrogen relative to that of other nutrients, subsidies can promote the inefficient and ineffective use of nitrogen. New evidence finds that regions like East and South Asia are using nitrogen fertilizers well beyond what is considered efficient, exacerbated by subsidies.

- This inefficient use is leading to diminishing crop productivity and increased nitrogen runoff in waterways. Globally, the inefficient use of input subsidies adversely affected the quality of freshwater supplies and was responsible for up to 17 percent of all nitrogen pollution in the past 30 years.

- Nitrogen pollution can have significant impacts on health. In areas of the world where input subsidies make up the largest share of the value of agricultural production in the global sample, subsidy-induced increases in water pollution are enough to reduce labor productivity by up to 2.7–3.5 percent.

- While these estimates are difficult to translate into monetary terms, their magnitude suggests that the marginal loss of health and later-life productivity can be substantial. This loss is especially critical in places where subsidies can induce nitrogen use to exceed optimal levels for plant growth.

- Coupled producer support subsidies are also drawing down global supplies of groundwater for irrigation. New evidence finds that, at the mean level of subsidy exposure, agricultural areas around the world risk losing up to 13.2 cubic kilometers of water per year, roughly equivalent to the total amount of water lost in California between 2011 and 2014 at the height of the drought.

Better policies offer considerable scope to limit or reduce the damage to water resources from the intensive use of inputs:

- Unlike coupled support, evidence suggests that decoupled support does not lead to harmful environmental spillovers, making it an important tool for achieving policy objectives without incurring costs that are counterproductive to policy goals.

- There is massive potential to reduce water pollution without affecting crop yields. The importance of efficiency gains points to a direct role for public policies to facilitate the uptake of better management practices and new technologies as well as to stimulate innovation tailored to site-specific conditions.

Introduction

This chapter examines the water-based externalities associated with agricultural subsidies. There is growing anecdotal and statistical evidence that modern agricultural practices adversely affect the quantity and quality of freshwater supplies. Policy design is key since it formulates the incentives for micro decisions made by farmers that aggregate to have a macro impact on the environment. For instance, as Tilman et al. (2002, 671) highlight, "New incentives and policies for ensuring the sustainability of agriculture and ecosystem services will be crucial if we are to meet the demands of improving yields without compromising environmental integrity or public health." At the heart of this issue is the mix of input and output subsidies that influence crop choices and farming practices.

Most, if not all, subsidies are well intentioned, but the externalities they have come to impose on the environment have been broadly ignored in policy design, perhaps because impacts often emerge with a lag, may be cumulative, and are less visible. This chapter explores the nature and extent of these externalities on water resources and provides new evidence of the way in which subsidies in the agriculture sector influence a resource that is critical to it.

Nitrogen legacies and the role of subsidies

Few innovations have transformed the world as much as nitrogen. Since the start of the 20th century, humans have been successful at making *"brot aus luft" or* "bread from air" by transforming atmospheric nitrogen in the air into ammonia, a form of reactive nitrogen that plants can use. A hundred years since the ingenious experiment in nitrogen fixation by Haber and Bosch, nitrogen has been poured into the ground as fertilizer. One result of the experiment is already clear. It has more than doubled the global rate of nitrogen fixation, enabled a 30–50 percent increase in yields, and supported the lives of several billion people who otherwise might have died prematurely or never been born (Erisman et al. 2008; Stewart et al. 2005).

The ability to produce fertilizers at scale, coupled with large fertilizer subsidies, has led to a steady rise in the consumption of nitrogenous fertilizers. Since the 1960s, nearly all of the growth in fertilizer use has been in Asia, particularly in China and India. This growth coincides with the onset of the green revolution, when government policies actively began to support a system of domestic price controls by way of large subsidies that distorted market prices.

In many countries, fertilizer subsidies are some of the largest expenditure items in government budgets. India spends a staggering US$10 billion to US$11 billion a year on fertilizer subsidies (Chatterjee et al. 2022), roughly five times more than what was spent 15 years earlier (Gulati and Banerjee 2015). Nearly 70 percent of this amount is allocated to nitrogen, causing a large gap between global and Indian domestic prices. In China, subsidies to the fertilizer industry averaged almost US$7 billion per year from 2008 to 2010, substantially depressing fertilizer prices.[1] In Mexico, the government created the Programa de Fertilizantes (Fertilizers Program) in 2019 with an initial budget of around US$75 million to support and supply smallholder farmers with up to 600 kilograms of fertilizer per year (Ding et al. 2021). In regions like Africa, which accounts for just 1.5 percent of the world's consumption of nitrogen, governments of at least 10 countries that account for more than half of the region's population spend nearly US$600 million to US$1 billion annually on subsidies, representing about 14–26 percent of their national budgets for agriculture (Jayne and Rashid 2013; Jayne et al. 2018).[2] These magnitudes, however, obscure the mixed track record in implementation, the vast disparities of fertilizer use across regions and between small and big farmers, and their uncertain effectiveness due to differences in soil composition (box 8.1; see also box 7.8 in chapter 7).

The rationale for devoting large resources to fertilizer subsidies is often to stimulate agricultural production, benefit poor rural households, stabilize food prices, and boost food security. At the same time, doing so provides politicians with a demonstrable way to support their constituents.[3] In the developing world, the economic case often rests on perceived market failures that might cause farmers to use inefficiently low levels of fertilizer. For instance, studies suggest that fertilizer markets can be prone to market failure

---

BOX 8.1
A divided world: Fertilizer feast and famine

Providing subsidized fertilizers may seem like a straightforward way to boost yields and make inputs more accessible and affordable to the poorest farmers, but the reality is far removed from this ideal. The distribution of fertilizers is uneven. Large areas of Africa and smaller but significant regions of Asia and Latin America continue to experience delays in access to affordable nitrogen fertilizers, causing farmers to miss the appropriate timing of fertilizer application (Austin et al. 2013; Houlton et al. 2019; Namonje-Kapembwa, Jayne, and Black 2015). As a result, many subsistence farmers in these parts rely on depleted soil nitrogen capital to grow food (Austin et al. 2013; Houlton et al. 2019). Even in countries like India where the consumption of nitrogen fertilizer is high, only 35 percent of subsidies reach their intended beneficiaries: small and marginal farmers (Chatterjee et al. 2022).

Moreover, because nitrogen is much more heavily subsidized than other fertilizers, it is also more prone to diversion for nonagricultural uses (Government of India 2016). In countries like Vietnam with relatively more progressive input subsidy programs, only 62 percent of the subsidy is captured by the bottom 50 percent of households in terms of income (see chapter 7). And in Malawi, an input subsidy program that uses community-based targeting with the precise objective of improving the targeting of the poor still fails to reach 46 percent of poor households while allocating inputs to 54 percent of nonpoor households (Houssou and Zeller 2011; Jayne et al. 2018). Studies also show that, while Zambia spends five times as much on farm subsidies as it does on cash transfers to the poor, a third of the subsidized fertilizer fails to reach the intended beneficiaries and is often resold commercially, with middlemen pocketing the subsidy (Economist 2017; Mason, Jayne, and Mofya-Mukuka 2013; Tesliuc, Smith, and Sunkutu 2013). Much of the rest is consumed by larger—presumably richer—farmers. Ironically, even as subsidies persist in many places, fertilizer application can still be low in some areas and fail to reach the intended beneficiaries due to improper implementation and leakage (see chapter 7 for additional discussion).

due to the high sunk costs of fertilizer producers, high transaction costs associated with poor market infrastructure, low demand by farmers because of liquidity constraints, imperfect information, beliefs about the quality of inputs, and uncertainty about the returns to fertilizers (Abay, Blalock, and Berhane 2017; Carter, Laajaj, and Yang 2013; Duflo, Kremer, and Robinson 2011; Michelson, Gourlay, and Wollburg 2022). In such contexts, economically rational farmers respond to the subsidy-induced price incentives by applying more fertilizer.

With exceptionally low prices of fertilizers, subsidies may also encourage farmers to deviate from optimal levels, resulting in the overuse of fertilizers beyond recommended rates (Schultz 1964). They may also lead to a failure to supply the right amount of fertilizer at the right time—a problem that seems to be pervasive in low- and middle-income countries where technical know-how is low (Duflo, Kremer, and Robinson 2011).[4] This failure is especially salient for nitrogen fertilizer, which, when compared to other fertilizers, calls for greater information to determine the appropriate timing and scale of application. In part, this level of information is needed because nitrogen in forms available for crops is highly volatile and can leave the soil very quickly.

Subsidized or distorted prices in combination with the inability of farmers to gauge the precise amount of application needed may result in excessive application as well as incorrectly timed application (Islam and Beg 2021). Indeed, several studies show that fertilizer subsidies have increased the intensity of fertilizer use as well as overuse (Huang, Gulati, and Gregory 2017).[5] Oftentimes, this balance is tilted on the side of nitrogen, which is subsidized much more heavily than other fertilizers.[6] Figure 8.1 shows that, while phosphorus and potassium fertilizer production rates have been increasing, their increase pales in comparison to that of nitrogen.

But like people, plants need a variety of nutrients to thrive. The right mineral fertilizers applied appropriately can alleviate nutrient deficiencies in soils and increase crop yields. While mineral fertilizers provide higher and more plant-accessible nutrients, organic minerals also provide carbon, which contributes to healthy soils and better crop productivity (Barrett and Bevis 2015; Gram et al. 2020). Neither mineral nor organic inputs can provide both of these properties on their own, and applying them in combination often creates added benefits (Vanlauwe 2015). However, subsidy-driven applications of nitrogenous fertilizer appear to be causing nutrient imbalances in many regions of the world, as farmers apply significantly more nitrogen than other primary nutrients like potassium and phosphorous or other secondary nutrients and micronutrients (Gautam 2015; Kurdi et al. 2020) (box 8.2).

The imbalanced use of fertilizers has created widespread deficiency of secondary and micronutrients such as zinc, sulfur, iron, and manganese in the soil that ultimately limits the ability of farmers to use fertilizers profitably (Giné et al. 2019; Goyal and Nash 2017; Jayne and Rashid 2013; Kishore, Alvi, and Krupnik 2021) and can also affect the nutritional content of crops and food consumed by people (De Groote et al. 2021).[7] This issue is of salience in Africa, where the variation in soil quality is particularly high (Carter, Lybbert, and Tjernström 2015) and where the presence of acidic soils requires secondary and micronutrients, organic amendments, and lime supplements for better soil management (Smale and Thériault 2018). Blanket recommendations and subsidy programs that focus heavily on nitrogen without paying careful attention to soil or agronomic conditions can thus fail to address the limiting factors for plant growth specific to local contexts (Smale and Thériault 2018). For example, studies show that Mali's subsidy program, which was heavily focused on urea, failed to address the more limiting factor for crop growth, which in some cases was

FIGURE 8.1 Nitrogen fertilizer consumption, by region, and total fertilizer production, by nutrient, 1961–2014

a. Consumption of nitrogen fertilizer, by region

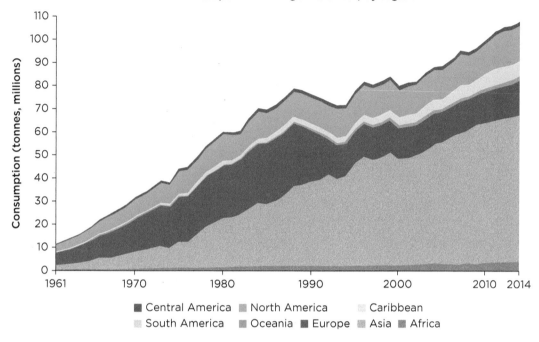

b. Total production of fertilizer, by nutrient

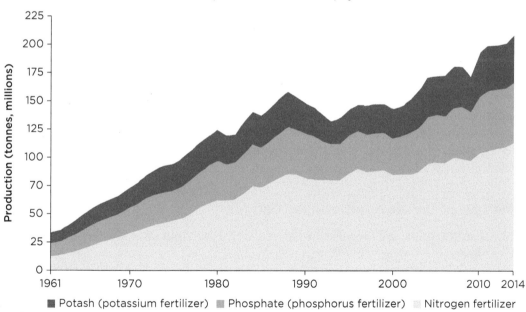

Sources: United Nations Food and Agriculture Organization; Ritchie, Roser, and Rosado 2013.
Note: The figures are sourced from ourworldindata.org and show the consumption of nitrogen fertilizer in tonnes across different regions (panel a) and the global production of nitrogen, phosphate, and potash fertilizer in tonnes (panel b).

NPK application: Skewed and distorted

In general, the optimal application of fertilizers requires a balanced proportion of nitrogen, phosphorus, and potassium (in its water-soluble form, potash) (NPK). While the correct proportions are context specific, measuring deviations from the accepted rule of thumb of N:P:K in the ratio 4:2:1 is revealing. Table B8.2.1 presents deviations from the optimal N:P and N:K ratios (with respect to N).

Fertilizer use clearly is tilted disproportionately toward the use of nitrogen. These distortions indicate extremely low applications of phosphorus and potash. In many of these countries, the structure of fertilizer subsidies contributes heavily to such a skew, which is particularly striking in China, India, and the United States.

TABLE B8.2.1 Deviations from the optimal N:P and N:K fertilizer ratios in select countries

Location	Optimal amount (kilograms per hectare)			Distortion (%)	
	N	P	K	N:P	N:K
Bangladesh	140.7	53.9	33.9	61	4
Brazil	42.6	48.8	56.1	−113	−81
China	267.6	91.0	41.2	94	62
India	85.8	28.9	10.8	97	100
United States	74.1	24.8	26.7	99	−31
Vietnam	127.5	66.7	43.1	−9	−26
World	69.5	26.3	17.9	64	−3

Source: Damania et al. 2019.
Note: K = potassium; N = nitrogen; P = phosphorus.

phosphorus rather than nitrogen (Smale and Thériault 2018). Similarly, a Tanzania field study conducted soil testing and found that sulfur was deficient in all the plots and a critical factor limiting the growth of maize yields. Yet national fertilizer recommendations did not include sulfur (Harou et al. 2022). These examples highlight the need to align nutrients with crop needs and soil fertility conditions.

The law of unintended consequences

More than 200 years ago, Frédéric Bastiat, economist and thinker, made a keen observation in his famous 1850 essay, "That Which Is Seen and That Which Is Unseen." He pointed out that, to evaluate the consequences of any action, it is necessary to look at both its seen effects, which are often the rationale behind the action, and its unseen effects, which include unintended consequences and other ripple effects. These insights continue to be vital for policy making.

Yet despite the ubiquity of fertilizer subsidies as an agricultural policy tool and the magnitude of resources devoted to them, little is known about their long-term impacts or the externalities and waste imposed by their ineffective use (Gautam et al. 2022). As discussed in the previous section, lopsided subsidies for fertilizers that substantially lower the price of nitrogen relative to other nutrients can promote inefficient and ineffective use. Such excess use and inefficiencies in application mean that not all of the nitrogen

BOX 8.3
The nitrogen cascade beyond water

Nitrogen pollution is one of the most important environmental issues of the 21st century in large part because the nitrogen element cascades through many chemical forms, with many complex effects (Kanter 2018; Keeler et al. 2016). There are several sources of nitrogen pollution, including fossil fuel combustion, industry, energy, transport, biomass burning, and wastewater, but the dominant source is the agriculture sector. Fertilizer is a key culprit in nitrogen pollution, which fouls the air and water worldwide. For example, nitrogen applied as fertilizer may end up in the atmosphere, surface water, and groundwater either directly or through food supply chains. Some of this nitrogen will be volatized as ammonia, causing local or regional air pollution; some will be denitrified to nitrous oxide, contributing to climate change; some will be lost to surface water, causing hypoxia and eutrophication; and some will enter groundwater, potentially affecting drinking water. Science suggests that the world may have surpassed the planetary boundaries for nitrogen, and some believe that nitrogen is the world's largest externality, exceeding even carbon.

Also known as laughing gas or the forgotten greenhouse gas, nitrous oxide is the third most abundantly emitted greenhouse gas in terms of carbon dioxide equivalents. It is 300 times as potent as carbon dioxide at trapping heat, is relatively long-lived, spending an average of 114 years in the sky before disintegrating, and is responsible for depleting the ozone layer (Kanter 2018; Tian et al. 2020). During the last four decades, nitrous oxide emissions have risen by 30 percent, constituting roughly 6 percent of greenhouse gas emissions, of which about three-quarters come from agriculture (Tian et al. 2020). Other nitrogen compounds with shorter atmospheric lifetimes—notably ammonia and nitrogen oxides—are key causes of air pollution, as they contribute to the formation of fine particulate matter that adversely affects human health (Kanter 2018).

Given the unique chemistry of the nitrogen cycle and the nitrogen cascade, policies that inadvertently exacerbate the unbalanced and ineffective use of nitrogenous fertilizers could have rippling impacts on land, water, air, and climate.

---

applied on fields gets absorbed by crops; there is a limit to how much a plant can produce based on nitrogen alone.[8] Due to the unique chemistry of the nitrogen cycle, the excess of reactive nitrogen gets lost to the surrounding environment in its multiple chemical forms—as nitrites and nitrates, polluting the waterways; as anhydrous ammonia or nitrogen oxide, worsening air quality; and as nitrous oxide, exacerbating climate change and stratospheric ozone depletion (Kanter 2018) (box 8.3).[9]

The following discussion highlights the unintended but significant long-term costs of fertilizer subsidies. In doing so, the results focus on how the ineffective application of nitrogen fertilizer can hurt crop productivity as well as scar freshwater resources.

Diminishing returns to nitrogen use

Previous work has provided causal estimates of the impact of nitrogen fertilizer on yields. McArthur and McCord (2017) use the unique economic geography of fertilizer production and transport costs to countries' agricultural heartlands to construct a new time-varying instrument for fertilizer use globally. They find that a 1 kilogram per hectare increase in fertilizer use causes an 8–9 kilograms per hectare increase in yields, which translates to a 10 percent increase in nitrogen fertilizer use, boosting global yields by about 29–32 percent.

However, the terrible paradox is that, in some places, this stunning expansion in food production has been achieved in a way that cannot be sustained. Although fertilizer subsidies may have enabled fertilizer use and hence raised agricultural productivity, subsidies can incentivize farmers to apply more than they need such that the beneficial impacts of nitrogenous fertilizers on productivity begin to wane beyond a point. A continuous measure of nitrogen use can mask such heterogeneity in impacts on yields.

Using disaggregated grid cell–level data sets on nitrogen fertilizer and yields from 2000 to 2013 covering about 150 countries, this section demonstrates the nonlinear effects of nitrogen use on agricultural productivity at the global scale (see box 8.4 for details). To capture the nonlinear impacts of nitrogen use and assess the impacts across the global distribution of fertilizer, the continuous measure is replaced with indicators for whether the value lies within different quantile bins.

Figure 8.2 provides a striking visualization of the global response curve and the likely magnitude of these impacts. It shows that nitrogen fertilizer can have large and significant, but heterogeneous, effects on yields across the fertilizer distribution. The response

---

BOX 8.4
Technical spotlight: Diminishing returns to fertilizer use

To study the impact of nitrogen fertilizer on agricultural productivity, this chapter uses a grid cell–level data set covering the entire world. For the analysis, the land area is split into grid cells measuring 0.5 degree on each side, which is approximately 56×56 kilometers at the equator. The sample period extends from 2000 to 2013. The following equation is estimated at the global scale:

$$\ln\left(NPP_{it}\right) = \alpha_1 + \sum_{i=2}^{9} \beta_i Q_{it} + X'_{it}\lambda + f_c(t) + \theta_t + \gamma_i + \epsilon_{it}. \tag{B8.4.1}$$

Here NPP_{it} is net primary productivity in grid cell i in year t. NPP, which can be measured from satellite imagery, is used to measure agricultural performance since it provides a common unit of productivity across different types of crops (Zaveri, Russ, and Damania 2020). NPP is linearly related to the amount of solar energy that plants absorb over a growing season and is measured in grams of carbon per square meter. NPP is combined with a land cover data set developed by the European Space Agency's Climate Change Initiative, which provides information on 37 classes of land cover globally at a 300-meter grid. This approach ensures that plant productivity as measured by NPP is only captured in grid cells that contain significant amounts of agriculture and avoids attributing impacts to forests or other natural habitats. Data on average annual use of nitrogen fertilizer per hectare of cropland are from Lu and Tian (2017). The analysis replaces continuous measurements of nitrogen use with indicators, Q_i, to denote whether the log values of nitrogen use lie within different quantiles of the nitrogen distribution to examine nonlinear effects on yields. The quantile coefficients, β_i, are the coefficients of interest. The first quantile is omitted and becomes the reference quantile.

A plethora of other factors can also affect the relationship between nitrogen fertilizer and yields. Control variables, including rainfall, temperature, and various fixed effects and time trends, are included to isolate the impact of nitrogen fertilizer as much as possible from other factors: θ_t are year fixed effects, γ_i are grid cell fixed effects, $f_c(t)$ are country-specific time trends, X'_{it} is a vector of other control variables, including precipitation shocks, a quadratic term for mean annual temperature (°C), and log of population. These controls account for baseline differences in yield and other factors that vary by year. They are meant to control for changes in agricultural policies, development levels, input availability, technological levels, and time-invariant factors such as terrain slope and soil type.

To the extent that unobservable factors like farmers' agronomic know-how might be correlated with both yields and inputs, the estimation may underestimate the impact of increasing fertilizer use on yields; for this reason, these results should be interpreted as lower-bound estimates.

Source: World Bank.

---

of yields to increasing levels of nitrogen application rises gradually until it reaches a peak, after which it falls sharply, indicating diminishing productivity. This response means that yields may be increasing but require increasing amounts of nitrogen to achieve this growth. Going from quantile 1 to quantile 2 (or 7) of nitrogen use increases yields by about 5 (or 20) percent, but the increase slows down thereafter, such that going from quantile 1 to quantile 8 (or 9) increases yields by only 17 (or 13) percent, respectively. Put simply, at low and moderate levels of use, nitrogen fertilizer has the intended beneficial impact on yields, but there is a law of diminishing returns, whereby, at high applications, extra nitrogen has a diminishing effect on yields.

The figure also highlights the median level of nitrogen use for various regions, with regional response curves depicted in box 8.5. To obtain a more complete picture of the differences in the distribution of nitrogen across the world, region-specific quintiles are

FIGURE 8.2 **Change in global agricultural productivity due to the application of nitrogen fertilizer, by quantile of use and region**

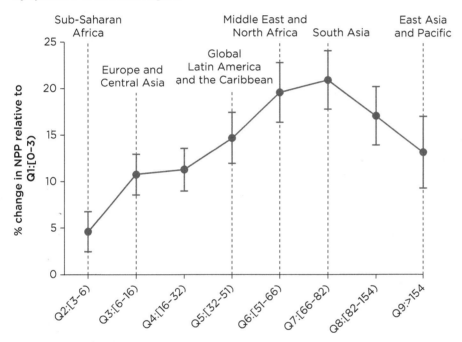

Source: World Bank.
Note: The figure shows point estimates and 95% confidence intervals of coefficients obtained for different quantiles of nitrogen fertilizer use from the second to the ninth quantile relative to the omitted first quantile. Vertical lines indicate where the median values of nitrogen fertilizer use lie for the global sample, and the different regions. The colored horizontal lines and dots below the graph indicate the bottom, middle, and top region-specific quintiles based on the regional distribution of fertilizer use. NPP = net primary productivity; EAP = East Asia and Pacific; ECA = Europe and Central Asia; LAC = Latin America and the Caribbean; MNA = Middle East and North Africa; SAR = South Asia; SSA = Sub-Saharan Africa.

Technical spotlight: Regional effects of fertilizer use

To account for the varying distribution of nitrogen use in each region, region-level response curves are also estimated based on region-specific quintiles following the methodology described in box 8.4. In the regional analysis, to examine nonlinear effects on yields, continuous measurements of nitrogen use are replaced with indicators to denote whether the values lie within different *quintiles*

FIGURE B8.5.1 Change in regional agricultural productivity due to the application of nitrogen fertilizer, by region-specific quintiles

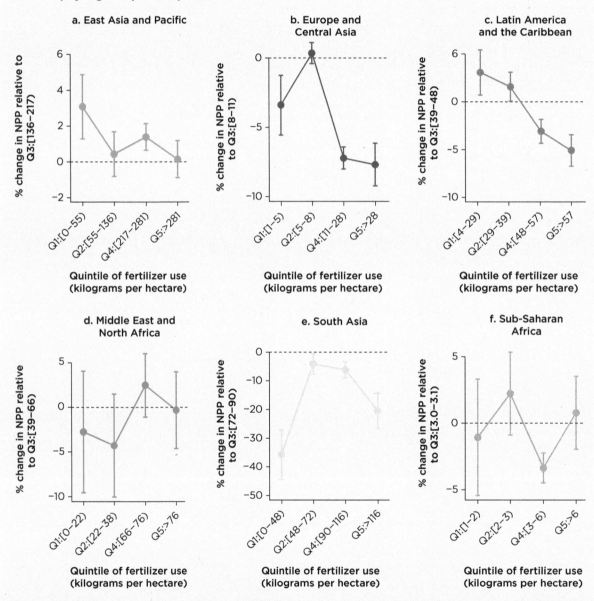

Source: World Bank.
Note: The figures show point estimates and 95% confidence intervals of coefficients obtained for different quintiles of nitrogen fertilizer use from the first to the fifth quintile relative to the omitted third quintile. NPP = net primary productivity.

(Continued)

of the nitrogen distribution. The middle, or third, quintile is omitted and becomes the reference quintile. The responsiveness of yield to the first, second, fourth, and fifth quintiles is therefore measured relative to quintile three and is depicted in figure B8.5.1. Depending on the spread of the distribution of fertilizer in each region, the third quintile may or may not represent the optimum level of nitrogen.

In regions like Latin America and the Caribbean and East Asia and Pacific, the amount of nitrogen use is high across most of the region, and the diminishing returns in the upper quintiles relative to returns in the middle quintile can be stark. In these regions, going from the middle third quintile to the lowest first quintile improves productivity by 3–4 percent. Further increasing the use of nitrogen fertilizer beyond the third quintile diminishes yields. In regions like South Asia, spatial disparities are significant, with areas of both low and high use of nitrogen fertilizer. Therefore, going from the middle third quintile to both the lowest first quintile and the highest fifth quintile of nitrogen use reduces productivity by about 35 percent or 20 percent, respectively. In Europe and Central Asia, both low and high quintiles of nitrogen use can reduce yields, relative to the middle quintile. The response curves in Sub-Saharan Africa and the Middle East and North Africa remain largely noisy. This response may reflect the vast uncertainty and heterogeneity surrounding yield responses across the African continent.

shown in figure 8.2. Regions like East Asia and South Asia are at the high end of the global distribution of nitrogen fertilizer, while Sub-Saharan Africa is at the low end of the distribution. South Asia, for example, is operating at the peak level of global response, indicating that about half of the areas within this region are on the decreasing return part of the global response curve. The median level of nitrogen use in East Asia is at the 95th percentile of the global level of nitrogen use and is situated on the decreasing part of the global response curve, indicating that much of the region is already facing the brunt of diminishing returns. This result is not surprising given the long history of nitrogen fertilizer use in these regions.

In contrast, the median level of nitrogen use in Sub-Saharan Africa is only a fraction of the global median. Many countries in this region get low yields *and* apply only small amounts of nitrogen to their crops. While a decline in overall fertilizer use would lead to significantly lower productivity for the region, with potentially serious consequences for food security, blanket recommendations and a sole focus on the application of nitrogenous fertilizers warrant further scrutiny. As discussed in the previous section, the responsiveness of yields to nitrogenous fertilizers, in particular, is still uncertain due to differences in the composition and acidity of African soils.[10] Balanced, soil-specific fertilizers can go a long way in helping African farmers to maximize their returns on investment.

Overall, these results demonstrate that crop production is becoming less efficient at using nitrogen fertilizer. This finding is qualitatively consistent with field-level agronomic data that also show decreasing returns to nitrogen fertilizer use. Underlying these results is the critical metric of nitrogen use efficiency, an indicator that describes how much of the fertilizer being used reaches a harvested crop. Various studies have sought to evaluate the exact value of nitrogen use efficiency. Studies suggest that only 32 percent is absorbed by plants in India, compared with 52 percent in Europe and 68 percent in Canada and the United States (Zhang et al. 2015). A recent global meta-analysis finds that nitrogen use efficiency has decreased by 22 percent since 1961 and remains stubbornly low at around 46 percent (Zhang et al. 2021). According to an average of 13 global databases, of the 161 teragrams of nitrogen applied to agricultural crops, only 73 teragrams of nitrogen reach the harvested crop (Zhang et al. 2021): almost two-thirds of all nitrogen applied to crops gets wasted. The European Union Nitrogen Expert Panel recommends nitrogen use efficiency of around 90 percent as an upper limit, indicating that a vast amount of the nitrogen that is poured into fields is wasted.

In sum, there is an optimum level of nitrogenous fertilizer application, which can vary with field conditions and combinations of other inputs. Policies that are intended to increase productivity but fail to consider local conditions can inadvertently exacerbate the unbalanced and ineffective use of nitrogenous fertilizers. Subsidizing the wrong type of fertilizer or subsidizing fertilizers in areas where fertilizer use is already well above levels that maximize its value can impose substantial costs on productivity and provide lower returns to farmers, leading to wasted fertilizers and government spending.

Ailing waters

The massive increase in nitrogen fertilizers has left a scar across many of the world's water bodies. As described earlier, because of excess use and inefficiencies in application, not all nitrogen applied on fields is absorbed by crops. Runoff of excess nitrogen increases concentrations of nitrate and nitrite in the waters. These concentrations can lead to cyanobacteria-related algal blooms. Conspicuous to the eyes, cyanobacteria can be deadly—they can emit neurotoxins and hepatotoxins such as microcystin and cyanopeptolin, which are toxic to humans, animals, and other aquatic life (Damania et al. 2019). Previous remote-sensing analysis has shown that between 2002 and 2012 the world experienced, on average, nearly two episodes of massive cyanobacteria-related algal bloom per year in 421 of the world's largest lakes, highlighting the recurring problem of fertilizer overuse, agricultural runoff, and subsequent deterioration in water quality (Damania et al. 2019).[11] Large algal blooms can devastate ecosystems, often resulting in hypoxia or dead zones, a condition that arises when water bodies lack sufficient oxygen. The legacy effects of nitrogen pollution on the environment can also endure decades after nitrogen inputs have ceased, with long time lags between the adoption of conservation measures and any measurable improvements in water quality (Basu et al. 2022; Van Meter, Van Cappellen, and Basu 2018). According to the Food and Agriculture Organization, agriculture—not cities or industry—is the biggest source of water pollution in many countries today, while worldwide, the most common chemical contaminant found in groundwater aquifers is nitrate from farming (Mateo-Sagasta et al. 2017).

A considerable body of literature has attempted to explain how human and environmental changes affect freshwater quality. Many of these pollution elasticities are available for individual countries or case-specific studies but have not previously been quantified at the global level for water pollution. This section employs a global gridded data set of nitrogen fertilizer and nitrogen pollution in water to calculate the change in water quality for every percentage change in nitrogen fertilizer use. Using a similar approach as in the previous section, global land is split into grid cells measuring 0.5 degree on each side. Data on nitrogen pollution from river monitoring stations in the GEMStat database are used to measure concentrations of nitrogen in water bodies. Nitrogen is measured using a combination of nitrates, nitrites, and ammonia. The methodology is described in box 8.6.

Technical spotlight: Nitrogen fertilizer use and water pollution

To estimate the impact of nitrogen fertilizer on water pollution, the setup in box 8.4 is modified to examine the total impact of nitrogen fertilizer use on water pollution spillovers. Like before, the analysis is conducted at the grid cell level, with grid cells measuring 0.5 degree on each side, which is approximately 56×56 kilometers at the equator. The sample period for this analysis extends from 1995 to 2013. The following equation is estimated at the global scale, where i denotes grid cell and t denotes year:

$$\ln(Q_{it}) = \alpha_1 + \beta_1 \ln(N_{it}) + X'_{it}\lambda + f_c(t) + \theta_t + \gamma_i + \epsilon_{it}. \tag{B8.6.1}$$

Water quality data are sourced from GEMStat, which is a globally harmonized database on freshwater quality developed by the United Nations Environment Programme's Global Environment Monitoring System (GEMS), maintained by the International Centre for Water Resources and Global Change, and hosted by the Federal Institute of Hydrology in Koblenz, Germany.[a] Data from river monitoring stations in the GEMStat database are used to measure concentrations of nitrogen in rivers. Total nitrogen in water, Q_{it}, is measured using a combination of nitrates and nitrites and aggregated to the 0.5 degree grid to match the resolution of nitrogen fertilizer data, N_{it}. A plethora of other factors can also affect the relationship between nitrogen fertilizer and water quality. Control variables, including various fixed effects and time trends, are included to isolate the impact of nitrogen fertilizer as much as possible from the impact of other factors: θ_t are year fixed effects, γ_i are grid cell fixed effects, and $f_c(t)$ are country-specific time trends. Finally, X'_{it} is a vector of other control variables, including annual rainfall, temperature, runoff, and log of population. In some specifications, the extensive or intensive margin of land use such as cropped area and yields is also included. Together, these controls account for baseline differences in water quality patterns and other factors that vary by year. They are meant to control for changes in agricultural policies, development levels, input availability, technological levels, and time-invariant factors such as terrain slope, soil type, and distance to coast or water bodies.

Source: World Bank.
a. For the Global Freshwater Quality Database, GEMStat, see https://www.gemstat.org.

The results show that an increase in agricultural fertilizer use leads to substantial deterioration in water quality such that there is a strong and positive impact of nitrogen fertilizer use on the concentration of nitrogen in water. These estimates are robust across multiple specifications, with elasticities ranging from 0.18 to 0.33, suggesting that a 10 percent increase in nitrogen fertilizer use leads to a 1.8–3.3 percent increase in the concentration of nitrogen pollutant. These pollution elasticities are remarkably consistent with estimates from prior econometric literature that finds adverse effects of nitrogen fertilizer use on water quality in country-specific settings (Paudel and Crago 2021).[12]

The subsidy toll

So far, this chapter has documented a link from subsidies on nitrogen fertilizer use to adverse impacts on productivity and water quality. To provide a better sense of the direct effect of such agricultural subsidies on water resources, analysis is undertaken for the sample of countries for which subsidy data are available.

A major challenge with quantifying support in agriculture is the difficulty of obtaining consistent measurements of such support across all countries, both low and middle income and high income. Agricultural support estimates are obtained from a combined database following Gautam et al. (2022), as discussed in chapter 6. It is important to distinguish between agricultural support that is coupled to input use and output levels and support that is decoupled from production and provided as direct payments to farmers. Coupled subsidies incentivize production and provide direct subsidies on output or inputs that create incentives to increase output. In contrast, decoupled supports are not linked to production and avoid altering incentives to change input or output levels. Instead, they provide direct income support to producers, acting as lump-sum subsidies, and are less distortionary. To separate producer support into coupled and decoupled payments, data from the Organisation for Economic Co-operation and Development's composition of producer support estimate (PSE) tables are used to construct coupled and decoupled support as a share of the total value of production.

In certain parts of the world, like South Asia and Sub-Saharan Africa, aggregate coupled support can be negative due to the inclusion of market price support (MPS), a variable in the database that accounts for trade measures and policies such as export bans, which can lead to a net tax on producers when global (free trade) prices exceed the domestic price. For this reason, three measures of coupled subsidies are used in the analysis. In one variant, MPS is removed to focus only on the portion of the subsidy that amounts to direct producer support. In the second variant, only support for inputs is included, and in the third variant, only MPS is included. Results in figure 8.3 indicate that coupled producer support has a positive and significant impact on levels of nitrogen pollution across various definitions of coupled support.

FIGURE 8.3 **Effect of subsidies on water pollution, by type of subsidy**

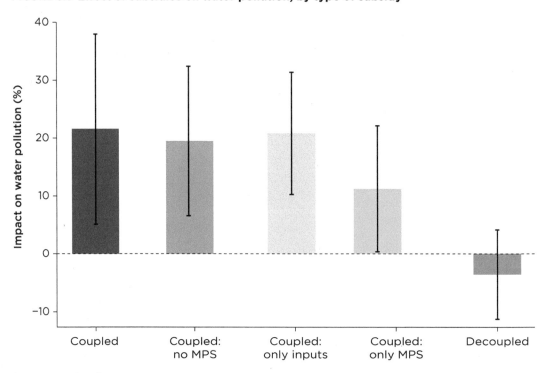

Source: World Bank.
Note: The figure shows point estimates of the impact of a 100 percentage point increase in the share of coupled or decoupled subsidy in total value of production on total nitrogen pollution in water with 90% confidence intervals. MPS = market price support.

Further, even when the definition of coupled subsidies is restricted to input support, the magnitude of the effect on pollution remains stable and similar to the magnitudes that correspond to broader definitions of coupled subsidies. A 100 percentage point increase in the share of coupled support for inputs leads to about a 20 percent increase in nitrogen pollution in water. These effects are quantitatively meaningful and amount to input support explaining approximately 17 percent of nitrogen pollution in the past 30 years across the global sample. However, when the definition of coupled support is restricted to MPS, the impacts on pollution are reduced by up to half. This result is consistent with the findings of chapter 7 that the effects of MPS on farm-level efficiency are not as deleterious as the effects of input support. Unlike the results of these coupled support measures, decoupled support has muted or statistically insignificant impacts.

17%: The approximate share of nitrogen pollution in water in recent years that can be attributed to the use of input subsidies

Following Paudel and Crago (2021), these elasticities can be used to estimate the effect of an increase in nitrogen water pollution on the size of the hypoxic zone to illustrate their economic significance. Applying the statistics reported in Hendricks et al. (2014) and Obenour et al. (2012),[13] a 100 percentage point increase in coupled input support, on average, can result in an increase of about 2,173 square miles in the size of the hypoxic zone globally. This area amounts to almost 30–40 percent of the measured size of the "dead zone" in the Gulf of Mexico, considered one of the largest dead zones in the world.

These spillover effects on water also have implications for human health. Although it is known that nitrogen in water is responsible for fatally inflicting what is known as blue baby syndrome, which starves infants' bodies of oxygen, studies have also shown that babies who do survive endure longer-term damage throughout their lives. Exposure to nitrogen pollution in early life can result in stunted growth and impaired development of infants, which could lead to poor productivity of future generations (Jones 2019; Zaveri et al. 2020). Using prior estimates, this exposure implies that, globally, a 10 percentage point increase in coupled input support and the subsequent release of nitrates into the water pose a risk large enough to wipe out up to 2.7 percent to 3.5 percent of labor productivity, especially in areas where input subsidies make up the largest share of the value of production in the global sample.

2.7% to 3.5%: The share of labor productivity that can be wiped out by subsidy-induced water pollution increases, where input subsidies are the highest

Understanding how subsidies affect water pollution can therefore also shine a light on the influence of subsidies on population health outcomes and a country's forgone human capital accumulation. Although these estimates are broad and imprecise, they suggest that fertilizer policies and vast fertilizer subsidies require careful scrutiny, particularly in places where nitrogen use exceeds optimal levels for plant growth. The significance of these findings and their ramifications for sustainable development cannot be overstated: subsidies can inadvertently lock in inefficiency for decades, or even longer, and can make the difference between success and failure.

Since levels of producer support are not randomly assigned, a potential concern is that subsidies themselves are endogenously determined and that other unobserved variables that are correlated with subsidies are really to blame. To alleviate some of this concern, it is instructive to devise a placebo check—that is, a test with a similar setup, but one in which a similar result is not expected. One such test is to check whether future levels of producer support are associated with higher levels of nitrogen pollution in water. Results show that the estimates of future producer support are small and insignificant, suggesting that the results are not simply picking up generic correlations between producer support and nitrogen pollution, providing further confidence in the result.

Drawing down the ocean underground

While the focus of this chapter is largely on environmental externalities related to water pollution, subsidies can also have an impact on water quantity. Along with fertilizer, groundwater is one of the core ingredients of what the Nobel Prize–winning economist Angus Deaton calls the "great escape" from scarcity (Deaton 2015). Vast quantities of groundwater have sustained the intensification of agriculture brought on by the green revolution in various regions of the world. To date, millions of farmers depend on groundwater irrigation to produce 40 percent of the world's agricultural production, including a large proportion of staple crops like rice and wheat (Jain et al. 2021). Yet groundwater reserves are perilously

---

BOX 8.7
Technical spotlight: Drawing down the ocean underground

Groundwater levels are being depleted at alarming rates in the world's arid and semiarid regions. These effects are most visible in India, where groundwater use has increased by an explosive 500 percent over the past 50 years (Garduño and Foster 2010), making India the largest user of groundwater in the world. While attention has been paid to the role of input subsidies such as cheap electricity for increasing groundwater extraction (Jessoe and Badiani-Magnusson 2019), recent work suggests that output subsidies that guarantee the purchase of rice and wheat at higher-than-market prices have contributed substantially toward declining water tables in the country (Chatterjee, Lamba, and Zaveri 2022; Devineni, Perveen, and Lall 2022). Overall, these procurement policies have led to a 30 percent overproduction of water-intensive crops. In the northwestern state of Punjab, which suffers from some of the largest increases in groundwater stress, output subsidies and rice procurement, in particular, account for 63 percent of the rise in groundwater declines in more than two decades (Chatterjee, Lamba, and Zaveri 2022).

To what extent are such subsidies implicated in the global drawdown of water? To provide a global assessment, a cross-country analysis is employed following the methodology used to assess the impacts of subsidies on water pollution. Global data on groundwater are drawn from the Gravity Recovery and Climate Experiment (GRACE) satellite mission, which enables measurement of changes in water mass distribution across the Earth's surface. GRACE data values capture changes in terrestrial water storage (TWS), which is an aggregate of changes in snow, surface water, soil moisture, and groundwater. To isolate the groundwater signal (GWS), output from the Global Land Data Assimilation System is used to subtract the non-groundwater components from the overall GRACE TWS value. GRACE monthly TWS (or GWS) data are represented as anomalies relative to a baseline gravity-field value rather than as changes in levels. The units are in terms of "centimeters of equivalent water thickness," which represents a change in gravity caused by a change in height (centimeters) of a water surface that is spread out over a given area. For the analysis, country-level changes in groundwater storage are measured across the entire land mass and in agricultural areas from 2002 to 2017.[a] Since GRACE is known to be effective at measuring

(Continued)

BOX 8.7
Technical spotlight: Drawing down the ocean underground (*continued*)

changes in groundwater storage for regions greater than 90,000 square kilometers, only countries that are greater than or equal to 90,000 square kilometers are included in the sample to respect the spatial limitations. Like before, data on coupled and decoupled subsidies are drawn from the database described in chapter 6.

The analysis leverages a suite of spatial and temporal fixed effects to account for all time-invariant effects unobservable at the country level as well as other annual shocks and trends to ensure that the result is not an artifact of trending variables. The regressions also control for weather variables, including contemporaneous temperature and precipitation along with cumulative precipitation over the past year. To rule out concerns that groundwater depletion is merely a reflection of general economic development, a control for GDP is included.

The statistical analysis measures the effect of coupled and decoupled support on changes in groundwater storage in cropped areas as well as across all areas. Given the focus on agricultural subsidies, it is expected that effects will occur largely in regions where agriculture dominates the landscape. Results in figure B8.7.1 show that the impacts of coupled subsidies are significant in cropped areas. On average, a 10 percentage point increase in coupled subsidies can lead to a water loss of about 2.5 centimeters of equivalent water height across cropland areas of countries over

FIGURE B8.7.1 Effect of subsidies on groundwater depletion, by type of subsidy

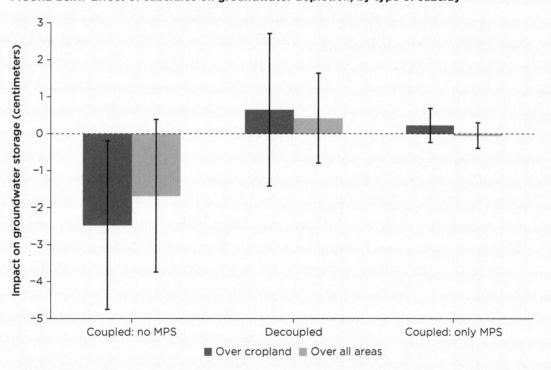

Source: World Bank.
Note: The figure shows point estimates of the impact of a 10 percentage point increase in the share of coupled or decoupled subsidy over the total value of production on groundwater storage with 95% confidence intervals. MPS = market price support.

(Continued)

the course of a year. In contrast, decoupled support or market price support alone has positive but statistically insignificant impacts. The impacts of coupled subsidies are economically meaningful and suggest that, at the mean level of coupled subsidy exposure, cropped areas across the globe could lose up to 13.2 cubic kilometers of water per year due to coupled agricultural subsidies. This amount is roughly equivalent to the total amount of water lost in California between 2011 and 2014 at the height of the drought (14.8 cubic kilometers). Although these estimates are broad, they suggest that coupled producer support subsidies have substantial implications for water resources and can lead to a perceptible drawdown of aquifers.

a. Cropland data from https://glad.umd.edu/dataset/croplands are used to identify cropland pixels where at least 20 percent of the share of total land area is denoted as cropland.

declining in many important agricultural regions across the globe. To what extent have agricultural subsidies influenced such declines in groundwater resources? Following a similar methodology used to assess the impacts of subsidies on water pollution, box 8.7 demonstrates the impact of coupled and decoupled subsidies on global groundwater depletion.

The way forward

Broadly, there are three ways to increase food production: greater use of land, greater use of inputs, and greater efficiency in how resources are used. Historically, most of the increase in food production came from increased land use, especially in low- and middle-income countries, as population growth led to the expansion of cropped area. But since the 1960s, global food production, on average, began to decouple from land use due to the intensive use of inputs such as fertilizer and irrigation water, which contributed to unprecedented growth in global yields.[14]

Over time, this intensification has brought new challenges due to inefficient and unbalanced input use. A recurring finding in this report is that poorly designed subsidies exacerbate these challenges. New global analyses presented in this chapter find that certain areas of the world are using egregiously large amounts of fertilizer that are well beyond efficient levels. This overuse is leading to diminished crop yields and increased nitrogen runoff into waterways, impairing the environment and human health. Subsidies are exacerbating these effects, incentivizing farmers to use excessive nitrogen that exceeds optimal levels for plant growth. Along with harmful impacts on water quality, producer support subsidies can also drive increases in the use of irrigation water and accentuate the risk of global groundwater depletion, in turn, compromising the long-term food and livelihood security of millions of people and the very objectives that the subsidies seek to achieve.

However, politicians often see coupled producer support subsidies as untouchable because farmers represent large and influential constituencies in many countries. But there are political and environmental win-wins in which these highly inefficient subsidies can be replaced with less distorting subsidies and better policies that are coupled with efficiency gains, such that nobody is made worse off. For instance, the analysis in this chapter shows that, unlike coupled support, decoupled support does not lead to harmful environmental spillovers, highlighting their importance in achieving policy objectives without incurring costs that are counterproductive to policy goals. At the same time, evidence from a global decomposition analysis of growth in agricultural production suggests that total factor productivity (TFP) has grown in importance as a driver of growth in recent decades.[15]

Growth in agricultural production has come increasingly from greater efficiency, not necessarily through bringing new land into agricultural production or intensifying input use (USDA 2021). In the most recent period, from 2011 to 2019, TFP grew by an annual rate of 1.31 percent, accounting for nearly two-thirds of the growth in global agricultural production (USDA 2021). While such global estimates can mask the vast variation across countries and regions, the overall trend implies that it is possible to produce more with fewer inputs. Instead of choosing between the "extensive" margin (that is, using more land) or the "intensive margin" (that is, using more inputs)—choices that also have the potential to cause adverse environmental spillovers—there is a third, often ignored, policy option of using resources more efficiently (World Bank, forthcoming).[16] Yet despite these important advancements, sustaining this growth in TFP is increasingly under stress from climate change. Anthropogenic climate change has reduced agricultural TFP by about 20 percent since 1961 (Ortiz-Bobea et al. 2021), suggesting that the sensitivity of the agriculture sector is likely to become even more pronounced in the decades ahead. This sensitivity could create greater pressure to increase the use of land and inputs, making *sustainable* productivity growth and closing efficiency gaps even more important policy objectives.

The importance of efficiency gains for sustainability points to a direct role for public policies to facilitate the uptake of more efficient and sustainable agricultural management practices as well as to stimulate innovation. Converting the fertilizer subsidy to cash transfer programs, investments in public goods, agricultural extension facilities, or other types of rural development programs may also be more beneficial from an efficiency standpoint. In the future, new technologies such as precision farming and new plant breeding might play an increasingly important role, which would require national government support and could vastly increase the efficient use of nitrogen. A broad view of the overall efficiency of the food system could also be transformational by incorporating solutions on the demand side. For example, the report of the EAT-Lancet Commission calls for changing the composition of our diets away from foods thought to be unhealthy for humans and the environment (Willet et al. 2019).

The current crisis in Ukraine and the fallout of war on rising fertilizer prices show the critical need to close efficiency gaps and transform our food systems (box 8.8). The response curve of crop yields with respect to nitrogen fertilizer in figure 8.2 and box 8.5 show that about 50 percent of the global calories produced are grown in areas where there is overuse of nitrogen fertilizers. Some countries and regions have room to bring fertilizer use closer to optimal levels with limited or benign impacts on food supplies. The recent increase in fertilizer prices highlights the importance of harnessing the most value out of any fertilizer that gets to the farm, which, in turn, can improve crop yields while also benefiting the environment. Indeed, recent research suggests that nitrogen pollution could be reduced by around 35 percent if polluting countries became as efficient as their neighbors. This reduction would have little impact on crop yields—increasing yield gaps by only 1 percent. In other words, closing yield gaps and mitigating nitrogen pollution need not

50%: The approximate share of global calories produced that are grown in areas where there is inefficient use of nitrogen fertilizers due to subsidies

BOX 8.8
The fallout of war

The recent war in Ukraine, the economic sanctions it triggered, and disruptions in the Black Sea trading routes have put a spotlight on rising fertilizer prices and the overall vulnerability of the global fertilizer market to supply shocks. A large amount of the three primary macronutrients used in fertilizers—38 percent of all nitrogen, 50 percent of all phosphorus, and 80 percent of all potassium produced—is traded on international markets, with only a few countries contributing to the majority of the traded share (Hebebrand and Laborde 2022a). The Russian Federation exports more than two-thirds of its production of each product (Smith 2022). One of the key ingredients in nitrogen fertilizers is ammonia. Since ammonia is produced primarily using natural gas or coal as feedstocks, the comparative advantage of manufacturing these ingredients lies with only a select group of facilities that are situated primarily in countries with relatively low gas or coal prices. In 2019, China, Russia, and Qatar together accounted for 33 percent (15, 13, and 5 percent, respectively) of traded nitrogen (Hebebrand and Laborde 2022a). The production of potash and phosphates is even more concentrated due to the uneven distribution of the source deposits.[a] For phosphates, the top three exporters—China, Morocco, and Russia—represent 57 percent of global trade, with Russia accounting for about 14 percent of the share (Hebebrand and Laborde 2022a). For potash, the market share of the top three exporters—Belarus, Canada, and Russia—reaches 80 percent, with Belarus and Russia jointly contributing a staggering 41 percent of globally traded potash.

Fertilizer prices have been climbing for more than a year due to rising prices of natural gas, coal, ammonia, and sulfur as well as disruptions in global supply chains due to COVID-19 restrictions. The fallout of the war has increased trade costs and uncertainty and shot up prices even further due to sanctions and supply disruptions. Adding to supply concerns, China, the largest producer of phosphate, has suspended exports of fertilizers through 2022 to ensure domestic availability. Together, these factors have led fertilizer prices to soar by nearly 30 percent since the start of 2022, following an 80 percent surge in 2021 (Baffes and Koh 2022). Three-quarters of the globe are dependent on imports for 50 percent or more of their fertilizer use.

Concerns about the affordability and availability of fertilizer are, therefore, growing, with worries that nutrient shortages will threaten future harvests and the global production and yield of a variety of crops. However, as the response curve of crop yields with respect to nitrogen fertilizer in figure 8.2 and box 8.5 demonstrate, reduced fertilizer supply can have vastly uneven impacts across regions. Regions that use large amounts of nitrogenous fertilizer but achieve very little in the way of additional yield are likely to be less vulnerable to rising fertilizer prices for the simple reason that increasing prices of fertilizers can provide an extra incentive to apply less fertilizer per unit of land without hurting farmers' returns. In fact, more than half of the calories produced are from areas that fall on the downward portion of the global response curve, suggesting that declining fertilizer use might not lead to a proportionate decline in crop production. Moreover, to the extent that farmers currently apply more than the recommended amount of fertilizer, reducing use can also provide an environmental benefit in the form of decreasing nutrients in water. In these cases, it can also provide an opportunity to experiment and learn whether the additional fertilizer is necessary as insurance against low yields (Smith 2022). In regions like Africa that use a limited amount of fertilizers, a decline in fertilizer use could lead to significantly lower productivity on farms reliant on synthetic fertilizers, with potentially serious consequences for food security. Moreover, its relatively smaller markets are likely to face a particularly difficult situation, as producers and traders might favor shipping limited supplies to larger markets (Hebebrand and Laborde 2022a). In countries like India, price shocks may be buffered as governments ramp up fertilizer subsidies, which will place tremendous fiscal pressure on budgets already stressed by substantial government outlays during the COVID-19 epidemic. India's fertilizer subsidy bill is likely to shoot up to US$30 billion for the fiscal year 2022–23, as the government provides additional funds to make up for the spike in cost from higher import prices (Bera 2022).

As the fertilizer crisis has shown, the world needs to produce fertilizer in ways that make environmental and geopolitical sense and to use what is available much more efficiently. In response to the crisis, encouraging best practices that squeeze the most value out of any fertilizer that gets to the farm will be vital. Applying knowledge of fertilizer use, efficient nutrient use, balanced fertilizers, and crop- and soil-specific fertilizers following sound agronomic principles will be critical.

a. Phosphorus fertilizers are typically produced by mining phosphate rock, while potassium fertilizers are created by mining potash from deep underground (Smith 2022).

be a necessary trade-off, and there is massive potential to employ country-specific policies to reduce nitrogen pollution without affecting crop yields (Wuepper et al. 2020). For instance, direct fertilizer subsidies, if applied, need to be temporary and targeted and need to encourage efficient and balanced plant nutrition. In the wake of the current crisis, governments and organizations should act quickly and judiciously to support the most vulnerable farmers while avoiding the pitfalls of trade restrictions and blanket high subsidies that eschew a tailored approach to nutrient management (Hebebrand and Laborde 2022b).[17]

Cultivating solutions

Nitrogen balances across the developing world run the gamut from acute deficiencies to extreme excesses. In general, farmers struggle to apply just the right amount of fertilizer. Even in high-income economies like those in the European Union, which has reduced the amount of nitrogen wasted over the past several decades, progress continues to stagnate (EEA 2018). How nitrogen is managed over the coming decades will determine whether humanity can return to being within the nitrogen planetary boundary—a level of human interference beyond which damage to ecosystems and human health could increase dramatically, perhaps permanently—without jeopardizing food security. Not surprising, improved nitrogen management is encapsulated in several of the Sustainable Development Goals (SDGs), including ending hunger (SDG 2) and protecting the environment and human health (SDGs 6, 12, 13, 14, and 15). Therefore, it is critical that better modes of nitrogen management be developed, deployed, and adopted globally both in areas that use too much and in areas that use too little fertilizer (Kanter, Chodos, et al. 2020). Well-designed policy that encourages experimentation can also help. For instance, research suggests that subsidies need not be permanent or universal to benefit farmers in substantial ways. Instead, temporary input subsidies can be useful for experimenting with and learning about best management practices (Carter, Laajaj, and Yang 2019).

On the farm, there is scope for improving the management of inputs using existing technology and resources by employing best management practices such as the four Rs to improve the efficiency of nitrogen use: using the right nutrients, at the right rate, at the right time, and in the right place. Overall, better fertilizer management through optimal quantity and timing of application can minimize waste, lower the direct fertilizer expense and associated environmental costs, and improve productivity by maintaining soil quality and ensuring that nitrogen is available when it is most beneficial for plant growth (Islam and Beg 2021). Methods that harness accumulated nitrogen legacies within the soil profile in areas where the availability of soil nitrogen is high could also lower fertilizer application rates without notable declines in crop yields and contribute to cost savings and environmental benefits (Basu et al. 2022). Since the ability of legacy nitrogen stores to sustain crop yields would vary spatially, tailored approaches are needed (Basu et al. 2022). Precision agriculture's focus on tailoring management decisions to site-specific conditions can help to refine these strategies such that they are more responsive and respectful of agricultural heritage and cultural practices (Kanter, Bell, and McDermid 2019). Low-cost, manual approaches such as seed priming and fertilizer micro-dosing can concentrate scarce resources in the vicinity of the plant, ensuring greater nitrogen uptake and leaving less nitrogen available to be lost to the environment. Other low-cost techniques suited to both low- and high-nitrogen smallholder systems include leaf color charts and chlorophyll meters (Kanter, Bell, and McDermid 2019).

However, the adoption of more efficient and sustainable intensification management practices is not inevitable.[18] Studies have pointed to various reasons for why the adoption is often slow or minimal (see box 7.1 in chapter 7). These reasons include perceived technological and financial barriers due to lack of information, farmers' reticence to overturn or augment long-running practices due to lock-in, as well as the inherent complexity in dealing with volatile nutrients like nitrogen that make it difficult to know the precise nitrogen needs of plants throughout the season (Jack and Tobias 2017; Kanter, Bell, and McDermid 2019). Changing farmers' practices surrounding the over- or under-application of nitrogen, therefore, requires combining the right information with the right incentives, training, and farmer education services.

China provides a notable example of success in educating and training farmers on good management practices. In a decade-long trial, researchers worked with 20.9 million smallholder farmers across the country to see if they could increase crop yields while also reducing the environmental impacts of farming. A combination of outreach program and workshops—about 14,000 workshops over 10 years—helped to convince farmers to adopt the recommendations (Cui et al. 2018). Results show that, in the decade from 2005 to 2015, average yields of maize, rice, and wheat increased by around 11 percent. At the same time, nitrogen fertilizer use *decreased* by around one-sixth, saving 1.2 million tons of nitrogen. By producing more crops and using less fertilizer, this experiment provided an economic return of US$12.2 billion (Cui et al. 2018). More recently, China also phased out nitrogen fertilizer subsidy and is instead funding improvements in nitrogen and manure management, with promising early results (Ji, Liu, and Shi 2020).

In contrast, a large-scale soil health card program launched in India in 2015 tested 23.6 million soil samples and distributed 93 million soil health cards to farmers, along with fertilizer recommendations, but it proved to be less successful. Without adequate education of what the data meant or how to apply the information, farmers failed to optimize the use of fertilizers (Fishman et al. 2016). However, when the soil health cards were simplified and made more user-friendly and farmers were given repeated access to extension services, there was a significant improvement in farmers' comprehension of soil health information along with a small, but significant, increase in the application of more balanced nutrients (Cole and Sharma 2017). Similarly, in Bangladesh, a simple rule-of-thumb training that deployed colored leaf charts to guide fertilizer application reduced fertilizer use by 8 percent without compromising yields. The behavioral intervention amounted to a savings of 180,000 metric tons of nitrogenous fertilizer, worth US$80 million or 14 percent of Bangladesh's input subsidy budget (Islam and Beg 2021).

Increasingly, new data through satellite technology coupled with a rapid rise in mobile phone penetration in rural areas[19] have opened new opportunities to connect farmers more easily to extension services and to empower them with timely and accurate information on fertilizer recommendations (Kanter, Bell, and McDermid 2019; Singh, Ganguly, and Dakshinamurthy 2018).[20] With greater availability and affordability, data from satellite sensors such as the European Space Agency's SENTINEL-2 Multi-Spectral Imager and the series of sensors from Planet's Dove satellites are being explored to bring satellite monitoring to the individual-field level. Recent research has shown that data from low-cost satellite sensors can help to predict crop yields at the individual-field scale, providing farmers with crucial information to inform their agricultural management practices and help them to make decisions about where and when to apply fertilizer (Jain et al. 2019). Building the capacity of local agricultural extension officers to appreciate the usefulness of these data sets and apply them in practice will be critical (Dash 2019).

Other initiatives and technologies, such as nitrogen-fixing bacteria, efficiency-enhancing fertilizers, or fertilizer deep placement that targets fertilizer at the source, also show promise. For example, the International Fertilizer Development Center has developed a method that uses fertilizer deep placement to enhance the efficiency of nutrient delivery to crops. The method buries nitrogen fertilizer into the soil, feeding nitrogen directly into a plant and reducing losses. However, the potential to scale such technology in regions like Africa may be more limited due to the nature of the soils, which can hinder the pathways of nutrient distribution (Cox, Kwon, and Koo 2015). Expanding frontier agricultural technologies such as insect and hydroponic farming that have a minimal land and water footprint and employ circular economy principles are also increasingly being seen as attractive options to enhance food security and livelihood opportunities, especially in Africa and countries affected by fragility, conflict, and violence. Waste from insects can also be fed back into the food system as organic fertilizers to improve soil health (Verner et al. 2021).

At the same time, even as countries focus attention on best management practices on-farm, there is an increasing need to extend attention to off-farm actors that are capable of influencing farm-level nitrogen management as well as off-farm initiatives that can mitigate nutrient pollution in water bodies (Basu et al. 2022; Kanter, Bartolini, et al. 2020). For example, since 2015, the government of India has made efforts to improve the efficient use of nitrogen in agriculture and has mandated the manufacture of urea, a type of nitrogen fertilizer that is commonly applied in India, to produce neem-coated urea. Neem-coating helps to reduce leakages by making it more difficult for black marketers to divert urea to industrial consumers (Government of India 2016). It also has the potential to benefit farmers. Since neem inhibits nitrification, it allows a more gradual release of nitrogen into the soil, thereby improving the efficiency of nitrogen use. Neem, however, is one of many possible compounds, and the responses of different crops under different conditions can be highly variable (Searchinger et al. 2020). Moreover, as large amounts of nitrogen in subsurface soil and groundwater can accumulate over several decades, more attention needs to be paid to off-farm land use policies and spatial planning that can protect water supplies from legacy nitrogen. Forests and wetlands act as natural buffers that absorb excessive nutrients that would otherwise pollute waterways. Evidence from the United States suggests that the targeted restoration of fluvial wetlands that are connected to rivers and streams are cost-effective measures that can remove both legacy and new nitrogen (Cheng et al. 2020). A strategic combination of off-farm watershed conservation measures with on-farm nutrient management can therefore mitigate the stubborn problem of nitrogen pollution faster than focusing solely on on-farm solutions whose benefits may take longer to realize (Basu et al. 2022).[21]

More research on these new technologies and initiatives on- and off-farm are needed to understand their efficacy across different locales and to quantify the environmental and economic consequences of such measures. Several countries have already taken steps to reduce problems related to fertilizer use using a mix of instruments. For instance, since the early 1990s, Denmark has reduced its nitrogen balance by 56 percent, although its agricultural productivity has risen over this period. Policy makers used a portfolio of strategies, including targets for reducing nitrogen discharges, fertilizer accounting systems, nitrogen quota systems to regulate use, bans on manure application on bare fields, fertilizer taxes for nonagricultural uses, as well as agricultural environmental schemes and advisory services (OECD 2021).

Ultimately, there is no easy solution for curbing nitrogen waste, given the diversity of agricultural, climatic, and political systems across the world. Nevertheless, as the challenge mounts and world population grows, it is urgent to close efficiency gaps and explore policy options on- and off-farm that increasingly make it possible to decouple agricultural production from its environmental impacts. Central to this effort is the design of agricultural subsidies that profoundly influence the structure, function, and trajectory of agriculture-food systems. As this report has demonstrated, even when eminently justifiable, poorly designed subsidies are enmeshed in trade-offs, suggesting the need to analyze the benefits and costs of whether, what, and how to subsidize.

Notes

1. In 2015, China took steps to phase out production subsidies for fertilizer by 2017. In 2017, the Chinese Ministry of Agriculture initiated a pilot program to replace chemical fertilizers with organic fertilizers for the production of fruits, vegetables, and tea in 100 counties. Under this program, subsidies are provided to organic fertilizer manufacturers to increase affordability and uptake (Huang, Gulati, and Gregory 2017; Searchinger et al. 2020).

2. Input subsidy programs in Africa were largely phased out in the 1980s and 1990s but were revitalized in the 2000s. Today, 10 countries implement second-generation subsidy programs. Despite recent evidence casting significant doubts on its touted success, Malawi's Agricultural Input Subsidy Program, known as the "Malawi miracle," sparked a resurgence of input subsidies in the other African countries in the middle to late 2000s (Jayne and Rashid 2013; Jayne et al. 2018).

3. Globally, input subsidies have become instruments of political expediency (Chatterjee et al. 2022; Takeshima and Liverpool-Tasie 2015).

4. In addition, critics argue that subsidies can discourage product innovation by fertilizer companies, crowd out commercial market purchases of fertilizers, and crowd out productive investments in agricultural research and development (Gulati and Banerjee 2015). Less common criticisms include misdiagnosed market failures—for example, using fertilizer subsidies to solve a transport cost problem that would be better addressed by investment in infrastructure (Gautam 2015; Smale and Thériault 2018).

5. For example, studies have found that maintaining low and stable prices for fertilizers in China contributed to its overuse in the past to a "moderate" or a "significant" extent (Cassou, Jaffee, and Ru 2018).

6. For example, subsidy regimes in countries like Bangladesh, India, Nepal, and Sri Lanka have been found to incentivize excessive application of urea and underapplication of phosphorus and potassium fertilizers (in India and Nepal), micronutrients, and organic inputs (Gulati and Banerjee 2015; Huang, Gulati, and Gregory 2017; Islam and Beg 2021; Kishore, Alvi, and Krupnik 2021).

7. Zinc deficiency in soils, for example, is an important constraint to crop production and the most ubiquitous micronutrient deficiency in crops worldwide. At the same time, zinc deficiency is one of the most common micronutrient deficiencies in humans, with more than 2 billion people estimated to be at risk worldwide (Hotz and Brown 2004). In Malawi, low-zinc soils and maize put semi-subsistence farming families at risk of zinc deficiency (Bevis 2018; Chilimba et al. 2012). In Bangladesh, rice farmers with low-zinc soil and low-zinc rice have lower-zinc status themselves (Bevis 2018; Mayer et al. 2007). A recent working paper by Bevis, Kim, and Guerena (2022) finds national-level large-scale evidence in Tarai, Nepal, of a bounded causal relationship between soil zinc status and child stunting, which is the primary clinical symptom of zinc deficiency, highlighting the enigma of South Asian micronutrient malnutrition driven partially by soil zinc deficiency.

8. Fertilizers are essential for plant growth, but beyond a point, adding more fertilizer may not be able to boost yield.

9. For this reason, the fallout of nitrogen pollution is considered one of the most important environmental issues of the 21st century (Kanter 2018). Recent estimates suggest that nitrogen may be the world's largest global externality, surpassing even carbon (Keeler et al. 2016).

10. Studies suggest that in acidic soils root growth can be stunted, and essential nutrients can be strongly bound in the soil solution, rendering them unavailable to plants. However, with adequate soil management (for example, liming), fertilizer use can be profitable on soils that would otherwise be acidic (Jayne and Rashid 2013).

11. Although the dominant source of nitrogen pollution in the water is the agriculture sector, many other sources also contribute to its proliferation, including livestock waste, fossil fuel combustion, and untreated wastewater. As cities grow and become denser, the threats of nitrogen leaching from below-ground septic tanks, human sewage, urban wastewater, and urban stormwater runoff are also expanding.

12. Paudel and Crago (2021) construct an empirical model on determinants of ambient water quality using more than 2.9 million pollution readings and find that a 10 percent increase in the use of nitrogen fertilizers (kilograms) leads to a 1.47 percent increase in the concentration of nitrogen (milligrams per liter) across US water sites.

13. Hendricks et al. (2014), using regression estimates from Obenour et al. (2012), report that a 0.23 percent increase in nitrogen concentration translates to an increase of 25 square miles in the size of the hypoxic zone, with a standard error of 5.45 square miles.

14. There is a great deal of heterogeneity in the pathways to growth of agricultural output across different world regions. For example, South Asia has followed an intensification path, whereby most of the gains in food production stemmed from increasing yields. At the other extreme, Sub-Saharan countries followed the extensification path in which increasing demand for food required expanding into new agricultural land. Latin American agriculture encroached on new land until the 1980s and then shifted toward intensification (Szerman et al. 2022).

15. Global agricultural TFP was about 76 percent higher in 2015 than in 1961 (Ortiz-Bobea et al. 2021).

16. Efficiency gains do not automatically guarantee sustainability. Higher yields, even when due to efficiency gains instead of intensification, do not by themselves protect forests and other valuable areas from an expansion of agricultural land. Simulations show that growth in agricultural productivity in Sub-Saharan Africa could increase the profitability of farming in the region and lead to more production on the continent and less production in other regions (Hertel, Ramankutty, and Baldos 2014). Since yields would still be lower in Africa than in the other regions, the net results may well be an increase in global agricultural land use and environmental emissions, at least in the short run. Therefore, the key takeaway is that efficiency gains create the *potential* to use fewer resources, but additional proactive and conservation policies are needed to ensure that the dividends are used wisely (OECD 2021).

17. While blanket subsidies for nutrients that are not fine-tuned to site-specific conditions can be detrimental, mineral fertilizers do play a crucial role in agricultural productivity. The renewed interest in organic fertilizers in the wake of the crisis, therefore, requires careful consideration because these fertilizers cannot replace mineral fertilizers altogether. Instead, crops need both organic and mineral fertilizers. Radical bans on mineral fertilizers, like the one seen in Sri Lanka in April 2021, can be devastating for yields.

18. Broadly, three assumptions are often made concerning adoption: (1) smallholder farmers will understand the benefits of improved practices, (2) they will trust the quality and reliability of information, and (3) they will be able to act on their altered preferences without being constrained by other factors that may affect their choices (Fishman et al. 2016).

19. Even among the poorest 20 percent in low- and middle-income countries, who tend to live overwhelmingly in rural areas, 70 percent have access to mobile phones—more than the share who have access to improved sanitation or electricity in their homes.

20. For example, UjuziKilimo, a Kenyan company, uses simple ground sensors to provide nitrogen recommendations via text messaging. A Nigerian company, Hello Tractor, connects farmers with tractor services via apps and text messaging (Kanter, Bell, and McDermid 2019).

21. Van Meter, Van Capellen, and Basu (2018) estimate that, even if runoff of nitrogen from cropland were fully stemmed, it would still take 30 years to realize the 60 percent decrease in load needed to reduce eutrophication in the Gulf of Mexico.

References

Abay, K. A., G. Blalock, and G. Berhane. 2017. "Locus of Control and Technology Adoption in Developing Country Agriculture: Evidence from Ethiopia." *Journal of Economic Behavior and Organization* 143 (November): 98–115.

Austin, A. T., M. M. C. Bustamante, G. B. Nardoto, S. K. Mitre, T. Pérez, J. P. H. B. Ometto, N. L. Ascarrunz, et al. 2013. "Latin America's Nitrogen Challenge." *Science* 340 (6129): 149. doi:10.1126/science.1231679.

Baffes, J., and W. C. Koh. 2022. "Fertilizer Prices Expected to Remain Higher for Longer." *World Bank Data* (blog), May 11, 2022. https://blogs.worldbank.org/opendata/fertilizer-prices-expected -remain-higher-longer?cid=SHR_BlogSiteTweetable_EN_EXT.

Barrett, C. B., and L. E. Bevis. 2015. "The Self-Reinforcing Feedback between Low Soil Fertility and Chronic Poverty." *Nature Geoscience* 8 (12): 907–12.

Basu, N. B., K. J. Van Meter, D. K. Byrnes, P. Van Cappellen, R. Brouwer, B. H. Jacobsen, J. Jarsjö, et al. 2022. "Managing Nitrogen Legacies to Accelerate Water Quality Improvement." *Nature Geoscience* 15 (2): 97–105.

Bera, S. 2022. "The Ripple Effects of the Global Fertilizer Crisis on India." *Mint*, April 20, 2022. https://www.livemint.com/industry/agriculture/can-india-brave-the-global-fertilizer-shock-1165038885 6854.html.

Bevis, L. 2018. "Invisible Heterogeneity in Crop Zinc Concentration and Child Zinc Intake in Rural Uganda." HarvestPlus Working Paper 30, HarvestPlus Program, International Food Policy Research Institute, Washington, DC.

Bevis, L. E., K. Kim, and D. Guerena. 2022. "Soil Zinc Deficiency and Child Stunting: Evidence from Nepal." Forthcoming in *Journal of Health Economics*. https://leahbevis.files.wordpress.com/2022/10 /bevis-kim-and-guerena-oct-2022.pdf.

Carter, M. R., R. Laajaj, and D. Yang. 2013. "The Impact of Voucher Coupons on the Uptake of Fertilizer and Improved Seeds: Evidence from a Randomized Trial in Mozambique." *American Journal of Agricultural Economics* 95 (5): 1345–51.

Carter, M., R. Laajaj, and D. Yang. 2019. "Subsidies and the Green Revolution in Africa: Direct and Social Network Effects of Randomized Input Subsidies in Mozambique." NBER Working Paper 26208, National Bureau of Economic Research, Cambridge, MA.

Carter, M., T. Lybbert, and E. Tjernström. 2015. "The Dirt on Dirt: Soil Characteristics and Variable Fertilizer Returns in Kenyan Maize Systems." Working Paper, USAID Feed the Future Innovation Lab for Assets and Market Access, Davis, CA.

Cassou, E., S. M. Jaffee, and J. Ru. 2018. *The Challenge of Agricultural Pollution: Evidence from China, Vietnam, and the Philippines.* Washington, DC: World Bank.

Chatterjee, S., D. Kapur, P. Sekhsaria, and A. Subramanian. 2022. "Agricultural Federalism: New Facts, Constitutional Vision." *Economic and Political Weekly*, September 3, 2022.

Chatterjee, S., R. Lamba, and E. Zaveri. 2022. "The Role of Farm Subsidies in Changing India's Water Footprint." Research Square PPR: PPR520912. doi:10.21203/rs.3.rs-1766947/v1.

Cheng, F. Y., K. L. Van Meter, D. K. Byrnes, and N. B. Basu. 2020. "Maximizing US Nitrate Removal through Wetland Protection and Restoration." *Nature* 588 (7839): 625–30.

Chilimba, A. D., S. D. Young, C. R. Black, M. C. Meacham, J. Lammel, and M. R. Broadley. 2012. "Agronomic Biofortification of Maize with Selenium (Se) in Malawi." *Field Crops Research* 125 (January 18, 2012): 118–28.

Cole, S., and G. Sharma. 2017. "The Promise and Challenges of Implementing ICT in Indian Agriculture." *India Policy Forum* 14: 173–240.

Cox, C. L., H. Y. Kwon, and J. Koo. 2015. *The Biophysical Potential for Urea Deep Placement Technology in Lowland Rice Production Systems of Ghana and Senegal.* IFPRI Discussion Paper 1448. Washington, DC: International Food Policy Research Institute. https://ssrn.com/abstract=2631732.

Cui, Z., H. Zhang, X. Chen, C. Zhang, W. Ma, C. Huang, W. Zhang, and G. Mi. 2018. "Pursuing Sustainable Productivity with Millions of Smallholder Farmers." *Nature* 555 (7696): 363–66.

Damania, R., S. Desbureaux, A. S. Rodella, J. Russ, and E. Zaveri. 2019. *Quality Unknown: The Invisible Water Crisis*. Washington, DC: World Bank.

Dash, J. 2019. "Satellites and Crop Interventions." *Nature Sustainability* 2 (10): 903–04.

Deaton, A. 2015. *The Great Escape: Health, Wealth, and the Origins of Inequality*. Princeton, NJ: Princeton University Press.

De Groote, H., M. Tessema, S. Gameda, and N. S. Gunaratna. 2021. "Soil Zinc, Serum Zinc, and the Potential for Agronomic Biofortification to Reduce Human Zinc Deficiency in Ethiopia." *Scientific Reports* 11 (1): 1–11.

Devineni, N., S. Perveen, and U. Lall. 2022. "Solving Groundwater Depletion in India While Achieving Food Security." *Nature Communications* 13 (3374). https://doi.org/10.1038/s41467-022-31122-9.

Ding, H., A. Markandya, R. Barbieri, M. Calmon, M. Cervera, M. Duraisami, R. Singh, J. Warman, and W. Anderson. 2021. "Repurposing Agricultural Subsidies to Restore Degraded Farmland and Grow Rural Prosperity." World Resources Institute, Washington, DC.

Duflo, E., M. Kremer, and J. Robinson. 2011. "Nudging Farmers to Use Fertilizer: Theory and Experimental Evidence from Kenya." *American Economic Review* 101 (6): 2350–90.

Economist, The. 2017. "Why Fertiliser Subsidies in Africa Have Not Worked: Good Intentions, Poor Results." *The Economist*, July 1, 2017. https://www.economist.com/middle-east-and-africa /2017/07/01/why-fertiliser-subsidies-in-africa-have-not-worked.

EEA (European Environment Agency). 2018. "Agricultural Land: Nitrogen Balance." AIRS Briefing, November 29, 2018. EEA, Copenhagen. https://www.eea.europa.eu/airs/2018/natural-capital /agricultural-land-nitrogen-balance.

Erisman, J. W., M. A. Sutton, J. Galloway, Z. Klimont, and W. Winiwarter. 2008. "How a Century of Ammonia Synthesis Changed the World." *Nature Geoscience* 1 (10): 636–39.

Fishman, R., A. Kishore, Y. Rothler, P. Ward, S. Jha, and R. Singh. 2016. *Can Information Help Reduce Imbalanced Application of Fertilizers in India? Experimental Evidence from Bihar*. IFPRI Discussion Paper 1517. Washington, DC: International Food Policy Research Institute.

Garduño, H., and S. Foster. 2010. "Sustainable Groundwater Irrigation: Approaches to Reconciling Demand with Resources." GW-MATE Strategic Overview Series 4, World Bank, Washington, DC.

Gautam, M. 2015. "Agricultural Subsidies: Resurging Interest in a Perennial Debate." *Indian Journal of Agricultural Economics* 70 (1): 83–105.

Gautam, M., D. Laborde, A. Mamun, W. Martin, V. Pineiro, and R. Vos. 2022. *Repurposing Agricultural Policies and Support: Options to Transform Agriculture and Food Systems to Better Serve the Health of People, Economies, and the Planet*. Washington, DC: World Bank.

Giné, X., S. Patel, B. Ribeiro, and I. Valley. 2019. "Efficiency and Equity of Input Subsidies: Experimental Evidence from Tanzania." *American Journal of Agricultural Economics* 104 (5): 1625–55.

Government of India. 2016. *Economic Survey 2015-16*, Vol. I. New Delhi: Department of Economic Affairs, Economic Division, Ministry of Finance.

Goyal, A., and J. Nash. 2017. *Reaping Richer Returns: Public Spending Priorities for African Agriculture Productivity Growth*. Africa Development Forum. Washington, DC: World Bank.

Gram, G., D. Roobroeck, P. Pypers, J. Six, R. Merckx, and B. Vanlauwe. 2020. "Combining Organic and Mineral Fertilizers as a Climate-Smart Integrated Soil Fertility Management Practice in Sub-Saharan Africa: A Meta-Analysis." *PloS One* 15 (9): e0239552.

Gulati, A., and P. Banerjee. 2015. "Rationalizing Fertilizer Subsidy in India: Key Issues and Policy Options." Working Paper 307, Indian Council for Research on International Economic Relations, New Delhi. https://www.academia.edu/14970315/Rationalizing_Fertilizer_Subsidy_in_India-Key _Issues_and_Policy_Options.

Harou, A. P., M. Madajewicz, H. Michelson, C. A. Palm, N. Amuri, C. Magomba, J. M. Semoka, K. Tschirhart, and R. Weil. 2022. "The Joint Effects of Information and Financing Constraints on Technology Adoption: Evidence from a Field Experiment in Rural Tanzania." *Journal of Development Economics* 155 (March): 102707.

Hebebrand, C., and D. Laborde. 2022a. "High Fertilizer Prices Contribute to Rising Global Food Security Concerns." *IFPRI Blog: Issue Post*, April 25, 2022. https://www.ifpri.org/blog /high-fertilizer-prices-contribute-rising-global-food-security-concerns.

Hebebrand, C., and D. Laborde. 2022b. "Short-Term Policy Considerations to Respond to Russia-Ukraine Crisis Disruptions in Fertilizer Availability and Affordability." *IFPRI Blog: Issue Post*, June 8, 2022. https://www.ifpri.org/blog/short-term-policy-considerations-respond-russia -ukraine-crisis-disruptions-fertilizer.

Hendricks, N. P., S. Sinnathamby, K. Douglas-Mankin, A. Smith, D. A. Sumner, and D. H. Earnhart. 2014. "The Environmental Effects of Crop Price Increases: Nitrogen Losses in the US Corn Belt." *Journal of Environmental Economics and Management* 68 (3): 507–26.

Hertel, T. W., N. Ramankutty, and U. L. C. Baldos. 2014. "Global Market Integration Increases Likelihood That a Future African Green Revolution Could Increase Crop Land Use and CO_2 Emissions." *Proceedings of the National Academy of Sciences* 111 (38): 13799–804.

Hotz, C., and K. H. Brown. 2004. "Assessment of the Risk of Zinc Deficiency in Populations and Options for Its Control." International Nutrition Foundation for the United Nations University, Tokyo.

Houlton, B. Z., M. Almaraz, V. Aneja, A. T. Austin, E. Bai, K. G. Cassman, and X. Zhang. 2019. "A World of Cobenefits: Solving the Global Nitrogen Challenge." *Earth's Future* 7 (8): 865–72.

Houssou, N., and M. Zeller. 2011. "To Target or Not to Target? The Costs, Benefits, and Impacts of Indicator-Based Targeting." *Food Policy* 36 (5): 627–37.

Huang, J., A. Gulati, and I. Gregory, eds. 2017. *Fertilizer Subsidies: Which Way Forward?* IFDC/FAI Report. Washington, DC: IFDC. https://ifdcorg.files.wordpress.com/2017/02/fertilizer -subsidieswhich-way-forward-2-21-2017.pdf.

Islam, M., and S. Beg. 2021. "Rule-of-Thumb Instructions to Improve Fertilizer Management: Experimental Evidence from Bangladesh." *Economic Development and Cultural Change* 70 (1): 237–81.

Jack, K., and J. Tobias. 2017. "Seeding Success: Increasing Agricultural Technology Adoption through Information." IGC Growth Brief 12, International Growth Centre, London.

Jain, M., R. Fishman, P. Mondal, G. L. Galford, N. Bhattarai, S. Naeem, U. Lall, Balwinder-Singh, and R. S. DeFries. 2021. "Groundwater Depletion Will Reduce Cropping Intensity in India." *Science Advances* 7 (9): eabd2849.

Jain, M., P. Rao, A. K. Srivastava, S. Poonia, J. Blesh, G. Azzari, A. J. McDonald, and D. B. Lobell. 2019. "The Impact of Agricultural Interventions Can Be Doubled by Using Satellite Data." *Nature Sustainability* 2 (10): 931–34.

Jayne, T. S., N. M. Mason, W. J. Burke, and J. Ariga. 2018. "Taking Stock of Africa's Second-Generation Agricultural Input Subsidy Programs." *Food Policy* 75 (February): 1–14.

Jayne, T. S., and S. Rashid. 2013. "Input Subsidy Programs in Sub-Saharan Africa: A Synthesis of Recent Evidence." *Agricultural Economics* 44 (6): 547–62.

Jessoe, K., and R. Badiani-Magnusson. 2019. "Electricity Prices, Groundwater, and Agriculture: The Environmental and Agricultural Impacts of Electricity Subsidies in India." In *Agricultural Productivity and Producer Behavior*, 157–84, edited by W. Schlenker. Chicago: University of Chicago Press.

Ji, Y., H. Liu, and Y. Shi. 2020. "Will China's Fertilizer Use Continue to Decline? Evidence from LMDI Analysis Based on Crops, Regions, and Fertilizer Types." *PloS One* 15 (8): e0237234.

Jones, B. A. 2019. "Infant Health Impacts of Freshwater Algal Blooms: Evidence from an Invasive Species Natural Experiment." *Journal of Environmental Economics and Management* 96 (July): 36–59.

Kanter, D. R. 2018. "Nitrogen Pollution: A Key Building Block for Addressing Climate Change." *Climatic Change* 147 (1): 11–21.

Kanter, D. R., F. Bartolini, S. Kugelberg, A. Leip, O. Oenema, and A. Uwizeye. 2020. "Nitrogen Pollution Policy beyond the Farm." *Nature Food* 1 (1): 27–32.

Kanter, D. R., A. R. Bell, and S. S. McDermid. 2019. "Precision Agriculture for Smallholder Nitrogen Management." *One Earth* 1 (3): 281–84.

Kanter, D. R., O. Chodos, O. Nordland, M. Rutigliano, and W. Winiwarter. 2020. "Gaps and Opportunities in Nitrogen Pollution Policies around the World." *Nature Sustainability* 3 (11): 956–63.

Keeler, B. L., J. D. Gourevitch, S. Polasky, F. Isbell, C. W. Tessum, J. D. Hill, and J. D. Marshall. 2016. "The Social Costs of Nitrogen." *Science Advances* 2 (10): e1600219.

Kishore, A., M. Alvi, and T. J. Krupnik. 2021. "Development of Balanced Nutrient Management Innovations in South Asia: Perspectives from Bangladesh, India, Nepal, and Sri Lanka." *Global Food Security* 28 (March): 100464.

Kurdi, S., M. Mahmoud, K. A. Abay, and C. Breisinger. 2020. "Too Much of a Good Thing? Evidence That Fertilizer Subsidies Lead to Overapplication in Egypt." MENA RP Working Paper, International Food Policy Research Institute, Washington, DC.

Lu, C., and H. Tian. 2017. "Global Nitrogen and Phosphorus Fertilizer Use for Agriculture Production in the Past Half Century: Shifted Hot Spots and Nutrient Imbalance." *Earth System Science Data* 9 (1): 181–92.

Mason, N. M., T. S. Jayne, and R. Mofya-Mukuka. 2013. "Zambia's Input Subsidy Programs." *Agricultural Economics* 44 (6): 613–28.

Mateo-Sagasta, J., S. M. Zadeh, H. Turral, and J. Burke. 2017. "Water Pollution from Agriculture: A Global Review; Executive Summary." CGIAR Research Program on Water, Land and Ecosystems. Food and Agriculture Organization, Rome; International Water Management Institute, Sri Lanka.

Mayer, A.-M. B., M. C. Latham, J. M. Duxbury, N. Hassan, and E. A. Frongillo. 2007. "A Food-Based Approach to Improving Zinc Nutrition through Increasing the Zinc Content of Rice in Bangladesh." *Journal of Hunger and Environmental Nutrition* 2 (1): 19–39.

McArthur, J. W., and G. C. McCord. 2017. "Fertilizing Growth: Agricultural Inputs and Their Effects in Economic Development." *Journal of Development Economics* 127 (July): 133–52.

Michelson, H., S. Gourlay, and P. Wollburg. 2022. "Non-Labor Input Quality and Small Farms in Sub-Saharan Africa: A Review." Policy Research Working Paper 10092, World Bank, Washington, DC.

Namonje-Kapembwa, T., T. S. Jayne, and R. Black. 2015. "Does Late Delivery of Subsidized Fertilizer Affect Smallholder Maize Productivity and Production?" Selected paper presented at the Agricultural and Applied Economics Association and Western Agricultural Economics Association Annual Meeting, July 26–28, 2015, San Francisco, CA.

Obenour, D. R., A. M. Michalak, Y. Zhou, and D. Scavia. 2012. "Quantifying the Impacts of Stratification and Nutrient Loading on Hypoxia in the Northern Gulf of Mexico." *Environmental Science and Technology* 46 (10): 5489–96.

OECD (Organisation for Economic Co-operation and Development). 2021. *Making Better Policies for Food Systems.* Paris: OECD Publishing. https://doi.org/10.1787/ddfba4de-en.

Ortiz-Bobea, A., T. R. Ault, C. M. Carrillo, R. G. Chambers, and D. B. Lobell. 2021. "Anthropogenic Climate Change Has Slowed Global Agricultural Productivity Growth." *Nature Climate Change* 11 (4): 306–12.

Paudel, J., and C. L. Crago. 2021. "Environmental Externalities from Agriculture: Evidence from Water Quality in the United States." *American Journal of Agricultural Economics* 103 (1): 185–210.

Ritchie, H., M. Roser, and P. Rosado. 2013. "Fertilizers." *Our World in Data.* https://ourworldindata.org/fertilizers.

Schultz, T. W. 1964. *Transforming Traditional Agriculture.* New Haven, CT: Yale University Press.

Searchinger, T. D., C. Malins, P. Dumas, D. Baldock, J. Glauber, T. Jayne, J. Huang, and P. Marenya. 2020. *Revising Public Agricultural Support to Mitigate Climate Change.* Development Knowledge and Learning. Washington, DC: World Bank. https://openknowledge.worldbank.org/handle/10986/33677.

Singh, V., S. Ganguly, and V. Dakshinamurthy. 2018. "Evaluation of India's Soil Health Card from Users' Perspectives." CSISA Research Note 12, International Food Policy Research Institute, Washington, DC.

Smale, M., and V. Thériault. 2018. *A Cross-Country Summary of Fertilizer Subsidy Programs in Sub-Saharan Africa.* Feed the Future Innovation Lab for Food Security Policy Research Paper 169. East Lansing: Michigan State University.

Smith, A. 2022. "The Story of Rising Fertilizer Prices." *ARE Update* 25 (3): 1–4. https://giannini.ucop.edu/filer/file/1645718420/20317/.

Stewart, W. M., D. W. Dibb, A. E. Johnston, and T. J. Smyth. 2005. "The Contribution of Commercial Fertilizer Nutrients to Food Production." *Agronomy Journal* 97 (1): 1–6.

Szerman, D., J. Assunção, M. Lipscomb, and A. M. Mobarak. 2022. "Agricultural Productivity and Deforestation: Evidence from Brazil." Yale Economic Growth Center Discussion Paper 1091, Yale University, New Haven, CT.

Takeshima, H., and L. S. O. Liverpool-Tasie. 2015. "Fertilizer Subsidies, Political Influence, and Local Food Prices in Sub-Saharan Africa: Evidence from Nigeria." *Food Policy* 54: 11–24.

Tesliuc, C., W. J. Smith, and M. R. Sunkutu. 2013. *Zambia: Using Social Safety Nets to Accelerate Poverty Reduction and Share Prosperity.* Social Protection and Labor Discussion Paper 1413. Washington, DC: World Bank.

Tian, H., R. Xu, J. G. Canadell, R. L. Thompson, W. Winiwarter, P. Suntharalingam, E. A. Davidson, et al. 2020. "A Comprehensive Quantification of Global Nitrous Oxide Sources and Sinks." *Nature* 586 (7828): 248–56.

Tilman, D., K. G. Cassman, P. A. Matson, R. Naylor, and S. Polasky. 2002. "Agricultural Sustainability and Intensive Production Practices." *Nature* 418 (6898): 671–7.

USDA (United States Department of Agriculture). 2021. "International Agricultural Productivity Data Product, Data, and Methods as of October 2021." Economic Research Service, Washington, DC.

Vanlauwe, B. 2015. "Time to End the False Debate of Organic vs Mineral Fertilizer." *Devex,* April 27, 2015.

Van Meter, K. J., P. Van Cappellen, and N. B. Basu. 2018. "Legacy Nitrogen May Prevent Achievement of Water Quality Goals in the Gulf of Mexico." *Science* 360 (6387): 427–30.

Verner, D., N. Roos, A. Halloran, G. Surabian, E. Tebaldi, M. Ashwill, S. Vellani, and Y. Konishi. 2021. *Insect and Hydroponic Farming in Africa: The New Circular Food Economy.* Agriculture and Food Series. Washington, DC: World Bank.

Willett, W., J. Rockström, B. Loken, M. Springmann, T. Lang, S. Vermeulen, T. Garnett, et al. 2019. "Food in the Anthropocene: The EAT–Lancet Commission on Healthy Diets from Sustainable Food Systems." *The Lancet* 393 (10170): 447–92.

World Bank. Forthcoming. *A Balancing Act: Efficiency, Sustainability, and Prosperity.* Washington, DC: World Bank.

Wuepper, D., S. Le Clech, D. Zilberman, N. Mueller, and R. Finger. 2020. "Countries Influence the Trade-Off between Crop Yields and Nitrogen Pollution." *Nature Food* 1 (11): 713–19.

Zaveri, E., J. Russ, and R. Damania. 2020. "Rainfall Anomalies Are a Significant Driver of Cropland Expansion." *Proceedings of the National Academy of Sciences* 117 (19): 10225–33.

Zaveri, E. D., J. D. Russ, S. G. Desbureaux, R. Damania, A. S. Rodella, and G. Ribeiro Paiva de Souza. 2020. "The Nitrogen Legacy: The Long-Term Effects of Water Pollution on Human Capital." Policy Research Working Paper 9143, World Bank, Washington, DC.

Zhang, X., E. A. Davidson, D. L. Mauzerall, T. D. Searchinger, P. Dumas, and Y. Shen. 2015. "Managing Nitrogen for Sustainable Development." *Nature* 528 (7580): 51.

Zhang, X., T. Zou, L. Lassaletta, N. D. Mueller, F. N. Tubiello, M. D. Lisk, C. Lu, et al. 2021. "Quantification of Global and National Nitrogen Budgets for Crop Production." *Nature Food* 2 (7): 529–40.

CHAPTER 9

The Effects of Agricultural Subsidies on Forests and Their Spillovers

"Humanity is cutting down its forests,
apparently oblivious to the fact that we may not live without them."
—*Isaac Asimov*

CHAPTER AT A GLANCE

Global deforestation is heavily linked to and driven by agricultural subsidies:

- Agricultural price supports are responsible for the loss of 2.2 million hectares of forest cover per year, which is equal to approximately 14 percent of annual deforestation. New analysis presented in this chapter establishes a causal link between agricultural subsidies and global deforestation rates. Agricultural subsidies distort the decisions of farmers, which can lead to unintended spillovers into natural and human capital that can cross international boundaries. One such spillover is into natural areas, where subsidies can incentivize farmers to expand the area devoted to agriculture into forests and other natural habitats.

- The impact of subsidies is not constrained by national borders—agricultural subsidies in rich countries are driving tropical deforestation around the world. This chapter shows that live-stock subsidies in the United States drive deforestation in Brazil by increasing the demand for soybeans as feedstock—a relationship that is likely not isolated to these two countries.

- Deforestation linked to agricultural subsidies leads to the release of 4.3 billion metric tons of carbon over a 20-year period. Using the World Bank's shadow price of carbon, this cost is valued at between US$174 billion and US$348 billion. Although the sensitivity of deforestation to subsidies varies based on region and the commodity that is subsidized, overall, agricultural subsidies have a substantial impact on deforestation and the subsequent release of carbon.

Agricultural subsidies increase the spread of emerging infectious diseases through deforestation:

- Globally, an estimated 1.3 million to 3.8 million cases of malaria each year can be attributed to agricultural subsidies. Deforestation can cause spillovers into other ecosystem services—namely, pest control and the spread of zoonotic and vector-borne diseases. The chapter presents results from a new global study that examines all countries where malaria is endemic and finds that deforestation is linked to an increase in malaria infections. Furthermore, malaria cases linked to subsidy-induced deforestation led to a loss of more than 400,000 disability-adjusted life years (DALYs) to malaria-endemic countries, with annual losses of up to US$19 billion to the world economy.

Introduction

As discussed in chapter 1, agriculture is among the most distorted sectors in the world economy, with more than US$635 billion provided annually in public support to farmers. While government support for agriculture is often intended to increase food security and reduce rural poverty, if poorly designed it can come at the cost of sustainable development objectives. Indeed, the agriculture sector is a leading cause of environmental degradation, accounting for more than 70 percent of freshwater use, 30 percent of energy consumption, and the majority of deforestation globally (FAO 2017). Any policies that distort this sector without proper safeguards in place are likely to exacerbate these environmental concerns.

Most agricultural subsidies and tariffs are aimed at reducing the price of production for farmers, distorting farmers' decisions about where, what, and how much to produce. This process may accelerate environmental degradation by reducing the shadow price of natural resource use below its marginal cost to society. This chapter investigates the extent to which subsidies contribute to global deforestation. Somewhat surprising, despite the magnitude of subsidies and their ubiquity, this important issue has not been investigated empirically (Busch and Ferretti-Gallon 2017).

The analysis begins by finding a significant link between annual changes in commodity prices and annual changes in loss of forest cover, globally. As agricultural products become more profitable for farmers due to fluctuations in global prices, farmers are incentivized to expand their cropland to grow more of these products, with much of this expansion pushing into the forest frontier. The global findings are supported by two case studies that provide robust causal evidence. The first exploits a quasi-experimental setting arising from a spatial discontinuity in subsidy levels at the Côte d'Ivoire–Ghana border to analyze the effect of cocoa subsidies on deforestation in West Africa. The second examines the effect of livestock subsidies in the United States on soybean-driven deforestation in Brazil, establishing a strong link between the two. Both case studies provide additional evidence in support of the global findings.

This chapter also presents new evidence of how subsidy-linked deforestation promotes the spread of diseases. COVID-19 has served as a stark reminder of the close links between human health and the environment. Before COVID-19 shocked the world, there had been a host of other emerging infectious diseases—HIV/AIDS (human immunodeficiency virus/acquired immune deficiency syndrome), Nipah virus disease, avian influenza, Ebola virus disease, SARS (Severe Acute Respiratory Syndrome), and MERS (Middle East Respiratory Syndrome)—all emerging from disturbed forests and contact with wild animals. Emerging infectious diseases, often of zoonotic origins, will continue to increase as humans extend their footprint on the environment. Agricultural encroachment into forest areas not only brings wildlife in closer contact with human populations but also modifies the biological and ecological composition of the forest and of forest-dwelling species, which, in turn, influences the behavior of vectors and the spread of pathogens. Even when diseases from forests do not kill people or affect the economy at the scale of the COVID-19 pandemic,[1] they frequently deepen poverty, diminish or destroy livelihoods, and undermine food security.

Malaria is one such disease: it affects more than 200 million people each year, claiming billions of dollars in macroeconomic costs. This chapter investigates the connection between deforestation and malaria with a particular focus on the indirect role of agricultural subsidies in this link. The results indicate that an increase in global deforestation of 14 percent due to agricultural producer supports is associated with 1.3 million to 1.8 million additional cases of malaria each year, which may burden the world with a loss of close to 436,000 DALYs, adding billions of dollars globally to public health expenditures.

Global deforestation is sensitive to changes in commodity prices

The world's forests are shrinking at an alarming rate. Since the turn of the 21st century, global tree cover has declined an estimated 11 percent, equivalent to 176 gigatons of carbon dioxide emissions (Hansen et al. 2013). Figure 9.1 shows deforestation levels over time, by geographic region. Annual loss of forest cover has accelerated significantly over the 21st century, with the highest levels of deforestation in East Asia and Pacific, Europe and Central Asia, and Latin America and the Caribbean.

FIGURE 9.1 **Global forest cover loss, by region, 2001–17**

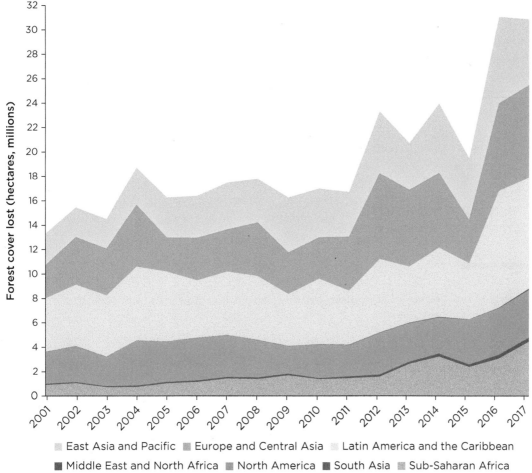

Source: Data on deforestation are obtained from the Global Forest Change data set (https://earthenginepartners .appspot.com/science-2013-global-forest) and Hansen et al. 2013.
Note: Deforestation levels were generally increasing over the 21st century, with the highest levels in East Asia and Pacific, Europe and Central Asia, and Latin America and the Caribbean.

The conversion of forests to agricultural land is by far the main cause of deforestation globally. Approximately one-third of the Earth's forests have already been depleted due to agricultural expansion (Ritchie and Roser 2021). Through the production of commodities, especially beef, soybeans, and palm oil, agriculture accounts for 27 percent of global forest loss (Curtis et al. 2018). An additional 24 percent is attributed to shifting agriculture, whereby small-scale forest conversion is followed by the eventual regrowth of forests to a degraded state. Furthermore, 26 percent of global deforestation can be linked to international demand for commodities, the bulk of which is exported to Asia and Europe (Pendrill et al. 2019).

The important connection to international demand suggests a relationship between commodity prices and forest cover loss. Indeed, a small literature focuses on this link (see Busch and Ferretti-Gallon 2017 for a meta-analysis). The overwhelming majority of these studies find a significant and positive association between the two. However, many of the studies linking agricultural prices and deforestation are limited in geographic scope—primarily to Brazil and Indonesia—and in the types of commodities examined. Another more technical limitation is that most studies do not assert or test a causal mechanism, instead examining commodity prices and deforestation as if they were not connected. However, it is possible that large-scale conversion of forests to agricultural lands could depress commodity prices because of increases in supply. For example, in the case of forest products, extensive logging could both increase deforestation and reduce wood prices, biasing estimated effects of wood prices on deforestation downward.

How much of global forest loss is due to the expansion of major agricultural commodities? Druckenmiller (2022), in a background paper for this report, estimates forest loss due to changes in commodity prices for the five major forest risk commodities: beef, soybeans, palm oil, sugarcane, and wood. Box 9.1 describes the data and methods. The relationship between agricultural commodity prices and deforestation is established at the global level, spanning the years 2000 to 2019. Estimates are then presented of how price elasticities differ across regions and by distance to the agricultural frontier. Together, these estimates can help policy makers to focus conservation efforts on areas where deforestation is most sensitive to fluctuations in agricultural prices and, by extension, subsidies that influence agricultural prices.

BOX 9.1

Technical spotlight: The effects of agricultural commodity prices and producer supports on global deforestation

Agricultural commodity prices and deforestation

Identifying the elasticity between commodity prices and loss of forest cover requires an instrument that changes prices in a way that is plausibly exogenous to other drivers of deforestation. Weather-induced yield shocks are a natural choice due to their economic exogeneity (humans cannot affect weather shocks) and near randomness (unpredictability). The empirical strategy therefore is to instrument for global prices using weather shocks in crop-growing regions of the world's top producing countries for each commodity. Because weather shocks generally occur in different regions of the world than observed deforestation and flexible controls for local weather are included, it is likely that the instrument only drives changes in land use through the influence on commodity prices.

(Continued)

Detox Development

BOX 9.1
Technical spotlight: The effects of agricultural commodity prices and producer supports on global deforestation (*continued*)

To establish the relationship between weather shocks and commodity prices, the natural log of price is regressed against lagged weather shocks in producing regions. The first-stage equation is given by

$$\log(P_t) = Z_{t-1} + W_{i,t-1} + X_{i,t} + f_r(t_r) + \alpha_i + \epsilon_{i,r,t}, \tag{B9.1.1}$$

where P is the commodity price and Z is the instrument. The instrument is defined as mean growing season temperature or precipitation for the commodity in crop-growing regions of the top five producing countries. Controls include a vector of socioeconomic covariates, X, including population and nighttime luminosity, and a vector of local climate controls, W, comprising quadratics in monthly temperature and precipitation. The equation also includes regional polynomial time trends and fixed effects for level-two administrative units. In equation B9.1.1, i indexes administrative units, r indexes region, and t indexes year. Standard errors are clustered by level-two administrative unit (that is, province or state) to account for spatial and temporal correlation in the disturbance terms.

The second stage estimates the elasticity between agricultural commodity prices and forest cover loss by substituting the predicted prices from the first stage, P, in place of actual prices and also including measures of annual deforestation, F, as well as the same controls for local weather, W, socioeconomic factors, X, regional time trends, and fixed effects that are included in the first-stage equation. The estimating equation is given by

$$\log\left(F_{i,r,t+k}\right) = \log\left(\hat{P}_t\right) + W_{i,t-1} + X_{i,t} + f_r(t_r) + \alpha_i + \epsilon_{i,r,t}. \tag{B9.1.2}$$

Forest cover loss today is modeled as a function of previous commodity prices (for example, one year in the past) in order to account for the lag between the prices farmers observe and the time it takes to clear land for additional planting. Ideally, a vector of lagged commodity prices would be included to estimate simultaneously the effect of prices one, two, and three years in the past on deforestation today. However, due to the high degree of collinearity in the data, the model instead follows a common approach in the literature (for example, Wheeler et al. 2011) and retains the single lagged value of the price variable that provides the best fit.

Producer supports and deforestation

To estimate the impact of agricultural support on deforestation, the following regression is run:

$$F_{c,t} = \log(coupled\_supports_{c,t}) + \log(decoupled\_supports_{c,t}) + \alpha_c + \delta_t + \epsilon_{c,t}, \tag{B9.1.3}$$

where $F_{c,t}$ is forest cover loss in country c and year t, *coupled_supports* and *decoupled_supports* are estimates of annual producer support from the Organisation for Economic Co-operation and Development, separated into estimates tied to production incentives and estimates decoupled from output levels. The equation includes country fixed effects to account for time-invariant differences between countries, such as geography and climate, as well as year fixed effects to account for common time trends. Standard errors are clustered at the regional level to account for spatial and temporal correlation in the disturbance terms.

7.2% to 8.1%:

The increase in deforestation that results from a 10% increase in the price of beef, sugarcane, or wood

The results show that agricultural commodity prices are strong drivers of deforestation. Globally, the findings suggest that deforestation is most sensitive to changes in beef, sugarcane, and wood prices. A 10 percent increase in these commodity prices increases forest cover loss by 7.2 percent, 7.3 percent, and 8.1 percent, respectively. Soybeans and palm oil have somewhat smaller price elasticities for forest loss of 3.9 percent and 1.7 percent, respectively.[2] These impacts typically emerge with a lag of one year for soybeans and sugarcane and a lag of two years for beef, palm oil, and timber before price changes induce forest clearing, reflecting the time it takes to clear land for additional plantings.

Substantial variation in deforestation elasticities across geographic regions highlights the need for a more nuanced, regional analysis. For example, deforestation elasticities are relatively high in Sub-Saharan Africa and North America and relatively low in Europe and Central Asia. This finding is not surprising, as countries differ in many ways, with different levels of integration into global commodity markets, different degrees to which government policy affects production decisions, and different availability of land and other inputs (Iqbal and Babcock 2018).[3]

Should policy makers target conservation efforts according to distance to the agricultural frontier? To answer this question, each 30-meter pixel of forest cover loss is classified by its distance to the boundary of current crop production (map 9.1 provides the example of South America). Separate deforestation elasticities are then estimated at different distances from the frontier (for example, within 1 kilometer, 1–2 kilometers, 2–5 kilometers, and so on). Consistent with the hypothesis that higher agricultural commodity prices induce forest clearing, a clear pattern is found of larger elasticities closer to current crop production. Elasticities are relatively large on the immediate boundary of the agricultural frontier (within 1 kilometer of current crop production), ranging from a low of 1.7 for palm oil to a high of 6.8 for wood for a 10 percent increase in prices. Deforestation elasticities decline for each additional distance band across all five commodities. This finding matches the intuition that forest clearing for new agricultural production will occur in areas nearby current croplands and can help policy makers to target conservation efforts toward areas with the highest likelihood of deforestation when subsidies distort agricultural commodity prices.

Having estimated the effects of changes in agricultural commodity prices on deforestation for major forest risk commodities, the next section tackles the question of whether subsidies have a direct impact on forest cover loss.

Assessing the effect of agricultural subsidies on deforestation

The complexities involved in understanding the relationship between commodity prices and forest loss are magnified when trying to link agricultural subsidies to deforestation. In essence, subsidies have proven particularly challenging to study because they are often hidden, come in complex packages, and are difficult to identify. Consequently, there is limited empirical evidence on the effect of specific agricultural subsidies or tariffs on deforestation. Yet if the links are significant, the risk of not addressing the issue may undermine global efforts to curb the extent of forest loss. This section brings this important topic to light through a direct assessment of producer supports on forest loss. The analysis addresses the shortcomings of the literature: it is global in scope and uses a novel approach to determine causal linkages between commodity prices, agricultural subsidies, and deforestation.

MAP 9.1 **Distance to the agricultural frontier in South America**

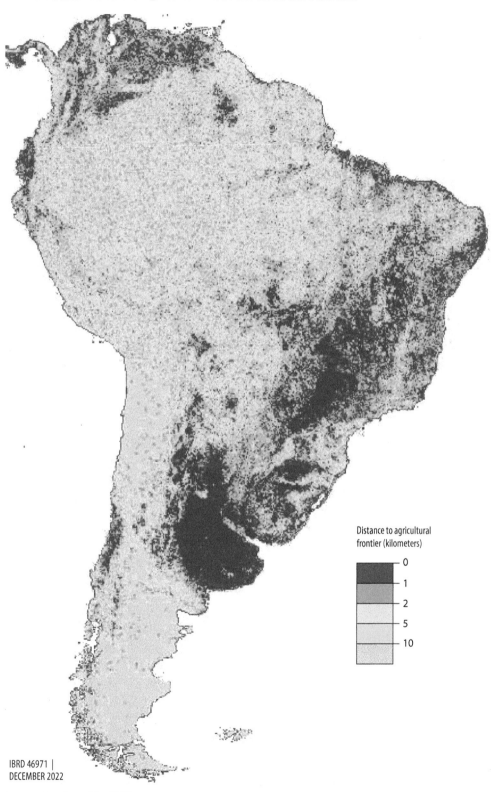

Distance to agricultural
frontier (kilometers)

0
1
2
5
10

IBRD 46971 |
DECEMBER 2022

Source: Druckenmiller 2022.
Note: Forest cover loss is measured by distance to the agricultural frontier, which is classified by 30-meter pixels.
Data on the extent of current crop production were obtained from the United States Geological Survey's Global
Croplands database (https://www.usgs.gov/apps/croplands/app/map?lat=0&lng=0&zoom=2).

MAP 9.2 **Average annual producer support estimates, by country, 2000–19**

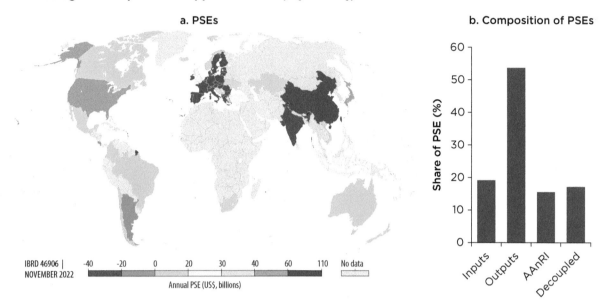

a. PSEs

b. Composition of PSEs

IBRD 46906 | NOVEMBER 2022

Annual PSE (US$, billions)

-40 -20 0 20 30 40 60 110 No data

Source: Druckenmiller 2022.

Note: PSEs were obtained from the Organisation for Economic Co-operation and Development (OECD) and are defined as the annual monetary value of gross transfers from consumers and taxpayers to agricultural producers, measured at the farm gate level, arising from policy measures that support agriculture. Negative values for PSEs arise when policies (like export taxes or bans) that reduce farmers' net revenues from agricultural products exceed the value of any support provided. The composition of PSEs (panel b) was also obtained from the OECD. AAnRI = subsidies where payments are based on area, animals, receipts, or income; PSE = producer support estimate.

Directly estimating the impact of agricultural supports on deforestation is of primary interest. Having established a statistical link between commodity prices and deforestation, the analysis is extended to agricultural subsidies by pairing country-level data on annual producer supports with forest cover loss. Producer support estimates (PSEs) are defined as the annual monetary value of gross transfers from consumers and taxpayers to agricultural producers arising from policy measures that support agriculture. Map 9.2 presents these data by country.[4] Additionally, agricultural supports that are "coupled" to input use and output levels and those that are "decoupled" from production incentives are examined separately. Finally, the relationship between producer support and deforestation is estimated. More details on the methodology are found in box 9.1.

2.2 million hectares: The size of forest cover loss per year attributable to coupled agricultural subsidies

Higher levels of coupled support are associated with higher levels of deforestation. The results suggest that a 1 percent increase in coupled PSE is associated with an 18,607 hectare increase in country-level loss of forest cover.[5] The elasticity estimates mean that producer supports explain approximately 2.2 million hectares of forest cover loss per year, equivalent to 14 percent of total and 37 percent of annual agriculture-driven deforestation. A negative but insignificant effect is found for decoupled PSE on forest cover loss.

These results indicate that agricultural subsidies have led to the release of 4.3 billion metric tons of carbon over the 20-year period of study. This impact can be monetized using a shadow price for carbon, which equates the estimated damage done per ton of carbon dioxide equivalent released into the atmosphere as a monetary figure. Using the World Bank's estimate of between US$40

Detox Development

and US$80 per ton, that carbon release would have a social cost of between US$174 billion and US$348 billion. However, accounting for the loss of biodiversity and ecosystem services, the total costs are surely much greater. Also important is the distribution of these costs, which often fall disproportionately on marginalized groups, including women who depend on forestry products for food, medicine, and water.

Box 9.2 presents two case studies as a further check on the results through deep dives on West Africa and the United States. Both examples support the link between agricultural subsidies and deforestation.

BOX 9.2
Technical spotlight: Two country case studies on the impact of subsidies on deforestation, 2000-10

Case study 1: Cocoa-driven deforestation in West Africa

The first case study evaluates the effect of cocoa subsidies on deforestation in Côte d'Ivoire and Ghana. Côte d'Ivoire and Ghana are the world's dominant producers of cocoa, together producing more than 60 percent of the global supply. Cocoa is a major source of income and employment in both countries and has historically been supported by large government subsidies. At the same time, cocoa production is a major driver of deforestation in the region.

Different levels of government support for cocoa production in Côte d'Ivoire and Ghana can be used as a quasi-experiment to evaluate the effect of agricultural subsidies on deforestation. Specifically, since these two nations share a common border along which much cocoa production occurs, the effect of government support on cocoa-driven deforestation can be evaluated using a spatial regression discontinuity design. While agricultural supports are significantly higher in Côte d'Ivoire than in Ghana (agricultural subsidies were 56 percent higher in Côte d'Ivoire than in Ghana during 2000 to 2010), other factors affecting deforestation rates—such as population levels, infrastructure density, and geographic terrain—are similar across the border. Therefore, different levels of deforestation in cocoa-suitable regions on either side of the border can arguably be attributed to differences in agricultural policy across the two nations.

Indeed, a sharp discontinuity in deforestation is found at the shared border in areas suitable for cocoa production (figure B9.2.1, panel a); deforestation between 2000 and 2010 was 3.1 percentage points higher on the Côte d'Ivoire side of the border. This estimate implies that a 10 percentage point increase in agricultural supports increases deforestation by 0.55 percentage point. In contrast, in areas not suited for cocoa production, as shown in panel b, there is no difference in deforestation rates on different sides of the border. This finding provides additional evidence that the higher levels of deforestation in Côte d'Ivoire can be attributed to higher levels of government support for cocoa production.

Are other differences at the shared border driving the observed difference in deforestation rates? Panels c–e of figure B9.2.1 demonstrate that several key drivers of deforestation—population, road density, and terrain ruggedness—are smooth across the border. Thus, it is unlikely that differences in geography or socioeconomic factors explain the difference in deforestation rates.

Case study 2: US livestock-driven deforestation in Brazil

The second case study evaluates the effect of US livestock subsidies on deforestation in Brazil. The United States is one of the world's largest producers of livestock, and the industry has received

(Continued)

BOX 9.2
Technical spotlight: Two country case studies on the impact of subsidies on deforestation, 2000–10 (*continued*)

FIGURE B9.2.1 Discontinuity in deforestation at the Côte d'Ivoire–Ghana border

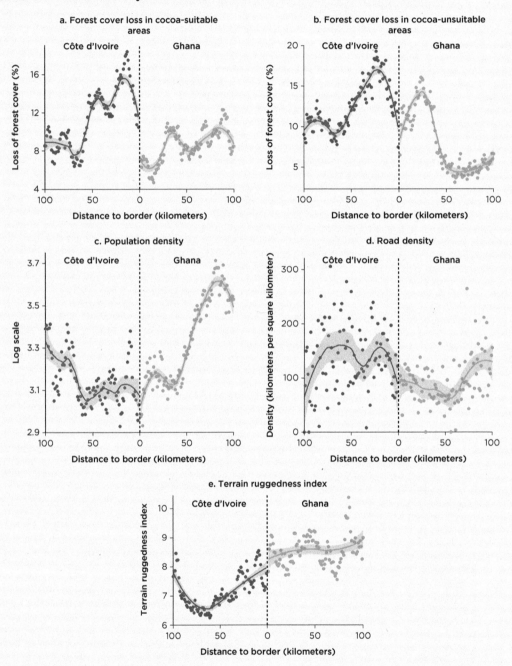

Source: Druckenmiller 2022.

Note: Agricultural supports are significantly higher in Côte d'Ivoire than in Ghana, whereas other drivers of deforestation—population, road density, and terrain—are similar across the border. Thus, the sharp discontinuity in forest cover seen in panel a at the shared border can arguably be attributed to higher levels of government support in Côte d'Ivoire for cocoa production.

(Continued)

approximately US$485 million annually in public support over the last two decades. Why might these subsidies affect deforestation in Brazil? Brazil is the world's largest producer of soybeans, and 75 percent of soybeans produced globally are used as animal feed. If livestock subsidies increase animal production, they are likely to increase demand for feed crops, affecting land use decisions in soybean-producing countries. This hypothesis is tested by linking US annual livestock subsidy payments with changes in soybean area planted, soybean production, and forest cover loss in Brazil.

The results show that soybean area planted and soybean production increase with livestock subsidies: a 10 percent increase in subsidy levels corresponds to an increase of approximately 1.5 percent in both outcomes. Forest cover loss also increases significantly with livestock subsidies: a 10 percent increase in subsidy levels is associated with a 0.4 percent increase in deforestation. However, the effect of livestock payments on forest cover loss is not due to the expansion of soybean area alone. Indeed, it is likely that these subsidies increase the incentives to produce other feed crops, such as maize, that also contribute to deforestation in Brazil. Much soybean expansion in Brazil has occurred in land currently used for beef production, with cattle then moving into forested land. This exercise does not capture these indirect effects.

Source: Druckenmiller 2022.

Agricultural subsidies and the emergence of infectious diseases: A focus on malaria

Through their impact on deforestation, agricultural subsidies and producer support policies unintentionally drive the emergence of infectious diseases. The reason is that forest edges are a major point of spillover for novel viruses from their natural hosts into humans or livestock (Dobson et al. 2020). Recent history provides many examples of the link between forest degradation, fragmentation, or loss of forests and emerging infectious diseases. For instance, the spread of Nipah virus in Malaysia during the late 1990s, the surge in malaria from the Brazilian Amazon to the Malaysian Borneo, and the propagation of Lassa virus in Liberia are all believed to be connected to changes in land use and the depletion of forests. Similarly, deforestation has been associated with the spike in Zika cases in Brazil during 2015–16 and is likely related to the resurgence of other pathogens such as chikungunya, dengue, and West Nile virus (Ali et al. 2017; Burkett-Cadena and Vittor 2018).

Despite the many possible spillovers, this section focuses on malaria. Although similar in prevalence to dengue, malaria is much deadlier and costlier in socioeconomic terms. In the 20th century alone, malaria was responsible for the deaths of 150 million to 300 million people, or 2–5 percent of all deaths during that time (Carter and Mendis 2002). Indeed, according to some estimates, malaria may have killed half of all humans who ever lived (Whitfield 2002). Despite a slowdown in deaths over the past few decades, as health care access and treatments improve, cases seem to be on the rise again, with the latest figure estimating 627,000 deaths attributable to malaria in 2020 (WHO 2021). This section focuses on malaria for these reasons, but also because the understanding of the links between deforestation and malaria is growing rapidly, with some evidence suggesting that an ecological mechanism is behind the deforestation-malaria link (Garg 2019).[6]

To the extent that malaria is associated with deforestation, agriculture and agricultural policies are connected to malaria incidence through their pressure on forests. Research over the past decade has brought to light a possible direct link between deforestation and malaria—that is, greater tree cover loss may increase the local risks of malaria transmission. If such a relationship indeed exists, it constitutes a prime example of the closely knit ties between the environment, the economy, and public health, suggesting complementarity in terms of natural capital and public health policy. The relationship is of particular concern in this section because agricultural subsidies may inadvertently be at the root of both deforestation and a higher number of malaria cases. This section reviews what is known and, just as important, what is not known about the malaria-deforestation relationship. It then delves into a new global analysis of this link and estimates a pass-through elasticity from agricultural subsidies to malaria, through deforestation.

Is deforestation a driver of malaria? The challenges and findings

Although substantial evidence points to a malaria risk rising with deforestation, the complexities of this relationship have given rise to contradictory results. While many studies find that malaria increases in deforested areas (Berazneva and Byker 2017; Chaves et al. 2018; Garg 2019; MacDonald and Mordecai 2019; Vittor et al. 2006, 2009), others fail to establish a statistically significant relationship between the two (Bauhoff and Busch 2020) or even conclude that forest conservation, not forest depletion, may increase malaria incidence (Valle and Clark 2013).

A close look at the literature reveals that the apparent contradictions are really manifestations of a fundamentally complex set of definitions, influences, and mechanisms that characterize the deforestation-malaria connection (figure 9.2). There are vast differences across countries, including the type of economic activities that drive deforestation and the confounding ecological, anthropogenic, and entomological elements that can lead to different outcomes, depending on which of the links from figure 9.2 dominate in a particular geography.

Gender is another factor that plays a role in the spread of malaria. The social and cultural context in which malaria spreads implies different levels of risk exposure for women and men. For example, women who work in agriculture, collect wood for fuel, gather other nontimber forest products, milk cows, or fetch water are often vulnerable to mosquito bites because of the early hours at which they carry out these outdoor tasks. However, men can be exposed to similar occupational risks through timber extraction, fishing, agriculture, and mining (UNDP 2015). Nonetheless, studies have shown that women tend to be at higher risk of mortality from severe falciparum than men. Pregnant women and children are at the highest risk, and nonpregnant women older than 40 years of age are at higher risk than younger nonpregnant women (Bauserman et al. 2019; Khadanga et al. 2014).

Despite the nuances and challenges, the overall literature agrees that deforestation has been a significant driver of malaria, possibly along with other emerging infectious diseases. Moreover, these studies are not limited to any specific region, but instead span

FIGURE 9.2 Deforestation and malaria: The confounding elements and mechanisms found in the literature

Ecological influences	Entomological influences	Anthropogenic influences

• Type of forest cover
  – Primary
  – Secondary
  – Canopy (heavily forested or initial deforestation)
 	
• Type of landscape
  – Hydrology (rivers, lakes, rainfall)
  – Level of land fragmentation (many divisions mean more forest fringes)
  – Highlands or lowlands
	• Dominant vector species
	  – Vector competence
	  – Speed of larval development (for given shade, sunlight, water, and temperature)
	  – Biting, resting habits (depends on vegetation and population)
	
	• Type of parasite
	  – P. falciparum, P. vivax mainly
		• Epidemiological
		  – Population immunity
		  – Health care and medicine access (different by gender)
		  – Eradication or control efforts
		
		• Social and economic
		  – Type of economic activity (large-scale or small-scale deforestation, gender roles)
		  – Migratory patterns of workers
		  – Type of settlements (urban or rural)
		  – Regulatory framework (protected areas)

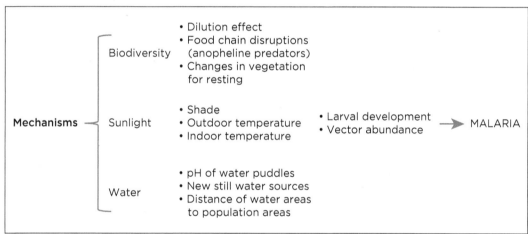

Source: World Bank.

the entire globe (online appendix I provides a detailed review of studies linking defor-estation to malaria in several African, Asian, and Latin American countries).[7] However, all of these studies are limited in geography and use different estimation strategies, mea-sures for deforestation, and data sources. Missing is a single analysis that uses consistent methods and data to examine this relationship at a global level in order to draw general-izable conclusions.

This study builds on the growing body of evidence that deforestation leads to greater incidence of malaria worldwide. The analysis takes a dual strategy approach. First, there is a detailed investigation in countries for which very high-quality, administrative data on malaria exist. This investigation is referred to as the Amazon-country analysis and includes Brazil, Colombia, Ecuador, and Peru. The second part of the analysis is global in scope, covering 73 malaria-prone countries in tropical and subtropical areas where climatic conditions are favorable to the anopheline vector and to larvae development.[8] The Malaria Atlas Project obtains, curates, and shares historical malaria incidence for all of these countries from a wide range of sources.[9] These data provide a consistent measurement that is comparable among all countries, which allows a comparison of the malaria effects with commodity price effects and producer support effects by geographic zones.

Deforestation in the Amazon leads to increasing transmission of malaria

The analysis in this subsection confirms the findings from other studies showing that deforestation is increasing the occurrence of malaria in Latin America. Box 9.3 describes the methodology used in this analysis. The results are shown in map 9.3.[10] When the entire country is considered, deforestation has an effect on malaria that is indistinguishable from zero. But when the analysis only includes states with partial or full rainforest cover, the magnitude and statistical significance of the relationship increase. Although the elasticity measures in map 9.3 cannot be taken as strictly causal, they consistently show that deforestation is strongly associated with the spread of malaria in the densely forested Amazonian regions of each country.[11]

BOX 9.3

Technical spotlight: Estimating the impact of deforestation on malaria transmission

For both the Amazon-country analysis and the global-scope analysis, the following fixed-effects finite distributed lag model is implemented to capture the immediate and lagged effects of deforestation on malaria cases or malaria incidence, Y. The regression is measured in logs to allow for an elasticity measure interpretation, where i,s,c index 2, 1, and 0 administrative levels, often city, state, and country.

$$lnY_{i,s,c,t} = \beta_0 + \sum_{j=1}^{4} \beta_j \ln(TCL_{i,t-j-1}) + W_{i,t} + X_{i,t} + \delta_s + \vartheta_t + \varepsilon_{i,s,c,t}. \qquad \text{(B9.3.1)}$$

Consistent with previous nomenclature, $W_{i,t}$ denotes weather controls—annual averages and squared averages of temperature and precipitation—$X_{i,t}$ is a control for population, in terms of either the log of nightlight intensity or the log of population density; finally, year and subnational level-two fixed effects are included to account for time-varying influences that are constant throughout regions or for regional particularities that are constant throughout time, respectively. The error terms are clustered at the subnational two level, which is the level of treatment. Subtle differences implemented for the global analysis are the inclusion of a country-year time trend and regional fixed effects at the subnational one level, s, and errors are clustered at the country level.

A useful trick available to finite distributed lag models is that the long-run propensity can be estimated as a simple sum of coefficients, which can be expressed with simple algebraic manipulations to provide standard errors for statistical inference. This study follows Berazneva and Byker (2017) in showing the cumulative impact of deforestation on malaria for each country and region studied. Up to five lags of annual deforestation are considered for the cumulative effect.

Source: World Bank.

Detox Development

MAP 9.3 Impact of deforestation on malaria transmission in Brazil, Colombia, Ecuador, and Peru

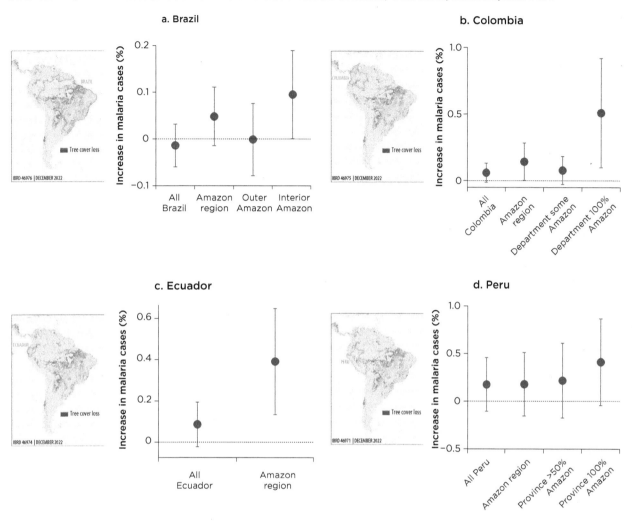

Source: World Bank.
Note: Point estimates for the coefficients of tree cover loss from estimating equation B9.3.1 in box 9.3 are shown with 95% confidence intervals. The point estimates can be interpreted as the percentage increase in malaria due to a 1% increase in tree cover loss. The different windows within each panel indicate a different level of forest cover, going from regressions sampling the entire country (left) to regressions only on densely forested areas (right).

Colombia, Ecuador, and Peru show very similar results, with a 1 percent increase in deforestation translating to anywhere between a 0.4 percent and 0.5 percent increase in malaria cases.[12] For Peru, the Departments of Amazonas, Madre de Dios, San Martín, Ucayali, and Loreto drive the positive correlation. For Ecuador, the entire Amazonian region composed of six provinces (Orellana, Sucumbíos, Napo, Pastaza, Morona Santiago, and Zamora Chinchipe) results in a positive association. And in Colombia, the densely forested southern departments with a significant relationship between tree cover loss and malaria are Amazonas, Caqueta, Guainia, Guaviare, Putumayo, and Vaupés.

The magnitude of the effect is smaller for Brazil than for the other countries considered. In Brazil, a 1 percent increase in tree cover loss is associated with a 0.09 percent increase in malaria. In the analysis, malaria data cover Amazonian and non-Amazonian states, but following MacDonald and Mordecai (2019), the data are divided further by "interior" and "outer" Amazon regions. The interior Amazonian states are Acre, Amapá,.

Amazonas, Pará, Rondônia, and Roraima. The outer Amazonian states are Maranhão, Mato Grosso, and Tocantins. Consistent with the results from Colombia, Ecuador, and Peru, in Brazil, regions that are deeper in the Amazon have a higher deforestation-malaria elasticity (map 9.3).[13]

Global analysis of deforestation's impact on malaria transmission

As deforestation increases, malaria incidence rises everywhere around the world where mosquito vectors thrive. The global analysis of 73 malaria-prone countries results in statistically significant associations between tree cover loss and malaria across four major geographic zones (figure 9.3). The elasticity coefficients range from 0.008 for Sub-Saharan Africa to 0.048 for East Asia and Pacific, whereas for the entire sample the elasticity measure is 0.037.[14]

FIGURE 9.3 Global impact of deforestation on malaria, by region

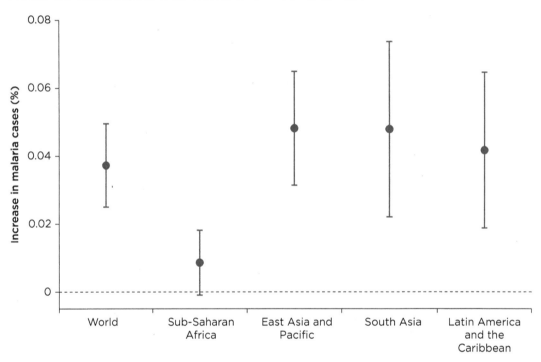

Source: World Bank.
Note: Point estimates for the coefficients of tree cover loss from estimating equation B9.3.1 in box 9.3 are shown with 95% confidence intervals. The point estimates can be interpreted as the percentage increase in malaria due to a 1% increase in tree cover loss.

The pass-through effect of agricultural subsidies on malaria through deforestation

Putting the results presented in this chapter together shows that a 10 percent increase in price for the agricultural commodities studied is associated with an increase in malaria incidence as high as 0.5 percent to 1.6 percent. Averaged over the five commodities analyzed, the immediate global impact of a 1 percent commodity price increase is a 0.021 percent increase in malaria incidence (figure 9.4). When considering cumulative tree cover loss, a 1 percent rise in commodity prices induces a 0.063 percent rise in malaria globally.

FIGURE 9.4 Global effect of agricultural commodity prices on malaria through deforestation, by commodity and region

a. Effect by commodity

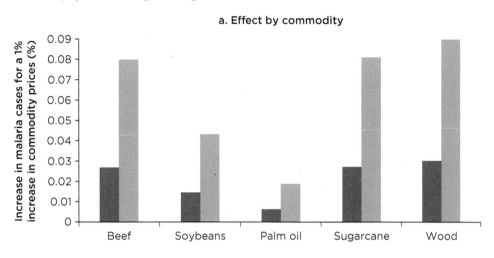

b. Effect by region

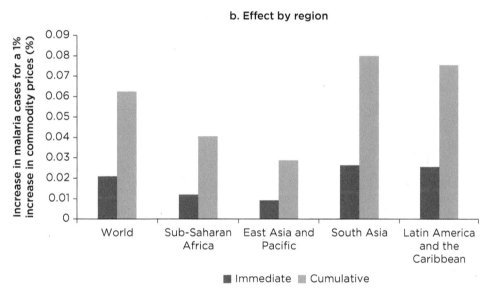

■ Immediate ■ Cumulative

Source: World Bank.

   However, responses vary by region and commodity. Deforestation for timber presents the greatest malaria risk given increases in market prices. This effect is driven mainly by the Latin America and the Caribbean region, where a 1 percent increase in the international price for wood leads to a 1.1 percent increase in deforestation, which correlates with a 0.042 percent increase in malaria. After several years of deforestation, this effect rises to 0.124 percent. South Asia is most vulnerable to malaria as a result of deforestation, driven by demand for beef, soybeans, and sugarcane. The East Asia and Pacific countries stand out for having the highest pass-through malaria elasticity for the palm oil industry. Figure 9.5 shows that, with very few exceptions, the malaria effect is increasing with years of tree cover loss, both for the different crops and across the different regions of the world.[15]

FIGURE 9.5 **Effect of agricultural commodity prices on malaria, by commodity and region**

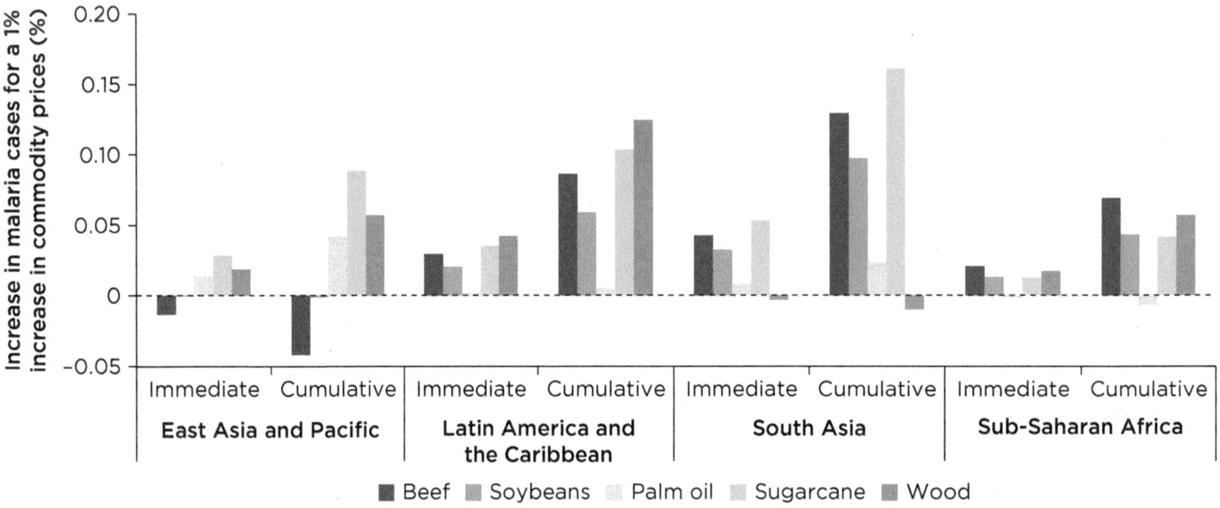

Source: World Bank.

The hidden health costs of agricultural subsidies

The hidden health losses and associated costs caused by subsidy-induced deforestation are significant in some regions. Producer supports are responsible for 14 percent of total annual deforestation globally. Using a global malaria elasticity of 0.037 translates to a 0.52 percent increase in malaria incidence or, for cumulative years of deforestation, an increase in incidence of 1.55 percent. For the 73 countries considered in the analysis, the population-weighted average incidence per 1,000 population is 45.02 for the year 2020. Using this value as a reference, a back-of-the-envelope estimate finds that the increase in incidence results in a range of 1.3 million to 3.8 million cases of malaria annually that are due to producer supports. Per country, this amounts to roughly 17,000 to 50,000 cases annually, on average, depending on baseline malaria cases and the rate of deforestation in each country.

1.3 million to 3.8 million:

The estimated annual increase in malaria cases that can be attributed to the use of agricultural producer supports

The potential costs of rising malaria linked to agricultural price supports is nontrivial. Table 9.1 presents several estimations and highlights the different methods that exist to assess these costs. The estimates vary widely, reflecting differences in what is considered as a cost as well as differences in methodology. Accordingly, the global economic burden of malaria due to agricultural subsidies is estimated to vary from a low of around US$3 billion a year to a high of around US$19 billion a year. Given that much of the burden is concentrated in low-income countries and in poor and rural areas, these impacts no doubt have significant consequences for development that are not fully reflected in cost estimates.

The direct financial, out-of-pocket cost of malaria treatment is based on estimates from White et al. (2011), who conclude that an episode of malaria costs US$5.84, if uncomplicated, and US$30.26, if severe, per sick person per year. Thus subsidy-induced deforestation entails an estimated cost of treatment

TABLE 9.1 Estimated economic impacts of increased malaria transmission due to subsidy-induced deforestation

Indicator	73-country sample	Average individual country
Increase in malaria	1.3 million to 3.8 million cases	17,400 to 51,900 cases
Out-of-pocket cost of treatment	US$23 million to US$68.3 million	US$314,500 to US$936,120
Cost to the macroeconomy	US$6 billion to US$18.6 billion	US$85.6 million to US$254.7 million
Burden in added DALYs	436,000 DALYs	6,000 DALYs
Health expenditure cost of DALYs	US$3 billion	US$40.6 million

Source: World Bank.
Note: DALYs = disability-adjusted life years.

between US$23 million and US$68.3 million globally. However, there is a well-known "malaria gap" whereby—owing to the many externalities of malaria on the community, such as lower tourism, less investment, and fewer labor market opportunities—the macroeconomic burden is much greater than the direct cost of treatment (Malaney, Spielman, and Sachs 2004). In a recent macroeconomic panel study of 180 countries, Sarma et al. (2019) find that a 10 percent decrease in malaria incidence is associated with a 0.27 percent increase in gross domestic product (GDP) per capita. Applying this result to the sample countries, the GDP per capita cost (purchasing power parity) of a 14 percent increase in deforestation would be anywhere from US$1.15 to US$3.44, which is equivalent to a total burden of US$6 billion to US$18.6 billion.

The subsidy-induced toll of malaria can also be expressed in terms of disability-adjusted life years lost. The DALY is a combined measure of the years lived with a disabling morbidity and the expected life years lost due to premature mortality from the disease.[16] A 14 percent producer support–driven increase in deforestation would add approximately 436,000 DALYs to the social burden of all malaria-affected countries. Conversely, the elimination of agricultural producer supports would avert an average 6,000 malaria DALYs per country per year, saving countries a global estimate of US$3 billion in health expenditures.[17]

Notes

1. In 2020, the global economy contracted 4.3 percent from the impacts of the COVID-19 economic shutdown, a loss of about US$3.6 trillion.
2. These estimated deforestation price elasticities are similar in magnitude to those reported in previous case studies for individual countries. For example, Scott (2014) estimates cropland acreage deforestation elasticities ranging from 2.4 to 6.5 in the United States for a 10 percent increase in commodity prices. In Indonesia, Wheeler et al. (2011) find deforestation elasticities with respect to sawlogs in the range of 5.9 to 13.5 and with respect to palm oil in the range of 3.6 to 9.0.
3. The results do show negative and significant deforestation elasticities for two commodities and regions: palm oil in Africa and beef in Europe and Central Asia. One potential explanation for the negative deforestation elasticity of palm oil is that the Global Forest Change data set classifies any area with vegetation taller than 5 meters in height as forest. Thus, it is possible that an increase in palm oil plantations is misidentified as an increase in forest cover in Africa, where the natural vegetation more closely resembles palm oil trees.
4. PSEs are available for 22 countries and the European Union. Map 9.2 plots the average annual PSEs in these nations over the period 2000 to 2019. The Organisation for Economic Co-operation and

Development's composition of PSE tables are used to separate producer support into coupled and decoupled payments.

5. The 90 percent confidence interval is the loss of 5,071 hectares to 28,920 hectares of forest cover. To ensure that no single country is driving the results, a "leave-one-out" sensitivity analysis reestimates the effect of PSEs on deforestation 25 times, each time dropping one country (or the European Union) from the analysis. All of the point estimates are between 13,592 and 22,052, and none is significantly different from the central estimate, suggesting that the association between coupled supports and deforestation is not driven by one country or policy in particular.

6. This is not to say that there are not significant challenges in identifying the deforestation-malaria relationship, but there are fewer for malaria than for other emerging infectious diseases.

7. Online appendix I can be found at https://openknowledge.worldbank.org/handle/10986/39423.

8. For a reference map of most of the countries included, see https://www.cdc.gov/malaria/about/distribution.html.

9. For information on the Malaria Atlas Project, see https://malariaatlas.org/.

10. Results for malaria incidence (per 1,000 population) were also tested and resulted in very similar coefficients and levels of significance for Brazil and Peru. Population data for Ecuador were found for only limited years and for Colombia for only a subgroup of cities.

11. Despite a lack of exogenous experimental evidence, the regressions focus on within-city changes through time, controlling for year fixed effects, precipitation, and temperature in terms of polynomials, but also with a specification controlling for ideal malaria temperature days. Different versions also control for nightlights or population density separately, obtaining similar results.

12. In Peru, Loreto, Amazonas, and San Martín are taken into account for the result shown in map 9.3 because Ucayali and Madre de Dios had very few cases of malaria. However, with Ucayali and Madre de Dios, the result is still higher, at 0.3 percent, although no longer statistically significant.

13. The long-run propensity of the effect of deforestation on malaria is tested for the four countries as described in box 9.3. The results show that the impact is persistent and increasing for Colombia, Ecuador, and Peru, with cumulative elasticities of 0.5, 0.6, and 0.8 percent, respectively. For Brazil, the cumulative effect in the interior fades over time. This finding may be consistent with smallholder, frontier settlements, which characterize deeper Amazon deforestation, since lower migration and eventual immunity could mitigate the initial impacts of deforestation.

14. The analysis again reveals that the effect of deforestation on malaria is growing with years of deforestation for all geographic zones. When accounting for the current year and five previous years, the impact suggests that a 10 percent increase in deforestation is related to a 1.1 percent increase in global malaria.

15. A related analysis consistent with the findings here concludes that about 20 percent of malaria risk in deforestation hotspots is driven by the international trade of crops such as timber, coffee, and cocoa (Chaves et al. 2020).

16. DALY estimates for the 73-country sample were obtained from the Global Burden of Disease Study 2019, provided by the Institute for Health Metrics and Evaluation (GBDCN 2021). A statistical analysis similar to that in box 9.3 reveals that, globally, a 1 percent increase in deforestation leads to a 0.07 percent increase in DALYs, which is significant.

17. This estimate is calculated using the results from Daroudi et al. (2021), which classifies the average cost per DALY averted for 176 countries according to different levels of their Human Development Index.

References

Ali, S., O. Gugliemini, S. Harber, A. Harrison, L. Houle, J. Ivory, S. Kersten, et al. 2017. "Environmental and Social Change Drive the Explosive Emergence of Zika Virus in the Americas." *PLoS Neglected Tropical Diseases* 11 (2): e0005135.

Bauhoff, S., and J. Busch. 2020. "Does Deforestation Increase Malaria Prevalence? Evidence from Satellite Data and Health Surveys." *World Development* 127 (March): 104734.

Bauserman, M., A. L. Conroy, K. North, J. Patterson, C. Bose, and S. Meshnick. 2019. "An Overview of Malaria in Pregnancy." *Seminars in Perinatology* 43 (5): 282–90.

Berazneva, J., and T. S. Byker. 2017. "Does Forest Loss Increase Human Disease? Evidence from Nigeria." *American Economic Review* 107 (5): 516–21.

Burkett-Cadena, N. D., and A. Y. Vittor. 2018. "Deforestation and Vector-Borne Disease: Forest Conversion Favors Important Mosquito Vectors of Human Pathogens." *Basic and Applied Ecology* 26 (February): 101–10.

Busch, J., and K. Ferretti-Gallon. 2017. "What Drives Deforestation and What Stops It? A Meta-Analysis." *Review of Environmental Economics and Policy* 11 (1): 3–23.

Carter, R., and K. N. Mendis. 2002. "Evolutionary and Historical Aspects of the Burden of Malaria." *Clinical Microbiology Reviews* 15 (4): 564–94.

Chaves, L. S. M., J. E. Conn, R. V. Mendoza López, and M. A. Mureb Sallum. 2018. "Abundance of Impacted Forest Patches Less Than 5 km² Is a Key Driver of the Incidence of Malaria in Amazonian Brazil." *Scientific Reports* 8 (1): 1–11.

Chaves, L. S. M., J. Fry, A. Malik, A. Geschke, M. A. Mureb Sallum, and M. Lenzen. 2020. "Global Consumption and International Trade in Deforestation-Associated Commodities Could Influence Malaria Risk." *Nature Communications* 11 (1): 1–10.

Curtis, P. G., C. M. Slay, N. L. Harris, A. Tyukavina, and M. C. Hansen. 2018. "Classifying Drivers of Global Forest Loss." *Science* 361 (6407): 1108–11.

Daroudi, R., A. A. Sari, A. Nahvijou, and A. Faramarzi. 2021. "Cost per DALY Averted in Low-, Middle- and High-Income Countries: Evidence from the Global Burden of Disease Study to Estimate the Cost-Effectiveness Thresholds." *Cost Effectiveness and Resource Allocation* 19 (1): 1–9.

Dobson, A. P., S. L. Pimm, L. Hannah, L. Kaufman, J. A. Ahumada, A. W. Ando, A. Bernstein, et al. 2020. "Ecology and Economics for Pandemic Prevention." *Science* 369 (6502): 379–81.

Druckenmiller, H. 2022. "The Effect of Agricultural Commodity Prices and Producer Supports on Global Deforestation." Background paper prepared for this report, World Bank, Washington, DC.

FAO (Food and Agriculture Organization). 2017. "Water for Sustainable Food and Agriculture." FAO, Rome. http://www.fao.org/3/i7959e/i7959e.pdf.

Garg, T. 2019. "Ecosystems and Human Health: The Local Benefits of Forest Cover in Indonesia." *Journal of Environmental Economics and Management* 98 (November): 102271.

GBDCN (Global Burden of Disease Collaborative Network). 2021. "Global Burden of Disease Study 2019 (GBD 2019) Reference Life Table." Institute for Health Metrics and Evaluation (IHME), Seattle, WA.

Hansen, M. C., P. V. Potapov, R. Moore, M. Hancher, S. A. Turubanova, A. Tyukavina, D. Thau, et al. 2013. "High-Resolution Global Maps of 21st-Century Forest Cover Change." *Science* 342 (6160): 850–53.

Iqbal, M. Z., and B. A. Babcock. 2018. "Global Growing-Area Elasticities of Key Agricultural Crops Estimated Using Dynamic Heterogeneous Panel Methods." *Agricultural Economics* 49 (6): 681–90.

Khadanga, S., P. K. Thatoi, B. N. Mohapatra, N. Mohapatra, and C. B. K. Mohanty. 2014. "Severe Falciparum Malaria—Difference in Mortality among Male and Nonpregnant Females." *Journal of Clinical and Diagnostic Research* 8 (12): MC01.

MacDonald, A. J., and E. A. Mordecai. 2019. "Amazon Deforestation Drives Malaria Transmission, and Malaria Burden Reduces Forest Clearing: A Retrospective Study." *The Lancet Planetary Health* 3 (S13).

Malaney, P., A. Spielman, and J. Sachs. 2004. "The Malaria Gap." *American Journal of Tropical Medicine and Hygiene* 71 (2): supplement: *The Intolerable Burden of Malaria II: What's New, What's Needed*.

Pendrill, F., U. M. Persson, J. Godar, T. Kastner, D. Moran, S. Schmidt, and R. Wood. 2019. "Agricultural and Forestry Trade Drives Large Share of Tropical Deforestation Emissions." *Global Environmental Change* 56 (May): 1–10.

Ritchie, H., and M. Roser. 2021. "Forests and Deforestation." OurWorldInData.org. https://ourworldindata.org/forests-and-deforestation.

Sarma, N., E. Patouillard, R. E. Cibulskis, and J.-L. Arcand. 2019. "The Economic Burden of Malaria: Revisiting the Evidence." *American Journal of Tropical Medicine and Hygiene* 101 (6): 1405.

Scott, P. 2014. "Dynamic Discrete Choice Estimation of Agricultural Land Use." TSE Working Paper 14-526, Toulouse School of Economics, Toulouse, France.

UNDP (United Nations Development Programme). 2015. *Gender and Malaria: Making the Investment Case for Programming That Addresses the Specific Vulnerabilities and Needs of Both Males and Females Who Are Affected by or at Risk of Malaria.* UNDP Discussion Paper, December 2015. New York: UNDP.

Valle, D., and J. Clark. 2013. "Conservation Efforts May Increase Malaria Burden in the Brazilian Amazon." *PLoS One* 8 (3): e57519.

Vittor, A. Y., R. H. Gilman, J. Tielsch, G. Glass, T. I. M. Shields, W. Sánchez Lozano, V. Pinedo-Cancino, and J. A. Patz. 2006. "The Effect of Deforestation on the Human-Biting Rate of *Anopheles darlingi*, the Primary Vector of Falciparum Malaria in the Peruvian Amazon." *American Journal of Tropical Medicine and Hygiene* 74 (1): 3–11.

Vittor, A. Y., W. Pan, R. H. Gilman, J. Tielsch, G. Glass, T. Shields, W. Sánchez-Lozano, et al. 2009. "Linking Deforestation to Malaria in the Amazon: Characterization of the Breeding Habitat of the Principal Malaria Vector, *Anopheles darlingi*." *American Journal of Tropical Medicine and Hygiene* 81 (1): 5.

Wheeler, D., D. Hammer, R. Kraft, S. Dasgupta, and B. Blankespoor. 2011. "Economic Dynamics and Forest Clearing: A Spatial Econometric Analysis for Indonesia." Center for Global Development Working Paper 280, Center for Global Development, Washington, DC.

White, M. T., L. Conteh, R. Cibulskis, and A. C. Ghani. 2011. "Costs and Cost-Effectiveness of Malaria Control Interventions—A Systematic Review." *Malaria Journal* 10 (1): 1–14.

Whitfield, J. 2002. "Portrait of a Serial Killer." *Nature*, October 3, 2002. https://doi.org/10.1038/news021001-6.

WHO (World Health Organization). 2021. *World Malaria Report 2021.* Geneva: WHO.

PART III
OCEANS

CHAPTER 10

The Economic, Social, and Environmental Effects of Harmful Fishery Subsidies

> "It has, I am afraid, been too common for vessels to fit out for the sole purpose of catching, not the fish, but the bounty."
> —*Adam Smith,* The Wealth of Nations

CHAPTER AT A GLANCE

Marine fish stocks and fisheries are crucial to meeting the food and nutritional needs of billions of people worldwide, while providing jobs and incomes to many more millions:

- *Fisheries and the oceans that contain them are critical drivers of long-term economic growth and environmental stability.* With 37 percent of the world's population living in coastal areas, and marine fisheries generating nearly half a trillion dollars in economic impacts each year, fisheries are a vital source of jobs, income, and food security.

- *Even though marine fish stocks are renewable, they are under serious threat.* The many contributing stressors include overfishing due to ineffective management and the perverse effects of government policies such as the provision of harmful subsidies; climate change, including through acidification, deoxygenation, and increased sea temperatures; and marine pollution from both marine and land-based sources, including plastics, oil spills and discharges, wastewater, and agricultural runoff, to name the most important.

According to the United Nations Food and Agriculture Organization (FAO), as of 2017, more than 30 percent of global stocks are overfished, while only 65.8 percent of fish stocks are within biologically sustainable levels (FAO 2020).

- *This chapter explores the effects of harmful subsidies on fisheries in three regional marine ecosystems.* These ecosystems—the Mauritanian exclusive economic zone (EEZ), the northern South China Sea, and the East China Sea—were chosen for their importance for food security, their size, and their diversity. This chapter examines the impact of subsidies in these fisheries to provide insights into the potential impacts of subsidy removal, reflecting ecological status and fishing pressures.

- *The results show that subsidies alone add to excess fishing capacity, reduced biomass, and lower fishing rents, even after fisheries that were previously managed with limited control over access have been closed.* The negative impact of distorting subsidies is even greater when fisheries are not managed sustainably, especially in situations where such stocks are already severely depleted. Consistent with economic theories of second best, repurposing subsidies without addressing open access may not resolve the problem in some situations.

- *Notwithstanding the benefits they may provide, fishery subsidy reforms must be carried out carefully, as their complete removal (as opposed to their repurposing) is likely to have adverse distributional outcomes.* While removing harmful subsidies will increase overall rents, small-scale artisanal fishers, in some instances, may be harmed. In contrast, when subsidies for large-scale (industrialized) operators are reformed while subsidies for some artisanal fishers remain in place, both large-scale and artisanal fleets stand to benefit.

The state of the world's oceans

At 70 percent of the Earth's surface, oceans are a crucial natural asset that, together with clean air, soils, and forests, make up the world's stock of foundational natural capital (Lange et al. 2021; World Bank 2017). Around 100 million tons of fish and invertebrates are caught annually from oceans (Pauly and Zeller 2016), generating total annual global gross revenues of US$150 billion and adding US$450 billion per year to the economy.[1] Marine fisheries are therefore vital for the livelihoods of hundreds of millions of people worldwide. They contribute directly to the food and nutritional security of billions of people (Golden et al. 2021), especially in the poorest coastal countries of the world, where, according to the FAO, fisheries supply up to 20 percent of the animal protein that people consume. They also serve as a source of vitally needed jobs and incomes, often in regions with few other opportunities for employment (Teh and Sumaila 2013). In addition to fisheries, marine ecosystems provide countless ecosystem services to sustain the planet and support the livelihoods of people around the world. Moreover, marine ecosystems can provide these benefits forever if managed wisely (Sumaila 2021).

The world's oceans are vital for climate change adaptation and mitigation, but climate change also puts them at risk. In addition to housing the world's fisheries, oceans regulate the climate in a significant way. In particular, oceans absorb 90 percent of the additional heat energy (von Schuckmann et al. 2020) and between 20 and 30 percent of carbon dioxide (Friedlingstein et al. 2021) generated from anthropogenic sources of greenhouse gas emissions. Coastal blue carbon ecosystems such as mangroves and seagrasses remove and store carbon. Vegetated marine habitats store up to 1,000 tonnes of carbon per hectare below ground, much more than most terrestrial ecosystems (Bindoff et al. 2019). In addition, marine organisms in upper waters, from phytoplankton to fish and marine mammals, contribute to the sequestration and storage of carbon in deeper waters. However, overexploitation, habitat disturbance, and losses from human activities release the carbon stored in marine and coastal ecosystems and weaken their capacity to sequester and store additional carbon (Gattuso et al. 2018).

Coastal ecosystems also play a critical role in adapting to climate change and building resilience. In addition to coastal protection and habitat provision, coastal ecosystems reduce the impact of storms and the inevitable sea-level rise that results from higher temperatures. Ensuring the effective management and sustainable growth of fisheries in oceans, while protecting and enhancing critical ecosystems, can thus provide global climate co-benefits as well as greater security to populations dependent on fisheries.

The increasing impacts of climate change—including rising sea levels, rising ocean temperatures, acidification, deoxygenation, and changes in the pattern of oceanic currents—are therefore an additional source of uncertainty for global fisheries (box 10.1). These processes will all have tremendous impacts on global fish stocks and related ecosystems in ways that are not yet fully understood. Fisheries reforms are long overdue and becoming increasingly urgent because climate change threatens the ability of depleted stocks to recover from overexploitation, which they were able to do in the past.

BOX 10.1
Managing the many stressors facing global fisheries

Recent assessments of the impacts of climate change, both observed and modeled, are increasingly demonstrating that the effects on marine ecosystems, fisheries, and the millions of fishers, processors, and traders who depend on them are likely to be more severe than originally expected. This concern arises where the intensity of climate impacts, combined with the limited adaptation capacity of many in the fisheries sector, contributes to the vulnerability of affected communities.

The effects of climate change on oceans are becoming better understood as climate impacts continue to increase in scale and severity. Some of these effects are already being felt acutely. Climate change is causing, for instance, changes in the patterns of fish migration, mostly from tropical areas toward the poles, as well as changes in fish-stock biomass as a result of higher temperatures, acidification, and deoxygenation. These changes, in turn, have economic impacts on the often-impoverished and vulnerable communities sustained by fisheries. Ocean acidification is affecting the distribution and survival of marine species, putting fisheries and aquaculture sectors under significant pressure. Sea-level rise and the increased frequency of extreme weather events are also affecting coastal communities around the world.

Besides climate change, demographic and economic growth are placing greater pressure on ocean resources and further straining the ecosystem services provided by oceans. Globally, the increasing migration of populations from inland areas to the coasts is adding to the growing pressure on coastal ecosystems and resources. Women are particularly exposed. For example, all along the fisheries value chain, women comprise a majority of the workforce but have historically been excluded from support programs that aim to reduce fishing capacity.

These challenges have thus far prevented oceans from reaching their full potential as engines of global development, and there is a pressing need to replace the "business-as-usual" model. Too often, oceanic sectors are managed separately by different ministries or other government entities that do not share or coordinate their efforts. This silo-based approach results in two important consequences. First, externalities generated by one sector are typically borne by others and are therefore not monitored, measured accurately, or mitigated. Second, a purely sectoral approach ultimately hampers the growth of the ocean economy because the development of one sector often prevents the full and optimal development of others.

The following conditions are needed to realize the potential of oceans as economic engines for development:

- The exploitation of marine living resources must be sustainable.
- Climate change must be fully incorporated in investment decisions.
- The impact on ocean health of all other activities that depend on oceans must be managed and restricted to sustainable limits.
- The management of different oceanic sectors must be better integrated to account for the externalities of one sector on others.

Now more than ever, the challenges are clear, as the threat of persistent overexploitation of ocean resources is exacerbated by growing demand. Furthermore, the effects of climate change, marine pollution, and litter from sea- and land-based sources combine to amplify the harmful impacts of such overexploitation on marine and coastal resources and ecosystems. Taken together, these challenges threaten to deplete fisheries and degrade the marine and coastal ecosystems and natural assets that drive growth in many coastal and island countries, undercutting the very foundation on which these countries rely.

The impacts of climate change on marine fisheries will make it difficult for many countries that depend on these fisheries to achieve several of the Sustainable Development Goals (SDGs)—in particular, SDG 1 (no poverty), SDG 2 (zero hunger), and SDG 3 (good health and well-being). Poor and remote fishing communities, in particular, are especially vulnerable to climate change because they depend on fisheries for their livelihoods and for food and nutrition security. More broadly, fisheries are at the heart of SDG 14 (life below water), which seeks to conserve, sustainably develop, and sustainably use the oceans, seas, and marine resources.

Ensuring that global fisheries remain economically and environmentally viable is of crucial importance, but overexploitation and pollution are degrading the ability of oceans to provide for humanity, causing widespread biodiversity loss, habitat damage, and destruction (box 10.1; Brondizio et al. 2019). Overfishing, destructive fishing practices, direct habitat damage, climate change, and pollution are among the major anthropogenic threats to ocean health and marine resources (Diaz et al. 2019; Lau et al. 2020). The consequences of these multiple stressors are illustrated in the finding by the FAO that 34 percent of the world's fish stocks are overfished today (figure 10.1).

The literature has advanced several reasons for the dismal state of global fisheries, including overcapacity; ineffective management; harmful fishery subsidies; illegal, unreported, and unregulated (IUU) fishing; pollution (including from land-based sources,

FIGURE 10.1 Global trends in the state of the world's marine fish stocks, 1974–2019

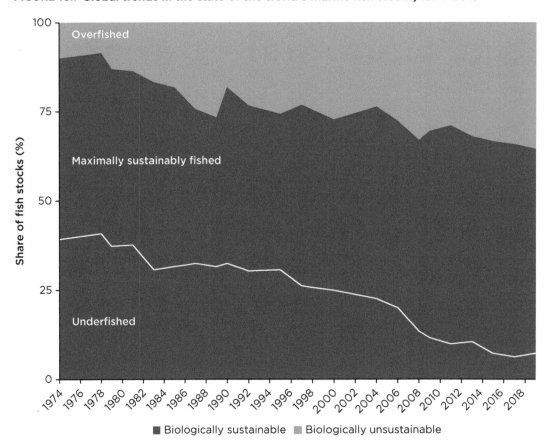

Source: United Nations Food and Agriculture Organization.

Detox Development

plastics, marine debris); insufficient coverage and ineffective management of marine protected areas; and the common-property, open-access nature of fish stocks (Gordon 1954; Pauly et al. 2002; World Bank 2017).

Here, the focus is on harmful subsidies that reduce the private cost of fishing through public support and artificially inflate profits beyond the point that is biologically sustainable and economically efficient. Such subsidies generate overcapacity and excess fishing effort, which result in both biological and economic waste, weaken the sustainability of fisheries, and reduce the global net benefits that could be generated by fisheries (World Bank 2017). Furthermore, global excess capacity also contributes to the prevalence of IUU fishing (Agnew et al. 2009; Sumaila et al. 2020).

Despite general recognition of the urgency of the situation, global action to address harmful fishery subsidies had remained elusive, as reflected in the slow progress of the World Trade Organization (WTO) negotiations on fishery subsidies, which started in 2001 and led to only partial agreement in June 2022.

The dual challenges of open access and direct subsidies

In addition to the effects of global climate change, two major issues harm the productivity and sustainability of global fisheries. The first is the lack of effective fishery management, which is often, but not always, driven by the quasi open-access nature of most fisheries and leads to unsustainable levels of effort and fishing. Following conventional use of the term in economics, open access defines a situation where access to a given fishery is either unrestricted or inadequately controlled. Resource rents (that is, revenues minus costs, including a normal market rate of return) that are untaxed can be considered an "implicit subsidy" to fishers, similar to an untaxed externality, since the social costs of fishery management are borne by all, but the benefits are entirely private. The open-access nature of fisheries, particularly on the high seas, which are beyond national jurisdiction (EEZs), leads to a free-for-all in the absence of effective regional fishery management organizations, with a race to the bottom that can push fish stocks beyond self-sustaining levels.

The second issue is the large explicit subsidies in the fisheries sector, which incentivize and fuel the expansion of fishing fleets or lower their cost of operation. Generally, and officially, fishery subsidies tend to be deployed to ensure food security, support employment and industry, or support low-income fishing communities. The unintended consequence, however, is that these subsidies often end up driving fishing efforts above the already unsustainable levels that result from ineffective management and open access. While *implicit* subsidies are studied at the global level in *The Sunken Billions Revisited* (World Bank 2017), this chapter focuses on *explicit* subsidies and examines the distortions they create, both on their own and when combined with different management scenarios.

The literature defines fishery subsidies in different ways—for instance, see the definitions of the World Trade Organization (WTO 2021), the Organisation for Economic Cooperation and Development (Cox 2006), the Food and Agriculture Organization (FAO 1995; Westlund 2004), the United Nations Environment Programme (Abaza and Fellus 2002; Porter 2004), Asia-Pacific Economic Cooperation (APEC 2000), the World Bank (Milazzo 1998), and the wider academic community (Sakai, Yagi, and Sumaila 2019; Sumaila et al. 2010). The common thread across these definitions is that a subsidy is a direct or indirect financial transfer from public entities that benefits the fisheries sector (box 10.2).

BOX 10.2
The economics of fisheries

The seminal assessment of the economics of fisheries is that of Gordon (1954), who observed that, in open-access fisheries or common-pool fisheries that are not effectively managed, effort will continue to increase even when revenues per unit of effort are declining. Ultimately, fishing and overfishing occur until there is breakeven where revenues equal costs.

The point at which total revenue equals total cost is commonly regarded as the bionomic equilibrium, where both industry profits and resource rents are completely dissipated (figure B10.2.1). The shape of the total revenue curve in this model reflects biological assumptions about how fish stocks grow. In this case, a peak is determined by the point beyond which further fishing begins to undermine the sustainability of fish stocks and hence revenue too. It reflects a logistic "density-dependent" growth process—when fish stocks are low, there is ample food, and growth rates (reproduction) are high. When fish stocks are high, there is growing competition for food, and reproduction slows. Between these states lies the peak, which is referred to as maximum sustainable yield (and labeled as MSY in figure B10.2.1). The model describes a situation where competition for fish leads to overexploitation and lower economic rewards than would be attainable with less fishing effort. It would make more sense to stop fishing at this point, but in open-access regimes or regimes with insufficient management in place, fishing will continue until revenue equals cost. Subsidies exacerbate the problem by artificially reducing the cost of fishing (lowering the total cost-of-fishing curve) and thus leading to even greater exploitation (that is, fishing effort shifts from E3 to E4 in figure B10.2.2).

FIGURE B10.2.1 **Gordon-Schaefer bioeconomic model**

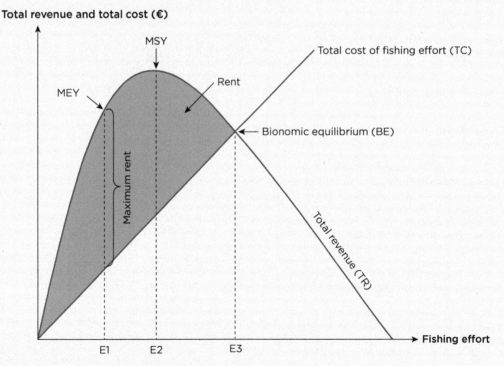

Source: Sumaila et al. 2013, 23.
Note: This model describes the different parameters commonly used in bioeconomics to study the performance of fisheries. MEY = maximum efficient yield; MSY = maximum sustainable yield.

(Continued)

BOX 10.2
The economics of fisheries *(continued)*

FIGURE B10.2.2 **Effect of cost-reducing subsidies on fishing effort**

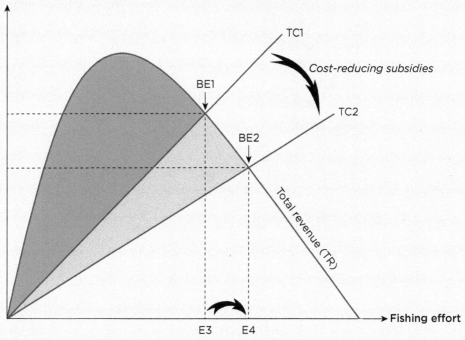

Source: Sumaila et al. 2013, 24.
Note: Subsidies that lower the total cost from TC1 to TC2 will also lower the bionomic equilibrium from BE1 to BE2, encouraging the growth of fishing effort from E3 to E4—hence the term "capacity-enhancing" subsidies.

The latest global estimate of fishery subsidies suggests that US$35.4 billion were provided by flag states in 2018, of which capacity-enhancing, or harmful, subsidies constituted US$22.2 billion (Sumaila, Ebrahim, et al. 2019). The top five providers of subsidies (China, the European Union, Japan, the Republic of Korea, and the United States) contributed 58 percent (US$20.5 billion) of all subsidies (harmful and otherwise). For the three ecosystems studied in this chapter, the Mauritanian EEZ, northern South China Sea (NSCS), and East China Sea (ECS), the magnitude of harmful subsidies to vessels fishing in these ecosystems is estimated at US$264 million, US$990 million, and US$2.5 billion, respectively. These harmful subsidies include subsidies for fuel, fees paid under fishing access agreements, boat construction and renewal, fisheries development projects, fishing port development, tax exemptions, and marketing and storage infrastructure.

US$35.4 billion: The amount of global fishery subsidies, of which **US$22.2 billion** are thought to be harmful

Figure 10.2 shows the 10 largest fishery subsidizers. China is by far the largest subsidizer, with subsidies totaling more than double the next largest subsidizer,

FIGURE 10.2 **Total fishery subsidies of the largest subsidizers, by effect of subsidy**

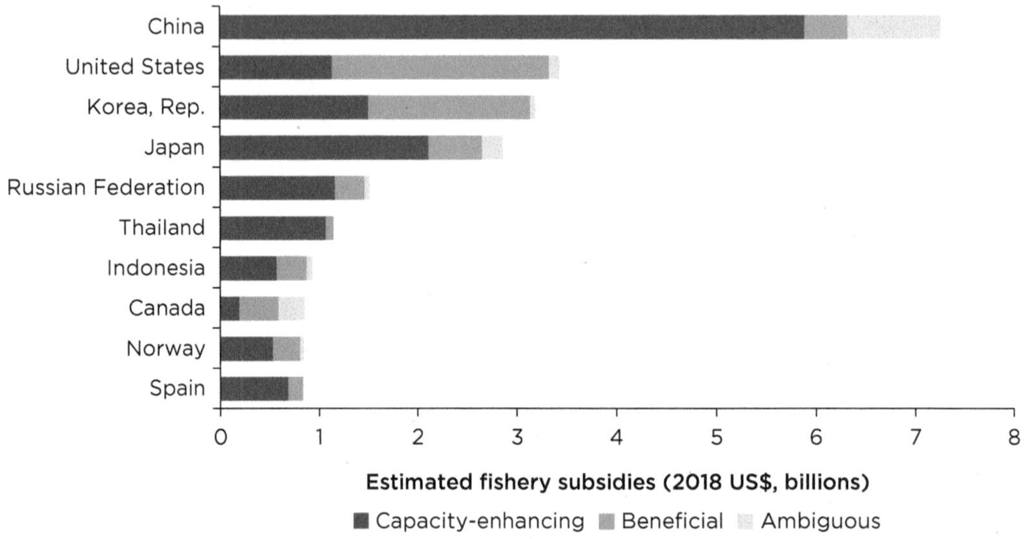

Estimated fishery subsidies (2018 US$, billions)

■ Capacity-enhancing ■ Beneficial ▨ Ambiguous

Sources: Data are from Sumaila, Ebrahim, et al. 2019; Sumaila, Skerritt, et al. 2019.
Note: The figure shows estimated total fishery subsidies for the 10 countries with the largest values in 2018 US dollars. Fishery subsidies include all direct and indirect transfers from the public sector to the private sector, divided into three broad categories: capacity-enhancing (boat construction, renovation, and modernization), beneficial (fishery management and research and development), and ambiguous (fisher assistance, vessel buyback, and rural fisher community development programs).

the United States. The nature of subsidies also matters, though, with Canada, Korea, and the United States devoting large percentages of their subsidies to activities like marine protected areas or other measures focused on the conservation of stocks, while more than 80 percent of subsidies in China, Japan, the Russian Federation, and Thailand drive capacity and effort and thus promote more fishing.

Although the countries in figure 10.2 provide large amounts of subsidies in absolute terms, when calculated as a *share* of total fisheries production, these countries are no longer outliers. As shown in figure 10.3, most countries subsidize fisheries at a rate of less than US$1,000 per metric tonne of catch. However, some significant outliers do exist, which are largely small coastal or island economies like the Comoros, Cyprus, Dominica, Eritrea, Kuwait, and Tonga.

Finally, figure 10.4 shows total subsidies by region and income group. Several patterns are evident. First, the vast majority of subsidies are paid in East Asia and Pacific, Europe and Central Asia, and North America, which coincides with both the location of wealthier countries, but also that of critical global fisheries in the Pacific and North Atlantic oceans. In addition, the figure highlights the considerable difference in subsidies between upper-middle-income and high-income countries versus low-income countries. Indeed, subsidies are largely a phenomenon in rich countries, but their impacts spread across the rest of the world. Even though the *share* of harmful, capacity-enhancing subsidies is lowest in high-income countries, those countries still spend considerably more resources in absolute terms than lower-middle-income and low-income countries.

FIGURE 10.3 Total fishery subsidies in fishery production, by GDP per capita, 2018

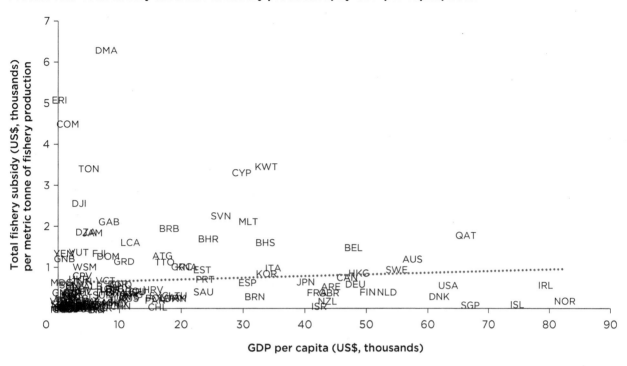

Sources: Fisheries data are from Sumaila, Ebrahim, et al. 2019; Sumaila, Skerritt, et al. 2019. Total fisheries production data are from the Food and Agriculture Organization's Fisheries and Aquaculture Statistics. GDP per capita is from the World Bank's World Development Indicators database.

FIGURE 10.4 Fishery subsidies, by effect and by region and income group

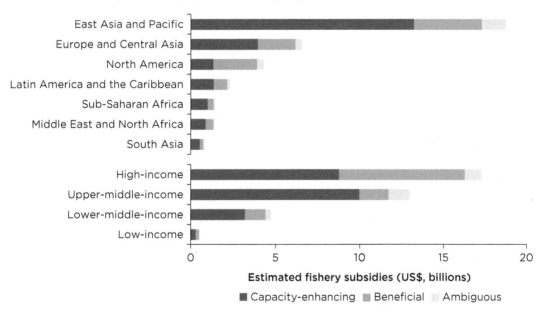

Sources: Data from Sumaila, Ebrahim, et al. 2019; Sumaila, Skerritt, et al. 2019.
Note: The figure shows estimated fishery subsidies, broken down by region and income group. Fishery subsidies include all direct and indirect transfers from the public sector to the private sector divided into three broad categories: capacity-enhancing (boat construction, renovation, and modernization), beneficial (fishery management and research and development), and ambiguous (fisher assistance, vessel buyback, and rural fisher community development programs).

The impact of subsidies in three critical fisheries

This section explores how different subsidy programs and management regimes can affect overall fish catch, revenue, and stock status. To this end, it presents new evidence from simulations of subsidy reform conducted using ecosystem models that reflect fisheries in three key regions around the world—the Mauritanian EEZ, the East China Sea, and the northern South China Sea. These fisheries were selected based on their importance to their associated countries and ecosystems as well as the availability of previously constructed ecosystem models (Pauly, Christensen, and Walters 2000) that were updated for the purposes of studying subsidies (Cheung 2007; Guenette, Meissa, and Gascuel 2014). Subsidies in each of these fisheries are very large, both in their absolute magnitude and as a proportion of gross fishery revenues (table 10.1), emphasizing the importance of ensuring that these fisheries remain economically productive. Box 10.3 describes the ecosystemic characteristics of each of these three fisheries.

TABLE 10.1 **Gross revenue, economic impact, and subsidy magnitude in three fisheries, 2018**

Fishery specifics	Mauritanian EEZ	Northern South China Sea	East China Sea
Total fish catch (tonnes, millions)	1.5	11	5
Harmful subsidies (US$, millions)	264	990	2,500
Gross revenue (US$, millions)	1,500	16,000	8,000
Estimated economic impact (US$, millions)	2,250	47,000	23,000

Source: Sumaila et al. 2021.
Note: Economic impact attempts to capture the multiplier effects of the fishery and is based on certain broad assumptions. EEZ = exclusive economic zone.

BOX 10.3

Ecosystems of the Mauritanian EEZ, the East China Sea, and the northern South China Sea

The Mauritanian exclusive economic zone (EEZ) extends to about 33,224 square kilometers with a depth of less than 200 meters. It includes a marine protected area, the Banc d'Arguin National Park, which covers about 6,450 square kilometers (Guenette, Meissa, and Gascuel 2014). The shelf ecosystem is enriched by the effects of an upwelling, with diverse fish species. The domestic fishery consists mainly of small-scale, artisanal fishing boats (Gascuel et al. 2007) that fish for mullets, especially in the Banc d'Arguin National Park, where only park residents operating small sailboats are allowed to fish (Guenette, Meissa, and Gascuel 2014). A larger proportion of the catch emanates from foreign vessels, both pelagic and demersal fleets, flagged in China, the European Union, the Russian Federation, and other countries (Gascuel et al. 2007).

The East China Sea (ECS) is an epicontinental large marine ecosystem covering about 770,000 square kilometers and bordered by China, Japan, and the Republic of Korea (Li et al. 2009). The ECS is rich in nutrients brought by the along-shore current, the Yellow Sea cold water mass, and the Kuroshio Current, which contribute to diverse fauna and flora. In the past decades, fish resources in the ECS have been degraded as a result of overexploitation. The main countries fishing in the ECS are China, Japan, and Korea. Landings in the ECS more than quadrupled from less than 10 million tons in the 1950s to 45 million tons in the 2000s (Li et al. 2009). The ECS ecosystem has been fished "down the food web," meaning that the biggest, most valuable fish were caught first and then smaller, less valuable species were targeted as the former become depleted (Pauly et al. 1998). Peak catch was reached in 2013 (Sumaila 2019), and the mean trophic level of landings declined from 3.5 to 2.8 between 1965 and 1990 (Chao et al. 2005). The proportion

(Continued)

To assess the impact of subsidies in the three ecosystems of interest, models are run under three different management policy scenarios:

- *Economic optimization,* where economic rents are maximized. This scenario most closely resembles a system where a permit or tenure system or a similar form of restriction limits the total catch at the economically optimal level, effectively avoiding unmanaged or open-access fishery problems. Results from simulations under economic optimization are thus indicative of outcomes that would prevail with and without a subsidy when entry into the fishery is controlled by adjusting fishing effort to levels that maximize economic rents.

- *Ecological optimization,* where the biomass of the ecosystem is maximized, with a particular focus on longer-lived species. In this regime, an attempt is made to maximize the ecological fitness of the fishery by controlling harvests.

- *Job growth or social objectives optimization,* where the number of jobs in the sector is maximized. This case mirrors a scenario where the aim is to expand employment and the fish catch in the short run. By extension, this scenario resembles management regimes where entry into the fishery is not managed to maximize either rents or ecological criteria and outcomes of a quasi open-access regime as conventionally defined.[2]

All three scenarios are run both with and without subsidies in order to isolate the effects of harmful subsidies on several key indicators of a fishery, including fishing effort,[3] biomass, economic rent, and catch. Box 10.4 briefly describes the Ecopath with Ecosim (EwE) models, while online appendix J provides a more in-depth description.[4] Most of the information employed was gathered from the works of the Fisheries Economics Research Unit and Sea Around Us (Pauly and Zeller 2016), including databases on fishing costs (Lam et al. 2011), ex-vessel prices (Tai et al. 2017), catch data (Pauly, Zeller, and Palomares 2020), and subsidies (Schuhbauer et al. 2020; Sumaila, Ebrahim, et al. 2019).

The analysis is computed for the net changes "without" and "with" subsidies, with the level of subsidy based on 2018 estimated amounts of subsidy provided to the fishing fleets active in each of the fisheries (Sumaila, Ebrahim, et al. 2019).

BOX 10.4
Technical spotlight: The Ecopath with Ecosim model

Mauritanian exclusive economic zone

The Mauritanian Ecopath with Ecosim (EwE) model applied in this study is adapted from a model developed by Guenette, Meissa, and Gascuel (2014). It is based on data from 1991 and contains 51 functional groups (including 27 fish, 1 marine mammal, 1 seabird, 18 invertebrate, 3 primary producers, and 1 detritus functional group). Fish functional groups are classified by habitat preferences, such as coastal, shelf, pelagic, and migratory. Fishing vessels are divided into artisanal, demersal, and pelagic fleets. The catch of the pelagic fleets is about 10.77 tonnes per square kilometer, more than 80 percent of the total catch in Mauritania. Pelagic fleets target mainly large pelagic fish, sardinellas, and horse mackerels. The Ecosim model is fitted using functional group biomass time-series data, spanning the period 1991–2006.

East China Sea

The East China Sea (ECS) model is adapted from a model developed by Li and Zhang (2012). The model has 38 functional groups, including major fishery resources and important groups in the ecosystem, such as marine mammals and seabirds. Fishing vessels are divided into 12 fleet groups. The landings and bycatch data were obtained from Sea Around Us (Pauly and Zeller 2016). The vulnerability index, V, values are taken from the estimates reported in Li and Zhang (2012), but V is limited to a maximum of 10 to avoid unrealistic population growth in the projections and scenario analysis. The time series of fishing effort is calculated based on initial data from the 1970s ECS model and data presented in Cao et al. (2017) on the number of Chinese fleet vessels and average horsepower from 1995 to 2014. The growth rate of fishing effort is estimated to be 0.65 percent per year (2000–14).

Northern South China Sea

The northern South China Sea (NSCS) model used in this analysis is from Cheung (2007) using data from the 2000s. The model covers the continental shelf (less than 200 meters in depth) of the NSCS (106°53′–119°48′ E to 17°10′–25°52′ N), which falls mainly in China's exclusive economic zone. The species are aggregated into functional groups based on their commercial importance, body size, ecology, and available species data (Cheung 2007). There are 38 functional groups, including 2 primary producers, 10 invertebrates, 21 fishes, 2 marine mammals, 1 marine turtle, and 1 seabird group. The fishing vessels are divided into six fleets: pair and stern trawls, shrimp trawl, purse seine, hook and line, gillnet, and others. The parameters of the NSCS model are evaluated based on government surveys, published literature, empirical equations, and global databases. The vulnerability parameters of the model are transferred from the 1970s NSCS model, which is fitted using data on the standardized catch per unit of effort from 1973 to 1988 for 17 commercially exploited taxa to determine the vulnerability parameters (Cheung 2007).

For all three fisheries, variable fishing cost and profits are computed as the percentage of total revenues (or landed values) for each type of fishing fleet in each of the models. These percentages are then used as input parameters for the economic conditions of the model. Fishing cost, ex-vessel price, landed value, and fishery subsidies data are based on databases developed by Sea Around Us (Pauly, Zeller, and Palomares 2020) and the Fisheries Economics Research Unit

(Continued)

Technical spotlight: The Ecopath with Ecosim model (*continued*)

(Lam et al. 2011; Lange et al. 2021; Schuhbauer et al. 2020; Sumaila, Ebrahim, et al. 2019; Tai et al. 2017) at the University of British Columbia. The variables for different species or types of gear in the model were extracted for each of the ecosystems being studied.

Online appendix J provides a more complete description of the EwE modeling framework and further details about economic models and data deployed in the current analysis. More details on the theoretical basis of EwE and its applications can be found in Cheung (2007); Cheung and Sumaila (2008); Pauly, Christensen, and Walters (2000); Sumaila (2004); and Sumaila et al. (2021).

Removing harmful subsidies and changes in fishing effort

Fishing effort, a term used to measure the aggregate inputs used for fishing, is a general proxy for the total amount of fishing. Increases in fishing effort are the main driver of overexploitation and have direct impacts on how rents and biomass change over time. Level of effort depends not only on the costs of fishing, but also on any established restrictions. Therefore, the subsidy amount and the management regime are both critical for determining fishing effort. Figure 10.5 shows how fishing effort changes in two scenarios—with and without subsidies—in each of the three fisheries and under different management regimes.

Under all management regimes (economic, ecological, and social objectives optimization), fishing effort declines when subsidies are removed. However, there are differences that reflect varying management objectives. For instance, in the Mauritanian EEZ and in the East China Sea, there are marginal upticks in fishing effort on removing

FIGURE 10.5 Changes in fishing effort with subsidy removed in three select areas, by management scenario

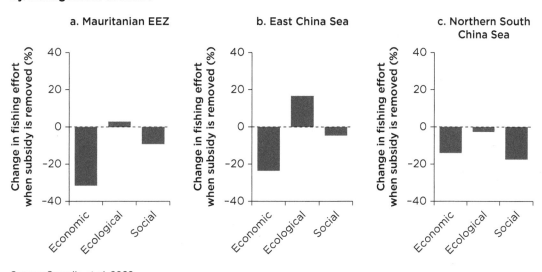

Source: Sumaila et al. 2022.
Note: EEZ = exclusive economic zone.

the subsidy. Under the ecological optimization scenario, the fishery is managed in ways that protect ecological objectives, meaning that subsidy removal may influence both the kind of fishing and the species targeted, but it does not have a major bearing on the level of effort. For example, in Mauritania and in the East China Sea, there is a decrease in the level of effort for pelagic fleets that benefits the biomass of long-lived pelagic species, but there is also an increase in demersal fishing that targets fast-growing invertebrates, which, in effect, compensates for revenues lost as pelagic fishing declines.

These results highlight the two-way interactions between fishery management and ecological factors. Management regimes affect species composition and trophic cascades along the marine food web, and these effects, in turn, affect the fish that are caught. These interactions lead to changes until there is an eventual convergence to a steady state (Beattie et al. 2002). Marine ecosystems are complex, and a small change in the predator-prey equilibrium can drive a large change in the aggregate biomass of the fishery, which is why ecosystem biomass does not directly follow changes in fishing effort. Box 10.5 explores the impact of subsidies and management regimes on biomass, where fishery management has a greater impact than subsidies.

BOX 10.5

Technical spotlight: Impact of subsidies and management regimes on biomass

To reveal how subsidies are likely to affect the biomass in an ecosystem, table B10.5.1 measures changes in biomass separately for invertebrates and for fish. For instance, in the case of the northern South China Sea, the biomass of invertebrates decreases by 1 percent, while the biomass of fish species increases by 8 percent when jobs are optimized. This decrease in the biomass of invertebrates is due to increases in predation once the biomass of fish species increases. A key observation from these results is that, when economic rent and jobs are optimized, subsidies reduce fish biomass overall but, in some cases, do not affect invertebrates and fish in the same way. When the objective is to optimize the ecology, subsidies do not have as large an impact on biomass. This finding is not surprising since protecting ecology remains the overarching goal whether or not subsidies are available.

The near constancy of fish stocks under ecological management with and without subsidies has an important policy implication, suggesting that management regimes may matter as much as, if not more than, subsidies reform.

TABLE B10.5.1 **Effects of harmful subsidies on biomass in three select areas, by management scenario**
net % change in biomass "without" relative to "with" harmful subsidies

Location	Optimizing economic rent		Optimizing jobs		Optimizing ecology	
	Invertebrates	Fish	Invertebrates	Fish	Invertebrates	Fish
Mauritanian EEZ	+2	+1	+6	+7	~0	~0
Northern South China Sea	−5	+27	−1	+8	~0	−1.2
East China Sea	+4	−2	+4	+13	~0	~0

Source: Sumaila et al. 2022.
Note: EEZ = exclusive economic zone.

The distributional impacts of subsidy removal, however, are unlikely to be neutral. Although removing the subsidies causes a general decline in effort (by increasing the cost of fishing), the decline is not felt equally across all types of fishing. The simulations highlight the significance of carefully considering these distributional impacts within a fishery because results for the fishery as a whole can mask an underlying heterogeneity. In the Mauritanian EEZ, where results are disaggregated between smaller, mostly poor artisanal fleets and larger commercial pelagic and demersal fleets, the steepest decline in fishing effort occurs among the artisanal fleets. Under the economic maximization scenario, pelagic and demersal fleets see a reduction in effort of 6 percent and 7 percent, respectively, but the effort is reduced by a striking 80 percent for the artisanal fleets. This finding suggests that artisanal fleets are overwhelmingly dependent on the current level of subsidies and that their removal could potentially have disastrous distributional impacts for the poorest fishers.

Removing harmful subsidies and changes in resource rents

The change in resource rents that would be gained or lost from subsidy removal is a key economic indicator to consider (box 10.6 considers fishery ecology). All things being equal, increasing effort in most industries leads to higher revenues. However, in the absence of fishery management measures, the opposite may be true: increased level of effort negatively affects biomass, fish stocks are not able to reproduce and replenish as rapidly, and rents decline overall. The results in table 10.2 suggest that the removal of harmful fishery subsidies leads to large increases in rents—in some cases. Under an economic optimization scenario, the removal of harmful subsidies results in a general increase in economic rents, depending on the ecosystem and the management regime. In the absence of effective management under the jobs-optimizing scenario, rents decline by 3 percent in fisheries with large-capacity fleets such as in the NSCS, suggesting that the removal of subsidies alone does not fully eliminate overfishing and the consequent depletion of rents. The state of overfishing is such that, even when subsidies are eliminated, the increase in fish stocks is so modest that, even though costs decrease, this decrease does not offset the loss from subsidy removal. In situations of overfished stock, more direct management of fish stocks and fishing effort is needed for rents to recover.

BOX 10.6

The importance of fishery ecology in determining the impact of harmful subsidies

Why do the impacts of subsidy removal vary so much from one fishery to another? While the direction of the impact remains largely the same (with a few exceptions), the magnitude can differ significantly.

The type of species targeted and the food web interactions between these species can matter significantly. For example, Fischer and Mirman (1996) identify three types of interactions that would have implications for the kind of outcome possible when fishing effort changes for any reason, including due to the provision of harmful subsidies:

- When the interaction is symbiotic, where having more of one species is good for another species
- When species compete for resources or are mutual predators
- When there is a predator-prey relationship between the species.

In addition, the initial condition, state of the ecosystem, and status of the fish stocks at the beginning of the analysis also matter, as does whether a fishery is already effectively managed or not. Indeed, fisheries in all three ecosystems studied in this chapter are overfished.

TABLE 10.2 Effects of harmful subsidies on economic rent in three select areas, by management scenario

net % change in economic rent "without" relative to "with" harmful subsidies

Location	Optimizing economic rent	Optimizing jobs	Optimizing ecology
Mauritanian EEZ	+10	+52	+7
Northern South China Sea	+213	−3	+27
East China Sea	+9	+49	−5

Source: Sumaila et al. 2022.
Note: EEZ = exclusive economic zone.

Removing subsidies and distributional consequences

The amplitude of the distributional consequences of removing fishery subsidies can be difficult to assess. Typically, fishing fleets are not neatly organized by income, making it difficult to analyze differential impacts on the rich versus the poor, let alone to identify effects disaggregated by gender or ethnic groups. However, the Mauritanian fishery offers a unique opportunity for such an analysis because data are available for the three categories of fleets: artisanal, demersal, and pelagic. This information makes it possible to isolate the effects by comparing two scenarios. One scenario covers the removal of harmful subsidies for all three fleets, and the other covers the removal of subsidies for demersal and pelagic fleets, while keeping subsidies for artisanal fleets, which tend to sustain poorer fishers.

The results suggest that, under economic rent optimization, removing harmful subsidies across all fishery sectors (artisanal, demersal, and pelagic) leads to reductions in fishing effort in all sectors of between 6 percent and 75 percent, depending on the sector (table 10.3). However, the artisanal sector is the only sector that sees total rents decline, and they decline significantly, by more than 60 percent. In contrast, economic rents increase by 28 percent and 6 percent, respectively, for the demersal and pelagic sectors. This finding suggests that, while removing subsidies across all sectors would have a net positive impact on rents, doing so would be undesirable from a distributional standpoint, as artisanal fleets stand to lose much more than the other two.

TABLE 10.3 Impact of removing harmful subsidies for all three sectors in the Mauritanian exclusive economic zone, by fishing sector

net % change "without" relative to "with" subsidies

Management objective	Artisanal	Demersal	Pelagic
Optimizing economic rent			
Relative rent without subsidies	−62	28	6
Relative fishing effort without subsidies	−75	−7	−6
Optimizing jobs			
Relative rent without subsidies	14	3	25
Relative fishing effort without subsidies	−7	−6	−29
Optimizing ecology			
Relative rent without subsidies	57	3	8
Relative fishing effort without subsidies	46	−6	−2

Source: Sumaila et al. 2022.

TABLE 10.4 Impact of removing harmful subsidies only for demersal and pelagic fleets in the Mauritanian exclusive economic zone, by fishing sector

net % change "without" relative to "with" subsidies

Management objective	Artisanal	Demersal	Pelagic
Optimizing economic rent			
Relative rent	22	8	6
Relative effort	16	−15	−6
Optimizing jobs			
Relative rent	3	−22	7
Relative effort	~0	−3	~0
Optimizing ecology			
Relative rent	18	−11	7
Relative effort	17	−19	−1

Source: Sumaila et al. 2022.
Note: Subsidies are retained only for the artisanal fleet.

If, instead, subsidies are removed for the demersal and pelagic fleets but maintained for the artisanal fleets, the result is dramatically different. Table 10.4 shows that, under economic rent maximization, the net change in economic rent for the artisanal fleet increases from −62 percent when harmful subsidies for all fishing sectors are removed to +22 percent when only subsidies for demersal and pelagic fleets are removed. In addition, both the demersal and pelagic fleets still see higher economic rents under this subsidy adjustment scenario. The result is a triple win, where harmful subsidies are removed, all types of fleets become better off, and poorer, small-scale fishers benefit the most, as distributional outcomes improve. Most of the pelagic and demersal fleets in the Mauritanian EEZ are foreign-owned, suggesting positive results from the removal of subsidies specifically for foreign fleets operating in the Mauritanian EEZ.

Implications and caveats

This chapter describes the results of EwE models for three diverse ecosystems, reflecting the current condition and economics of the fisheries and the stocks on which they depend. The models incorporate different levels of harmful subsidies provided to fishing fleets and suggest some of the impacts that could occur as a result of subsidy removal. Allowing for differences in ecology and the effectiveness of management measures currently in place, these results confirm that, generally speaking, removing harmful subsidies could reduce fishing effort and overfishing—but not uniformly so. Neither is subsidy removal necessarily sufficient. Subsidy removal is most effective in a scenario when the objective is "optimizing for economic rent." By design, this ideal is hypothetical and eliminates the ubiquitous problem of overfishing when limitations on access are imperfect. More generally, these results illustrate what is suggested by the theory of second best in economics: eliminating one distortion may be insufficient in the presence of multiple distortions.

The most realistic and policy-relevant case involves the job maximization scenario, which shows that, if the subsidy is large enough, overfishing and resource depletion are intensive. In this case, subsidy removal is uniformly ineffective in stimulating a recovery of the fishery. This result is seen most clearly in table B10.7.2 in box 10.7, which shows

BOX 10.7
Technical spotlight: Subsidy removal and the recovery of fisheries under job optimization and ecological optimization

In addition to the management regime and the underlying complex ecology of a fishery, it is possible that the magnitude of the subsidy drives outcomes. Since, in practice, some fisheries may resemble quasi open-access regimes, it is instructive to compare whether the size of a subsidy affects the outcome when that subsidy is removed across different management regimes. This examination provides insight into whether subsidy removal per se is sufficient for rents and biomass to recover and whether the size of the subsidy matters.

Table B10.7.1 shows results when the initial subsidy is low (10 percent of landed value of the catch) under a job optimization scenario. As subsidies are removed, the level of effort declines in all three fisheries, and fish stocks recover to varying extent across the fisheries. As biomass increases, costs decline and rents rise in all three cases, suggesting that overexploitation is low enough to achieve recovery by increasing costs through the removal of subsidies.

Table B10.7.2 describes a scenario where subsidies are large (50 percent of landed catch value). In this case, eliminating the subsidy leads to a decline in rents despite some recovery of biomass (columns 3 and 4 in table B10.7.2), contrary to what conventional economic wisdom would suggest. It may be that, even though biomass recovers and costs decline, these benefits cannot compensate for the elimination of subsidies. As a result, rents decline despite a fall in fishing effort and a rise in biomass.

Such outcomes are not uncommon among a wide class of renewable resources where recovery and reproduction rates are low, which tends to happen when species are on the brink of endangerment.

If subsidy removal does not help a badly depleted fishery to recover, might other improvements in ecological management be more effective instead? Table B10.7.3, which provides results for subsidy removal under ecological management, suggests the answer. When the fishery is managed to optimize for ecology, rents recover in Mauritania and the northern South China Sea (NSCS). This finding contrasts with the findings in table B10.7.2, where there was no recovery in Mauritania and the NSCS, despite elimination of the subsidy. Not surprising perhaps, rents remain persistently negative in the East China Sea (ECS), reflecting the parlous state of that overexploited fishery and the fact that the ECS is home to one of the largest fishing fleets in the world, along with the largest subsidies in the world.

TABLE B10.7.1 Scenario optimizing job growth with initial subsidy of 10 percent of landed catch value
net % change without harmful subsidies

Location	Rent	Fishing effort	Biomass	
			Invertebrates	Fish
Mauritanian EEZ	+32	−10	+7	+5
Northern South China Sea	0	0	+2	−6
East China Sea	+171	−43	+19	+5

Source: Sumaila et al. 2022.
Note: EEZ = exclusive economic zone.

(Continued)

BOX 10.7

Technical spotlight: Subsidy removal and the recovery of fisheries under job optimization and ecological optimization (*continued*)

TABLE B10.7.2 Scenario optimizing job growth with initial subsidy of 50 percent of landed catch value
net % change without harmful subsidies

Location	Rent	Fishing effort	Biomass	
			Invertebrates	Fish
Mauritanian EEZ	−13	−7	−4	+4
Northern South China Sea	−18	−18	+15	−4
East China Sea	−58	−43	+19	+5

Source: Sumaila et al. 2022.
Note: EEZ = exclusive economic zone.

TABLE B10.7.3 Scenario optimizing ecology with initial subsidy of 50 percent of landed catch value
net % change without harmful subsidies

Location	Rent	Fishing effort	Biomass	
			Invertebrates	Fish
Mauritanian EEZ	+35	+3	+5	0
Northern South China Sea	+45	+25	−2	−0.2
East China Sea	−33	−201	+38	+27

Source: Sumaila et al. 2022.
Note: EEZ = exclusive economic zone.

that, when pressures on the ecosystem are significant, the removal of subsidies is not enough to restore rents and fish stocks. In contrast, the same simulation under ecological management largely leads to the recovery of rents and effort, except in the East China Sea, where resource depletion is so pronounced that neither ecological management nor subsidy removal can reverse the situation. The ECS is home to the largest fishing fleets in the world, which also benefit from generous subsidies.

Overall, these simulations suggest that management regimes are key and, while subsidy reform and repurposing are necessary to bring fishing to within reasonable limits, they alone cannot compensate for the complete absence of fishery management measures.

Some important qualifications need to be considered when interpreting these results. First, the models provide a stylized and simplified representation of the economic dimensions of fisheries. Model outcomes depend crucially on assumed functional forms and the accuracy of both calibration and parameterization of the equations. Results therefore need to be treated with caution and interpreted as indicative of trends in the model rather than as model projections or empirical forecasts. In these models, the representation of the fishing industry clearly is a simplification that abstracts from dynamic, large fixed and sunk costs, the role of expectations, and strategic interactions between large fleets that compete for the same fish. All of these factors make a material difference to the results and affect how fishers react to subsidies. In addition, as the simulations suggest, more complex and granular descriptions of fish ecosystems also alter the results.

Given the complexity of the issues, it is unlikely that all of these interactions and factors can be feasibly modeled. Ideally, the effects of subsidy reforms would be based on econometrically estimated empirical models. This approach eschews the need for modeling complex interactions by allowing for reduced-form estimates of the factors of interest. An advantage of such an approach is that robustness can be empirically verified, and the accuracy of the results can be tested against real-world data. Without statistical validation, margins of error can be difficult to determine a priori. Such developments may need to await the collection of more data on both biological indicators as well as economic indicators of the fisheries.

Notes

1. This figure assumes an average ex-vessel price per tonne of US$1,500 (Tai et al. 2017) and an average economic impact multiplier of three (Dyck and Sumaila 2010).
2. As in conventional definitions, (quasi) open access describes a situation where control over access to the resource is imperfect. The case of full open access is one of complete free entry and exit and describes a textbook example that is seldom observed. In practice, there may be restrictions, agreements, and other impediments to costless entry and exit. Where the level of effort exceeds the level of rent maximization, there is more access than is optimal from a strict economic efficiency perspective, which reflects what is often termed quasi open access. Other objectives suggest different levels of optimal entry and may be preferable based on distributional, ecological, or other criteria.
3. Fishing effort is the term used to measure the aggregate amount of inputs used for fishing and hence is a proxy for the amount of fishing. A composite indicator may be used for a given combination of inputs, such as the number of hours or days spent fishing, number of hooks used (in long-line fishing), kilometers of nets used, and so forth.
4. Online appendix J can be found at https://openknowledge.worldbank.org/handle/10986/39423.

References

Abaza, H., and E. Fellus. 2002. "Fisheries Subsidies and Marine Resource Management: Lessons Learned from Studies in Argentina and Senegal." United Nations Environment Programme, Geneva.

Agnew, D. J., J. Pearce, G. Pramod, T. Peatman, R. Watson, J. R. Beddingtonne, and T. J. Pitcher. 2009. "Estimating the Worldwide Extent of Illegal Fishing." *PloS One* 4 (2): e4570.

APEC (Asia-Pacific Economic Cooperation). 2000. *Study into the Nature and Extent of Subsidies in the Fisheries Sector of APEC Member Economies*. Singapore: APEC. https://www.apec.org/Publications /2000/10/Study-into-the-Nature-and-Extent-of-Subsidies-in-the-Fisheries-Sector-of-APEC -Member-Economies-2000.

Beattie, A., U. R. Sumaila, V. Christensen, and D. Pauly. 2002. "A Model for the Bioeconomic Evaluation of Marine Protected Area Size and Placement in the North Sea." *Natural Resource Modeling* 15 (4): 413–37.

Bindoff, N. L., W. W. Cheung, J. G. Kairo, J. Arístegui, V. A. Guinder, R. Hallberg, N. J. M. Hilmi, N. Jiao, M. S. Karim, and L. Levin. 2019. "Changing Ocean, Marine Ecosystems, and Dependent Communities." In *IPCC Special Report on the Ocean and Cryosphere in a Changing Climate*, 477–587. Paris: Intergovernmental Panel on Climate Change.

Brondizio, E. S., J. Settele, S. Díaz, and H. T. Ngo, eds. 2019. *Global Assessment Report on Biodiversity and Ecosystem Services of the Intergovernmental Science-Policy Platform on Biodiversity and Ecosystem Services*. Bonn: IPBES Secretariat.

Cao, L., Y. Chen, S. Dong, A. Hanson, B. Huang, D. Leadbitter, and D. C. Little. 2017. "Opportunity for Marine Fisheries Reform in China." *PNAS* 114 (3): 435–42.

Chao, M., W. M. Quan, L. C. Hou, and Y. H. Chen. 2005. "Changes in Trophic Level of Marine Catches in the East China Sea Region." *Marine Sciences* (Chinese edition) 29 (9): 54.

Cheung, W. W. 2007. *Vulnerability of Marine Fishes to Fishing: From Global Overview to the Northern South China Sea*. PhD dissertation, University of British Columbia.

Cheung, W. W., and Y. Sadovy. 2004. "Retrospective Evaluation of Data-Limited Fisheries: A Case from Hong Kong." *Reviews in Fish Biology and Fisheries* 14 (2): 181–206.

Cheung, W. W., and U. R. Sumaila. 2008. "Trade-Offs between Conservation and Socio-economic Objectives in Managing a Tropical Marine Ecosystem." *Ecological Economics* 66 (1): 193–210.

Cox, A. 2006. *Financial Support to Fisheries: Implications for Sustainable Development*. Paris: OECD Publishing.

Diaz, S., J. Settele, E. Brondizio, H. Ngo, M. Guèze, J. Agard, A. Arneth, et al. 2019. "Summary for Policymakers of the Global Assessment Report on Biodiversity and Ecosystem Services of the Intergovernmental Science-Policy Platform on Biodiversity and Ecosystem Services." IPBES Secretariat, Bonn.

Dyck, A. J., and U. R. Sumaila. 2010. "Economic Impact of Ocean Fish Populations in the Global Fishery." *Journal of Bioeconomics* 12 (3): 227–43.

FAO (Food and Agriculture Organization). 1995. "Code of Conduct for Responsible Fisheries." FAO, Rome.

FAO (Food and Agriculture Organization). 2020. *The State of World Fisheries and Aquaculture 2020: Sustainability in Action*. Rome. https://doi.org/10.4060/ca9229en.

Fischer, R. D., and L. J. Mirman. 1996. "The Compleat Fish Wars: Biological and Dynamic Interactions." *Journal of Environmental Economics and Management* 30 (1): 34–42.

Friedlingstein, P., M. W. Jones, M. O'Sullivan, R. M. Andrew, D. C. E. Bakker, J. Hauck, C. Le Quéré, et al. 2021. "Global Carbon Budget 2021." Earth System Science Data Discussion, begun November 4, 2021. https://doi.org/10.5194/essd-2021-386.

Gascuel, D., P. Labrosse, B. Meissa, M. T. Sidi, and S. Guenette. 2007. "Decline of Demersal Resources in North-West Africa: An Analysis of Mauritanian Trawl-Survey Data over the Past 25 Years." *African Journal of Marine Science* 29 (3): 331–45.

Gattuso, J.-P., A. K. Magnan, L. Bopp, W. W. L. Cheung, C. M. Duarte, J. Hinkel, E. Mcleod, et al. 2018. "Ocean Solutions to Address Climate Change and Its Effects on Marine Ecosystems." *Frontiers in Marine Science* 5 (October 4, 2018): 337.

Golden, C. D., J. Z. Koehn, A. Shepon, S. Passarelli, C. M. Free, D. Viana, H. Matthey, et al. 2021. "Aquatic Foods to Nourish Nations." *Nature* 598 (7880): 315–20.

Gordon, H. S. 1954. "The Economic Theory of Common-Property Resource: The Fishery." *Journal of Political Economy* 62 (2): 124–43.

Guenette, S., B. Meissa, and D. Gascuel. 2014. "Assessing the Contribution of Marine Protected Areas to the Trophic Functioning of Ecosystems: A Model for the Banc d'Arguin and the Mauritanian Shelf." *Annals of Geophysics* 46 (1): 1–26.

Jia, X., C. Li, and Y. Qiu. 2005. *Survey and Evaluation of Guangdong Marine Fishery Resources and the Measures for Sustainable Utilization*. Beijing: Chinese Oceans Press.

Lam, V. W. Y., U. R. Sumaila, A. Dyck, D. Pauly, and R. Watson. 2011. "Construction and First Applications of a Global Cost of Fishing Database." *ICES Journal of Marine Science: Journal du Conseil* 68 (9): 1996–2004. http://icesjms.oxfordjournals.org/content/68/9/1996.abstract.

Lange, G.-M., M. W. Beck, V. W. Y. Lam, P. Menéndez, and U. R. Sumaila. 2021. "Blue Natural Capital: Mangroves and Fisheries." In *The Changing Wealth of Nations 2021: Managing Assets for the Future*, ch. 6. Washington, DC: World Bank Group.

Lau, W. W., Y. Shiran, R. M. Bailey, E. Cook, M. R. Stuchtey, J. Koskella, C. A. Velis, et al. 2020. "Evaluating Scenarios toward Zero Plastic Pollution." *Science* 369 (6510): 1455–61.

Li, Y., C. Yong, D. Olson, N. Yu, and L. Chen. 2009. "Evaluating Ecosystem Structure and Functioning of the East China Sea Shelf Ecosystem, China." *Hydrobiologia* 636 (1): 331.

Li, Y., and Y. Zhang. 2012. "Fisheries Impact on the East China Sea Shelf Ecosystem for 1969–2000." *Helgoland Marine Research* 66 (3): 371–83.

Milazzo, M. 1998. *Subsidies in World Fisheries: A Re-examination*. Washington, DC: World Bank.

Pauly, D., V. V. Christensen, J. Dalsgaard, R. Froese, and F. Torres, Jr. 1998. "Fishing Down Marine Food Webs." *Science* 279 (5352): 860–63.

Pauly, D., V. Christensen, S. Guénette, T. J. Pitcher, U. R. Sumaila, C. J. Walters, R. Watson, and D. Zeller. 2002. "Towards Sustainability in World Fisheries." *Nature* 418 (6898): 689–95.

Pauly, D., V. Christensen, and C. Walters. 2000. "Ecopath, Ecosim, and Ecospace as Tools for Evaluating Ecosystem Impact of Fisheries. *ICES Journal of Marine Science* 57 (3): 697–706.

Pauly, D., and D. Zeller. 2016. "Catch Reconstructions Reveal That Global Marine Fisheries Catches Are Higher Than Reported and Declining." *Nature Communications* 7 (1): 1–9.

Pauly, D., D. Zeller, and M. L. D. Palomares. 2020. "Sea Around Us Concepts, Design, and Data (Seaaroundus.Org)." Sea Around Us, University of British Columbia, Vancouver.

Porter, G. 2004. *Incorporating Resource Impact into Fisheries Subsidies Disciplines: Issues and Options.* Economics and Trade Branch Discussion Paper. Geneva: United Nations Environment Programme.

Sakai, Y., N. Yagi, and U. R. Sumaila. 2019. "Fishery Subsidies: The Interaction between Science and Policy." *Fisheries Science* 85 (2019): 439–47.

Schuhbauer, A., D. J. Skerritt, N. Ebrahim, and U. F. Le Manach. 2020. "The Global Fisheries Subsidies Divide between Small- and Large-Scale Fisheries." *Frontiers in Marine Science* 7 (September). doi:10.3389/fmars.2020.539214.

Sumaila, U. R. 2004. "Intergenerational Cost-Benefit Analysis and Marine Ecosystem Restoration." *Fish and Fisheries* 5 (4): 329–43.

Sumaila, U. R. 2019. "Comparative Valuation of Fisheries in Asian Large Marine Ecosystems with Emphasis on the East China Sea and South China Sea LMEs." *Deep Sea Research Part II: Topical Studies in Oceanography* 163 (May): 96–101.

Sumaila, U. R. 2021. *Infinity Fish: Economics and the Future of Fish and Fisheries.* Amsterdam: Elsevier.

Sumaila, U. R., and W. W. Cheung. 2015. *Boom or Bust: The Future of Fish in the South China Sea.* Vancouver, BC: University of British Columbia. https://www.admcf.org/wp-content/uploads/2019 /11/4-Resource-a_-Boom-or-Bust_-The-Future-of-Fish-in-the-South-China-Sea-English-version -November-2015.pdf.

Sumaila, U. R., W. W. Cheung, L. S. L. Teh, A. H. Y. Bang, T. Cashion, Z. Zeng, J. J. Alava, S. le Clue, and Y. Sadovy de Mitcheson. 2021. *Sink or Swim: The Future of Fisheries in the East and South China Seas.* Hong Kong SAR, China: ADM Capital.

Sumaila, U. R., N. Ebrahim, A. Schuhbauer, D. Skerritt, Y. Li, H. S. Kim, T. G. Mallory, V. W. L. Lam, and D. Pauly. 2019. "Updated Estimates and Analysis of Global Fisheries Subsidies." *Marine Policy* 109: 103695.

Sumaila, U. R., A. S. Khan, A. J. Dyck, R. Watson, G. Munro, P. Tydemers, and D. Pauly. 2010. "A Bottom-Up Re-estimation of Global Fisheries Subsidies." *Journal of Bioeconomics* 12 (2010): 201–25.

Sumaila, U. R., V. Lam, F. Le Manach, W. Swartz, and D. Pauly. 2013. "Global Fisheries Subsidies." Report to European Parliament, October. https://www.europarl.europa.eu/RegData/etudes/note /join/2013/513978/IPOL-PECH_NT(2013)513978_EN.pdf.

Sumaila, U. R., D. Skerritt, A. Schuhbauer, N. Ebrahim, Y. Li, H. S. Kim, and D. Pauly. 2019. "A Global Dataset on Subsidies to the Fisheries Sector." *Data in Brief* 27: 104706.

Sumaila, U. R., D. Zeller, L. Hood, M. L. D. Palomares, Y. Li, and D. Pauly. 2020. "Illicit Trade in Marine Fish Catch and Its Effects on Ecosystems and People Worldwide." *Science Advances* 6 (9): eaaz3801.

Sumaila, U. R., Z. Zeng, V. W. Lam, and W. W. Cheung. 2022. "A Rich Analysis of the Economic, Social and Environmental Effects of Harmful Fisheries at the Ecosystem Level." IOF Working Papers 2022-04, Institute for the Oceans and Fisheries, University of British Columbia, Vancouver. https://fisheries.sites.olt.ubc.ca/files/2022/11/WP2022-04-Sumaila-et-al.Final_.pdf.

Tai, T. C., T. Cashion, V. W. Lam, W. Swartz, and U. R. Sumaila. 2017. "Ex-Vessel Fish Price Database: Disaggregating Prices for Low-Priced Species from Reduction Fisheries." *Frontiers in Marine Science* 4 (November 13, 2017): 363.

Teh, L. C., and U. R. Sumaila. 2013. "Contribution of Marine Fisheries to Worldwide Employment." *Fish and Fisheries* 14 (1): 77–88.

von Schuckmann, K., E. Holland, P. Haugan, and P. Thomson. 2020. "Ocean Science, Data, and Services for the UN 2030 Sustainable Development Goals." *Marine Policy* 121 (February): 104154. https://doi.org/10.1016/j.marpol.2020.104154.

Westlund, L. 2004. *Guide for Identifying, Assessing, and Reporting on Subsidies in the Fisheries Sector.* Rome: Food and Agriculture Organization.

World Bank. 2017. *The Sunken Billions Revisited: Progress and Challenges in Global Marine Fisheries.* Washington, DC: World Bank.

WTO (World Trade Organization). 2021. "Negotiations on Fisheries Subsidies." WTO, Geneva. https://www.wto.org/english/tratop_e/rulesneg_e/fish_e/fish_e.htm.

PART IV

FROM EVIDENCE TO ACTION

CHAPTER 11

Reforming Harmful Subsidies in a Complex Political Economy

"The changes we dread most may contain our salvation."
—*Barbara Kingsolver*

CHAPTER AT A GLANCE

Reforming subsidies brings both rewards and challenges:

- *If done incorrectly, subsidy reforms can have counterproductive economic, social, and political consequences.* This chapter highlights common political economy challenges associated with subsidy reform and proposes guiding principles for designing and implementing effective reforms.

More often than not, subsidies are a candidate for reform:

- *If a subsidy is not achieving its stated goals or if the unintended consequences outweigh the benefits, it is a candidate for reform.* Often established with good intentions, subsidies tend to be inadequate at contributing to sustainable and inclusive development. What is more, they cause a wide range of externalities and impose social costs that outweigh any benefit they may bring, as much of this report has established.

To devise effective subsidy reforms, first assess how things could go wrong:

- *Simply removing subsidies may not be enough to cause the behavioral or technological shifts needed to fix negative externalities.* People may face significant barriers, such as information, capacity, financial, or technical constraints or systemic risks and uncertainty. Ignoring such barriers could result in unnecessarily high transition costs and missed opportunities.

- *Subsidy programs are often intricately linked to political interests and influence.* Powerful interest groups can have outsized influence over policy processes, capture the message that is conveyed to the public, and mobilize formidable public opposition. Sometimes, second-best compromises are unavoidable in order to deliver the public benefits of subsidy reform.

- *Subsidy reform can have significant transition costs, particularly in the short term, while the economic system adjusts.* Policy makers need to anticipate and tackle potential impacts on the poor or vulnerable, impacts on the macroeconomy and competitiveness, employment effects, and adverse substitution effects.

It then is necessary to design and implement subsidy reforms that go beyond subsidy removal:

- *Subsidy reforms should consist of a package of measures that mitigate the downside risks of reform, while maximizing their contribution to sustainable development.* Subsidy reforms should consist of carefully timed and sequenced measures that include communication and consultation programs, effective compensation and social protection, as well as complementary measures and prudent reinvestment strategies for reform revenues. This way, reforms have the potential to minimize disruption, while contributing to countries' sustainable development objectives.

Introduction

Previous chapters of this report have made the case for the need to reform subsidies that pollute the air, land, and oceans, distort agricultural production, and contribute to unsustainable fishing. The detailed analyses find that most subsidies do not achieve their intended economic or social purposes. Indeed, many increase the incentives to exploit natural capital—land and water for agriculture, fossil fuel combustion, or fishing stocks—unsustainably, leading to a wide range of externalities with enormous costs and dwarfing any short-term benefits.

In an era when public coffers are empty and debts are reaching unsustainable levels, countries must reevaluate their spending programs and repurpose subsidies that are ineffective, inefficient, or counterproductive. Past experience shows that reform can be difficult and, if done hastily, can have unintended and perverse consequences.

How can policy makers evaluate whether reforms are warranted? And how can they design and implement reforms in a way that protects vulnerable households while also contributing to national sustainable economic development goals? This chapter presents an overview of best practices and lessons learned from the empirical evidence as well as from countries' experiences with implementing reform.

Of course, every country and every sector is different, and no single approach to subsidy reform will work in every context. The principles laid out here are intended to act as a guide. Countries' experiences with fishery, agriculture, and energy subsidies offer guiding principles for designing and implementing effective reforms and provide lessons for avoiding approaches that lead to resistance and backlash that stymie efforts at reform. At the same time, sectoral examples drill down from the general to the specific to provide sector-specific guidance to help governments to reform inefficient or harmful subsidy programs in a way that raises public revenues, supports the poor, and mitigates a wide range of negative externalities.

This chapter explores the importance of establishing the case for subsidy reform—anticipating the political challenges of reform—and outlines six overarching principles or imperatives that can help countries to navigate the political complexities and design and implement effective reforms.

Establishing the case for reform

Evaluating the rationale for existing subsidies

Subsidy programs—regardless of whether they are for the energy, agriculture, or fishery sector—are often well-intentioned policy schemes with defined development objectives. They usually aim to promote the productivity and competitiveness of certain sectors or to support low-income groups. Table 11.1 provides a detailed overview of the common policy objectives of environmentally harmful subsidies in different sectors. Yet subsidy programs are often inefficient or wholly inadequate to achieve their original policy objectives. As table 11.1 highlights, subsidies can have a host of unintended consequences, such that they exacerbate inequalities, inefficiencies, and environmental degradation.

Even when subsidy programs are initially implemented efficiently, it is not uncommon for the original purpose of such programs to be lost over time, as the sector, economy, or interest groups adjust to the new norm created by a subsidy. Indeed, many subsidy programs are associated with adverse environmental, distributional, or economic externalities, which become entrenched over time. Adverse environmental effects, for instance,

TABLE 11.1 **Subsidy programs: Objectives versus the common reality**

Sector	Objective	Reality
Energy	Boost industrialization and competitiveness of energy-intensive industries	Energy subsidies entrench inefficiency and wasteful practices (skilled labor and access to markets tend to be better drivers of competitiveness).
	Support fuel affordability for the poor	Subsidies increase affordability of fuels, but are highly regressive—that is, most subsidy payments benefit richer households for whom affordability is not a concern.
	Redistribute resource revenues	Subsidies empower powerful vested interest groups in extractive and energy-intensive industries, giving rise to corruption and lack of transparency.
Agriculture	Boost productivity	Subsidies may boost domestic production, but often lead to distorted and inefficient input use and extensification onto marginal lands.
	Provide food security	The increase in domestic production may improve food security, but import restrictions increase volatility in the case of domestic supply shocks, for example, due to weather.
	Support rural livelihoods	Subsidies tend to be regressive and disproportionately support groups at the upper end of the income distribution.
	Provide environmental benefits	Although well-designed, decoupled subsidies can lead to a virtuous cycle (see box 6.1 in chapter 6) by restoring landscapes and watersheds, most subsidies are coupled to production and result in harmful environmental impacts. These impacts can include increased deforestation due to agricultural land expansion, worsened water pollution due to excessive fertilizer use, and increased water scarcity due to incentivizing excessive water use.
Fisheries	Provide employment and social protection	As most fisheries are open-access, common-property resources, subsidies often lead to overfishing, depleted fish stocks, and ultimately a decline in employment. Although this effect is known and widely observed, subsidies continue due to fears of free riding on subsidy withdrawal efforts and lack of coordination.
	Provide food security and mitigate climate change	Overfishing diminishes food security. Global environmental security is compromised in ways that are poorly understood, since oceans and the biodiversity they host sequester more greenhouse gases than all terrestrial ecosystems do.

Source: World Bank.

may not be apparent immediately, but only materialize years later. Fishery subsidies may contribute to the excessive buildup of fishing fleets over time, which eventually exhaust the sustainable yield of fish stocks. Similarly, fossil fuel subsidies erode the incentives to invest in energy efficiency and technological modernization over time, so that countries eventually are locked into obsolete polluting technologies.

For these reasons, a constant and thorough reevaluation of subsidy programs is key to ensuring that their objectives remain relevant and continue to be met. The evidence from different sectors presented in this report shows that subsidies can undermine the very policy objectives they are meant to achieve. As chapters 7–9 show, agricultural input subsidies may be ineffective in promoting the productivity gains they are intended to achieve and instead contribute to water pollution, deforestation, and degraded soils. Similarly, chapter 10 shows that fishery subsidies can diminish fish stocks, catch, and revenues, even when they are intended to boost the productivity of fisheries. Evaluating a country's subsidy programs is key to understanding whether they are meeting their objectives, which can often differ from reality (table 11.1).

The evidence presented in previous chapters suggests that most often subsidy programs are falling short of their goals. And even when subsidies contribute to their policy objectives, it is often unclear whether they are the most cost-effective way of doing so. This fact alone is often enough to make a case for reforming environmentally harmful subsidies. Determining the balance of costs and benefits through careful empirical analysis at the program and country levels is crucial for informing policies.

Do the adverse effects of subsidies outweigh their benefits?

Even when subsidies make beneficial contributions to their intended policy objectives, they often do so at substantial cost. For instance, fossil fuel subsidies may make energy goods more affordable to low-income households, but at the same time contribute to air pollution challenges, reinforcing inequality and perpetuating inefficient and obsolete technologies. Do the benefits of continued subsidization programs justify their unintended externalities in terms of social, environmental, and economic impacts?

Conventional tools for assessing the balance of costs and benefits are often inadequate to the task. Many of the social and environmental costs of subsidies remain unmonetized and unmonetizable, such as the loss of biodiversity or further endangerment of rare and threatened species. Moreover, monetary measures—for instance, measures of lost ecosystem services—do not distinguish whether the household incurring the loss is poor or rich. Yet the same US$100 loss could be devastating for a poor household, while being a mere inconvenience for a rich one.

Yet even though reliably estimating all benefits and externalities of subsidy programs is challenging, the evidence points to a clear conclusion: the unintended adverse side effects of subsidies outweigh the benefits of keeping subsidies in place. The policy evaluations presented in this report and throughout the literature can help policy makers to evaluate whether existing subsidy programs are good value for money. When assessing the adverse side effects of subsidy programs, three dimensions are crucial to consider:

- *Environmental externalities.* Subsidies can directly incentivize behaviors and technologies that degrade the natural capital of countries, which can compromise development prospects.

- *Distributional incidence.* If ill-designed, subsidy programs may predominantly benefit the rich and other advantaged groups and thus aggravate preexisting inequalities.

- *Economic efficiency.* Subsidies influence the choices of producers and consumers alike and result in technical and allocative inefficiency.

Evaluating these issues is key to determining whether the unintended adverse side effects of externalities outweigh the benefits of subsidy programs (box 11.1). Table 11.2 presents an overview of common direct and long-run impacts of subsidy programs that run counter to policy objectives. But even when these perverse impacts are known and highly visible to policy makers and the public, lock-in effects and political economy forces can make subsidy reforms challenging to accomplish. The implication is that understanding the negative impacts of subsidies is necessary, but it is not enough to engender a sufficient policy coalition for reform.

BOX 11.1

The World Bank's Energy Subsidy Reform Facility

The World Bank—through the Energy Sector Management Assistance Program—is supporting countries seeking to reform energy subsidies that can threaten their economic, fiscal, and environmental health. The Energy Subsidy Reform Facility (ESRF) offers a concrete guide to analyzing energy subsidies and developing pragmatic and effective reform strategies. Based on a wealth of country case studies and empirical evidence, it holds lessons that can help policy makers to navigate complex political economy challenges in several key areas (figure B11.1.1). Its recommendations are relevant beyond the energy sector—in particular, those related to mobilizing public support and protecting livelihood risks associated with subsidy removal.

FIGURE B11.1.1 **Key areas of focus of the Energy Subsidy Reform Facility**

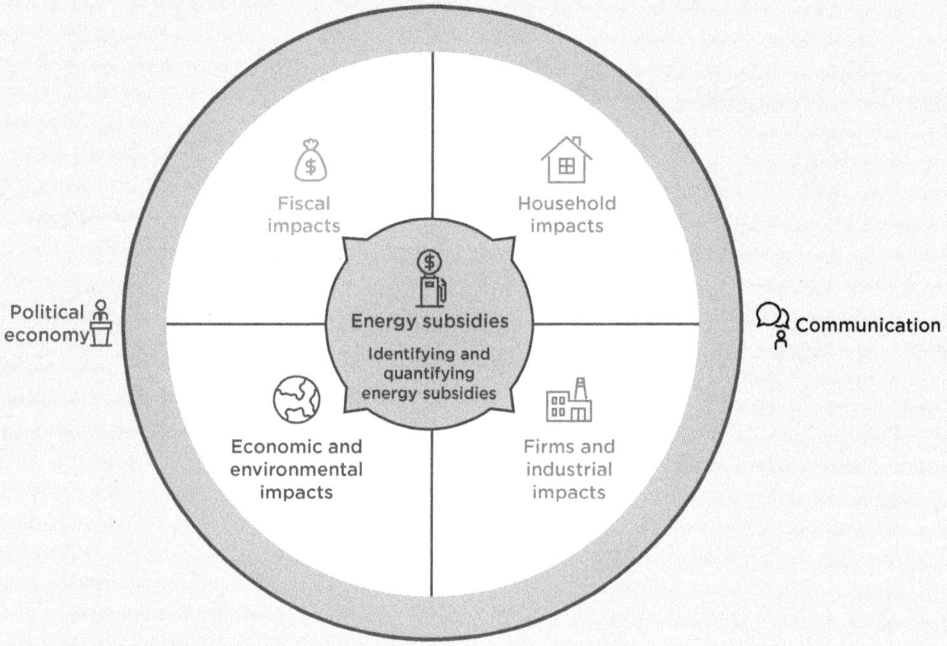

Source: Energy Subsidy Reform Facility.

The ESRF approach offers concrete guidance notes on specific topics that policy makers are likely to encounter when designing reform. These guidance notes cover the following key areas:[a]

- Identification and quantification of energy subsidies
- Assessment of the fiscal cost of subsidies and fiscal impact of reform
- Distributional impact on households
- Readiness of social safety nets to mitigate price shocks
- Impacts on firms and industrial competitiveness
- Macroeconomic impacts and global externalities
- Local environmental externalities—air pollution and health
- Political economy of energy subsidies reform
- Communication campaigns for energy subsidy reform.

a. Guidance notes are available at https://esmap.org/esraf.

TABLE 11.2 Adverse side effects of subsidy programs in the natural resource space

Effects	Distributional incidence	Environmental externalities	Economic efficiency
Direct effects	• Disproportionately benefit the rich and other advantaged groups • Reinforce spatial inequalities	• Land degradation • Air pollution • Water pollution • Overfishing • Climate change • Resource depletion	• Crowd out public funds from other productive uses (for example, health, education) • Compromise debt sustainability • Lock in inefficiency • Misallocate resources across sectors
Long-run impacts	• Reinforce inequalities and social disparities	• Affect human health, productivity, and livelihoods	• Reduce long-run growth and competitiveness

Source: World Bank.

Anticipating the political challenges to reform

Just because subsidy reforms are deemed necessary and beneficial, does not imply they are politically feasible. Experience from around the world has shown that, even when an existing subsidy system is found to be wholly harmful, unaffordable, and inefficient, reforming it and replacing it with an alternative policy framework are difficult tasks. And even if a subsidy program is successfully removed, there is no guarantee that the environmental externalities or economic inefficiencies will disappear automatically. Successful and durable changes in subsidy policy are mindful of three, often overlooked, challenges in the zeal for reform: (1) the barriers that prevent people from changing their behaviors or adopting new technology, (2) the influential interest groups and public opposition to reform, and (3) the potential adverse effects of reform that may subtract from a reform's success.

Identifying the barriers that prevent people and firms from adjusting behaviors and technology

Subsidy reform can have multiple objectives, from freeing up public budgets to reducing inefficiencies, achieving pro-poor redistribution, or addressing environmental externalities. Policy makers need to consider whether subsidy removal is enough to fix externalities and whether market frictions and transition costs need to be addressed. Because people and firms face a wide range of market failures, simply removing subsidies may not be enough to bring about the behavioral or technological shifts needed to fix negative externalities.

In a perfect market, price signals would trigger the behavioral and technological shifts required to achieve policy objectives, but in reality, households and firms may be unable or unwilling to adjust. In principle, two related issues may be at play. First, the presence of other market failures can act as transition barriers (table 11.3). Second, subsidy removal alone may not fully solve the underpricing of public goods, such as air, soil, or international fish stocks.

> Simply removing subsidies may not be enough to cause the behavioral or technological shifts needed to fix negative externalities.

TABLE 11.3 Transition barriers: Why prices alone may not change technology and behaviors

Level and type of constraint	Example
Investment barriers at the firm or government level	
Information constraints	• Limited information on scale and type of inefficiency • Limited information on modern technology and methods
Capacity constraints	• Technical capacity • Managerial capacity • Lack of awareness • Behavioral bias
Financial constraints	• Uncertain payoffs that hamper financing (for example, due to lack of information) • Competing investment opportunities • Inadequate credit markets
Market structures	• Lack of competition (for example, monopolies, oligopolies) • Protected industries • Trade protectionism
Fiscal mismanagement	• Subsidies to inefficient, polluting industries • Lack of enforcement (for example, carbon taxes, landfill tariffs)
Systemic barriers that can exacerbate existing barriers	
Risks and uncertainty	• Resource price volatility • Economic, political, and social instability • Lack of long-term credibility of policies

Source: World Bank.

Examples illustrate these barriers. Removing gasoline subsidies will create an incentive to invest in more fuel-efficient cars or to drive less and thus reduce air pollution. But, without access to credit to invest in a new car or efficient and affordable public transport as an alternative, people may face significant barriers to driving less. Avner, Rentschler, and Hallegatte (2014) show that carbon or energy taxes are only half as effective when dense public transport infrastructure is not available. They estimate that, in the absence of public transport options, the price elasticity of carbon dioxide emissions is halved in the short run—thus requiring higher (that is, less acceptable) energy prices to achieve the same reduction in commuting-related emissions.

In the case of agriculture, credit constraints can reduce poor farmers' ability to buy inputs such as fertilizers and seeds for high-yield production, even if their profits would be higher come the harvest. So, removing subsidies for such products without addressing credit constraints can leave farmers worse off. Evidence also suggests that subsidized irrigation can, in some contexts, reduce the pressure on land (Damania et al. 2017). So, removing irrigation water subsidies can lead to the expansion of agricultural land and increase the pressures on forests and natural areas, if no additional measures are taken. But the reverse may also be true in other circumstances. Similarly, policy makers need to consider whether removing fishery subsidies would address the fundamental problem of overfishing or would the unsustainable equilibrium continue, even after subsidies are reduced.

The bottom line is that, in theory, economic agents are assumed to adjust seamlessly to price signals. When environmentally harmful practices become more expensive because of subsidy reform, they will engage less in such practices. Yet while prices do indeed play this powerful role, it is crucial to recall the theoretical assumptions on which this seamless market response to prices rests: perfect information, no monopolies, no barrier to market entry or exit, perfect factor mobility, zero transaction costs, and absence of externalities.

Of course, reality may be more complex, and violations of these assumptions are likely ubiquitous. Information or capacity constraints can lead to inefficient decision-making even when price distortions are corrected. Likewise, other impediments, such as missing markets (for example, for credit), can constrain the implementation of efficiency-enhancing measures. A growing problem in some sectors and industries is the lack of competitive pressures to invest in efficiency gains, especially if protectionist trade policies are in place. Finally, with high fixed costs it may be costly to adjust capital stocks that may lead to long-term technology lock-in.

To facilitate the transition, policy makers need to address the financial, capacity, and information barriers that may obstruct the behavior or technological change that policy makers wish to achieve. Although it may be tempting to focus on a single distortion—such as a visible subsidy that carries a significant fiscal burden—it is vital also to consider complementary reforms that tackle these barriers. The long-established general theory of second best asserts that, in the presence of multiple distortions, correcting a single distortion will not necessarily improve outcomes (Bennear and Stavins 2007).

Understanding opposition from lobby groups and the role of public perception

Experience shows that political economy challenges create some of the most serious barriers to subsidy reform.[1] The tendency to design reforms to be technically sound and administratively feasible rather than politically viable aggravates these barriers. In reality, subsidy programs are often intricately linked to political interests and influence. This can make subsidy removal and reform complex because support was designed to confer a benefit to key stakeholders who command influence on policy decisions. Pivotal coalitions and public perceptions need to be considered in the design of feasible reforms.

Powerful interest groups can have outsized influence over policy processes and mobilize formidable public opposition, as is established in the literature on the political economy of policy. Even if a subsidy reform is a net positive gain for an economy, certain sectors or groups may lose out. This fact is limited not just to firms in the agricultural, energy, or fishery sectors, but can extend to their value chains, workers, or geographic regions. Ignoring these concerns and interests can undermine the feasibility and credibility of reform efforts. Hence, providing support or compensation for certain industries may be inevitable, even if deemed unfair (box 11.2).

This report has also highlighted how subsidy removal can represent significant losses to the rich (chapter 5). When subsidy programs are regressive, in absolute monetary terms, the highest income groups stand to lose the most when subsidy payments are stopped. In such cases, it comes as no surprise that high-income households have a strong interest in maintaining subsidy schemes and opposing reform or rendering the changes ineffectual. Indeed, survey evidence from Indonesia illustrates that the highest income group is twice as likely to oppose fuel subsidy removal as the lowest income group (figure 11.1).

BOX 11.2
When compensation paves the way for reform: Murray-Darling Basin, Australia

The water reform undertaken in the Murray-Darling Basin, Australia, since 2007 is considered a model response to water insecurity. In this agricultural heartland, the reform involved a major shift away from a command-and-control system of water allocation that assumed no environmental limits and toward a new market-based system that acknowledges sustainable resource limits.

At the heart of the reform was the 2007 National Plan. A critical first step included setting diversion limits for surface and groundwater. These limits effectively capped extraction from strained systems, which then enabled widespread water trading. This transformation required the conversion of existing, ill-defined water rights into secure, long-term, tradable entitlements and the gradual development of water trading rules and water markets.

Existing irrigators received compensation when entitlements were reduced. This compensation was facilitated through a generous funding package of $A 10 billion, which included provisions to compensate those who may have lost entitlements. Indeed, it went considerably further by paying for significant amounts of on-farm irrigation infrastructure with the stated aim of "addressing once and for all water overallocation." The upshot was wide acceptance of the reforms, despite many misgivings and concerns. It is more than likely that, without such generous compensation, resistance would have stymied reform efforts (irrespective of equity considerations). Regardless of whether the reform will achieve its long-term objectives, compensation made it possible in the first place.

FIGURE 11.1 **Opposition to subsidy reform in Indonesia, by household consumption decile**

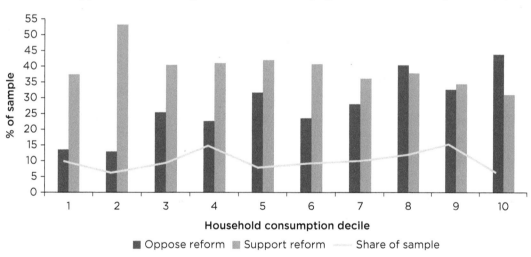

Source: Kyle 2018, © SAGE Publications. Reproduced with permission from SAGE Publications; further permission required for reuse.

The issue is exacerbated because subsidy schemes are often used to exert political influence by conferring favors—for instance, to key voters or constituents in politically critical regions. Even if these groups represent a minority, their political clout is likely to influence policy design considerations in practice. In such cases, policy makers may be forced to compromise on the efficiency of reform design in order to gain the support of these groups. A large literature identifies a link between spikes in deforestation and political cycles in Asia and Latin America and the Caribbean, as concessions in forests for beef

Opposition to
subsidy reform is
particularly high
when corruption is
perceived to be
high. Here,
promises of
compensation and
redistribution can
lack credibility.

and palm oil are conferred as gifts in exchange for votes (Cisneros, Kis-Katos, and Nuryartono 2021; Pailler 2018).

As with all economic rents, subsidies tend to breed rent seeking. In some cases, subsidy programs have given rise to opaque systems of public administration and misappropriation (Kojima 2016). In countries with poor governance and widespread corruption, citizens are far less trusting of public policy promises and the capacity of institutions to deliver in the public interest.

There is strong empirical evidence that opposition to subsidy reform is particularly high in regions where corruption is perceived to be high (Kyle 2018). When corruption levels are perceived to be low, poor households are more than two and a half times more likely to support rather than oppose fuel subsidy reform. However, when corruption is perceived to be high, support for the reform declines by 18 percentage points, and opposition increases by 14 percentage points. In short, public perceptions matter: to be able to support a subsidy reform, people need to have confidence that the proposed reform is indeed in their interests and that promised compensation payments will materialize when the subsidy is removed (box 11.3).

Moreover, understanding and mitigating the impacts of subsidy reform on the poorest and most vulnerable groups are crucial—not only to protect livelihoods and ensure a pro-poor reform, but also to galvanize support from these groups. Spatial inequalities, social marginalization, and low-income status may cause certain population groups to suffer disproportionately from subsidy reforms. For instance, subsidy removal could constitute significant shocks to the disposable income of low-income taxi drivers in cities or artisanal fishing communities. Identifying these groups and devising adequate compensation and social protection schemes are vital.

Failing to understand the needs and impacts of reform on each population group can spark dissent, which can cause reform to fail. Subsidies are often considered part of the social contract between government and the governed. Any perceived unfairness in the reform process can feel like a violation of that contract, leading to political unrest.

BOX 11.3
When winners feel like losers: Public perceptions can drive opposition to subsidy reform

The case of El Salvador illustrates that winning public trust is key for implementing subsidy reforms (Calvo-Gonzalez, Cunha, and Trezzi 2015). In 2011, the government implemented a reform to the gas subsidy that increased the welfare of households in all but the top two deciles of the income distribution. However, the reform turned out to be unpopular, especially among lower-income groups—the intended winners of the reform. Using household surveys, the study shows that misinformation (a negativity bias by which people with limited information infer negative consequences), mistrust of the government's ability to implement the policy, and political priors explain most of the dissatisfaction before implementation. Perceptions improved gradually—and significantly—over time when the materialized benefits of the reform induced households to update their initial expectations. Therefore, even well-designed pro-poor subsidy reforms may not be perceived as such by beneficiaries, and reform strategies need to address such information constraints and prior perceptions.

Nigeria's decision to remove subsidies on fossil fuel imports in 2012 illustrates the immense political challenges of subsidy reforms. After fuel prices more than doubled, extensive strikes and public protests prompted the government to reintroduce subsidies immediately (for further details, see Bazilian and Onyeji 2012; Siddig et al. 2014). Likewise, India's attempted agricultural subsidy reforms in 2020 ended in a similar reversal after public backlash (box 11.4). Rentschler (2016) illustrates how urban regions with high energy dependency can face particularly high poverty due to subsidy reform and thus become hotspots of public opposition and protests. Subsidy reforms are bound to fail if they ignore the needs of the poor.

BOX 11.4

Lessons from India: The complexities of reforming agricultural subsidies and protectionism

In September 2020, India's national government passed the "farm laws" to alter the degree of state regulation over the sale, movement, pricing, and storage of farm produce. The stated purpose was to deregulate agricultural markets and allow trade outside of regulated markets, facilitate free and unfettered interstate movement, and incentivize the entry of private corporations and investment. Although the assumption was that the private sector would pass the gains on to farmers, boosting their incomes, farmers feared that the reform would lead instead to elite capture by large corporates and destroy rural livelihoods (Chatterjee and Krishnamurthy 2021). As a result, some state governments passed their own resolutions against the laws, with some pronouncing protectionist measures for their own farmers against farmers from other states, which led to greater regulatory ambiguity and uncertainty.

Farmers and farmer organizations expressed deep mistrust and anxieties over a potential loss of familiar systems of trade and livelihood. These concerns stemmed from historical experiences of unmet promises by successive governments, including the promise to provide a floor price for 23 commodities, easier access to credit, insurance, technical assistance, marketing infrastructure, and protection against the market power of larger firms (Chatterjee and Krishnamurthy 2021; Chatterjee and Mahajan 2021). Farmers were also concerned that, as sales outside of regulated market sites were not subject to taxes or regulation, they would stand to lose the public goods provided by the regulated markets, such as dispute resolution, use of proper weights, and timely payments (Chatterjee and Mahajan 2021).

Against this backdrop, protests against the reform laws erupted in most parts of the country and were particularly large in Haryana and Punjab. In these states, farmers stood to lose the most from the new laws (Chatterjee and Mahajan 2021), as public procurement of paddy and wheat is nearly universal, and farmers have received the floor price for rice and wheat, reliable credit, and timely payments through commission agents for half a century. In states where public procurement is more limited, or farmers depend less on intermediaries for credit, or regulated markets are absent, the protests were fairly muted and revolved largely around misgivings related to the ability of large corporations to capture markets (Chatterjee and Mahajan 2021; Johari 2021). The "farm laws" were ultimately repealed in late 2021, ending one of the longest episodes of attempted regulatory overhaul in the country's history of agricultural law and policy (Krishnamurthy 2021).

The attempted reforms highlight the need to establish policy credibility and the consequences of ignoring the vast heterogeneity in the structural conditions and realities across regions. The result was to push a well-intentioned but poorly designed reform package that failed to overcome distrust and misgivings.

When fiscal pressures mount, subsidies become unaffordable public expenditures. Quickly removing subsidies can appear attractive as a quick fiscal relief measure, but policy makers need to consider the far-ranging consequences that subsidy reform may have. Reforms of subsidies and pricing mechanisms can have significant transition costs, particularly in the short term, while the economic system adjusts to a subsidy-free environment. Anticipating and tackling such effects can increase the chances of successful reform:

- *Employment impacts.* When subsidies are large and prop up entire sectors of the economy, removing them is bound to cause structural pain, induce sectoral shifts in the economy, and lead to job losses. Knock-on effects can affect firms along supply chain networks or force migration decisions in affected communities. It can also create jobs in the longer term—for example, as investments increase in energy systems, renewable energy, and energy efficiency—but these jobs may not match the workforce's prevailing skills profile.

- *Macroeconomic impacts.* Inflationary impacts and the associated affordability challenges of subsidy reforms can be significant, albeit usually temporary. Food price changes are salient and affect low-income households disproportionately, as they tend to spend a larger share of their income on food than higher-income groups. For most sectors, energy goods are critical inputs to production, and cost increases may be passed on to consumers, which can reduce aggregate demand.

- *Reduced competitiveness.* Subsidy removal can reduce competitiveness and cause firm exits, especially in industries that were artificially propped up by subsidies. For example, higher fuel and electricity prices may necessitate costly investments in energy and material efficiency, affecting manufacturing costs and output. However, subsidies can also reduce domestic efficiency and competitiveness on the international stage, so that subsidy removal is likely to strengthen competitiveness in the longer run (Mgeni et al. 2018).

- *Substitution with inferior goods.* Removing subsidies may also force people and firms to shift to less expensive, but inferior goods. For example, the reduction of liquefied natural gas may force poor households to shift toward cheaper, lower-grade energy sources, such as charcoal or kerosene for cooking and lighting, with adverse impacts on health and the environment. Similarly, removing agricultural input subsidies for high-yield seed varieties may force farmers to revert to unimproved seeds, reducing yields and productivity.

- *Hardship for the poor and vulnerable.* Subsidy reforms risk inflicting significant hardship on poor and vulnerable groups, who might depend heavily on subsidies, even when these subsidies are highly regressive. As seen in chapter 7, although the poor are often less likely to participate in input subsidy programs and derive disproportionately smaller benefits from them, those benefits still make up a significant share of income for participating households.

Crucially, these adverse effects (and their associated political challenges) may vary significantly, depending on the type of subsidy.

Navigating political complexity: Six principles for effective reforms

When subsidy programs fail to meet their objectives and entail large societal costs, the question is not *whether* to reform subsidies, but *how*. The collective experience of governments around the world can help to guide the design and implementation of subsidy reforms, even in challenging political economy contexts. This section presents six overarching principles for designing subsidy reform that have proven to enhance the chances of success, illustrated with evidence from previous chapters and past reform experiences.

What makes a subsidy reform successful? At the very minimum, *reform success* should entail the permanent repurposing of harmful subsidies in a way that mitigates major socioeconomic disruptions while protecting vulnerable livelihoods. A more extensive notion of *success* may include the successful removal of the subsidy alongside a comprehensive set of reform measures to ensure that reform contributes to a country's long-term sustainable development objectives and does not simply offer financial relief.

As summarized in figure 11.2, subsidy reform is not merely about removing subsidies; to ensure effectiveness and long-term sustainability, a comprehensive strategy encompassing six key principles is needed:

- *Comprehensive assessment of existing subsidies that goes beyond conventional cost-benefit analyses* is key to understanding their magnitude, adverse side effects (including environmental externalities), distributional incidence, and potential costs and benefits.

- *Building public acceptance* is a prerequisite for reform, especially when there are high risks of political interest groups derailing reform efforts. Effective communication and transparency are key to addressing the trust deficits, which may detract from the credibility of assurances to address the adverse consequences of reform.

- *Social protection and compensation* are an imperative, especially in the short run, in all contexts where subsidy removal may threaten the livelihoods of vulnerable groups and increase poverty.

- *Sound strategies for reinvesting reform revenues* can ensure that subsidy reforms deliver on development priorities, such as health and education. Even if reinvestment strategies are adjusted later on, formulating them early can lend credibility to the public good objectives of subsidy reform.

- *Targeted complementary measures* may be necessary when price-based instruments (such as subsidy reform) alone are insufficient to solve environmental externalities. Improving public transit can facilitate switching away from fossil fuels; laws and regulations can protect critically endangered natural capital; and capacity-building programs can enhance the efficiency of subsidy reforms.

- *Carefully sequenced and coordinated reforms* can reduce the disruption from large price shocks due to a one-off subsidy removal and enable households and firms to adjust gradually. Coordination between government agencies is crucial to ensure that different elements of reform are complementary and aligned.

FIGURE 11.2 Six principles for successful subsidy reform: A package of carefully planned measures

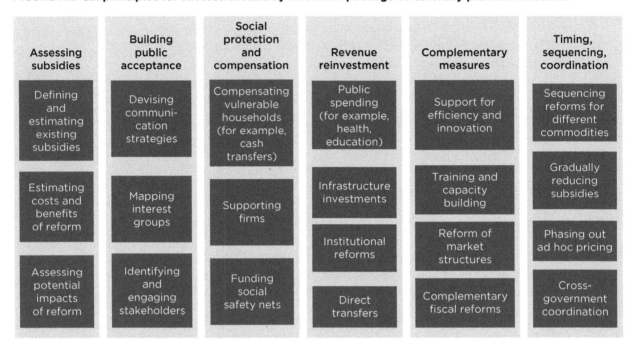

Assessing subsidies	Building public acceptance	Social protection and compensation	Revenue reinvestment	Complementary measures	Timing, sequencing, coordination
Defining and estimating existing subsidies	Devising communication strategies	Compensating vulnerable households (for example, cash transfers)	Public spending (for example, health, education)	Support for efficiency and innovation	Sequencing reforms for different commodities
Estimating costs and benefits of reform	Mapping interest groups	Supporting firms	Infrastructure investments	Training and capacity building	Gradually reducing subsidies
Assessing potential impacts of reform	Identifying and engaging stakeholders	Funding social safety nets	Institutional reforms	Reform of market structures	Phasing out ad hoc pricing
			Direct transfers	Complementary fiscal reforms	Cross-government coordination

Source: World Bank.

Assessing subsidies and pricing mechanisms: Defining, identifying, and evaluating reform

Subsidy programs often evolve into complex and opaque policy schemes that make it difficult for policy makers to identify and quantify existing subsidies precisely. As preceding chapters have shown for energy, agriculture, and fishery subsidies, the definitions and estimates of subsidies can vary significantly. However, conflating different types of government interventions into the same subsidy "basket" can result in generic and inadequate reform attempts.

For example, removing subsidies on different types of fuel—such as petrol and kerosene—can have significantly different distributional and environmental impacts (chapter 5). Temporary natural gas subsidies may even yield positive air pollution outcomes, as they can support the switch from inferior fuels like charcoal to cleaner ways of cooking and heating. Similarly, different types of agricultural output subsidies can have different impacts on economic efficiency (chapter 7). So, before designing a reform, policy makers should conduct a thorough assessment to understand the types and magnitude of subsidies and whether and to what extent subsidies are harmful and inefficient. For instance, Kojima (2018) offers a practical guide to defining and measuring energy subsidies. For agricultural subsidies, the Organisation for Economic Co-operation and Development (OECD) shows how to estimate a country's agricultural support schemes using public accounts (Melyukhina and Ilicic-Komorowska 2016).

In addition to estimating the magnitude of subsidies, it is crucial to simulate the likely impacts of subsidy removal on different groups—in particular, low-income households and other disadvantaged groups. A thorough assessment of the likely effects of reform is a central element of advance planning. Previous chapters have demonstrated various

MAP 11.1 **Spatially disaggregated simulations of the poverty impacts of subsidy reform in Nigeria, with and without compensation**

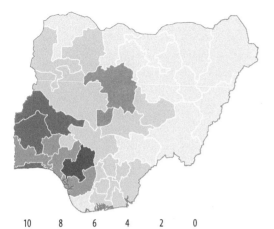
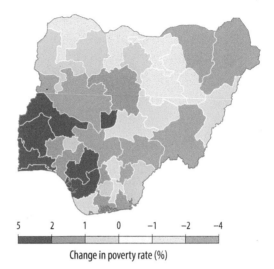

a. Without compensation

Uncompensated poverty rate increase (%)
10   8   6   4   2   0

b. With uniform cash transfer compensation

Change in poverty rate (%)
5   2   1   0   −1   −2   −4

IBRD 46907 | NOVEMBER 2022

Source: Rentschler 2016.

approaches to conducting ex ante impact assessments of reforms and illustrated that they can provide practical guidance for policy makers seeking to estimate the potential impacts of reform.

Econometric approaches based on household or firm survey data are suitable for obtaining a disaggregated and granular picture of micro-level effects. Spatially disaggregated approaches can help to unmask national averages, identify regions that would be particularly affected by subsidy reforms, and offer targeted compensation measures to avoid an increase in poverty (map 11.1).

Computable general equilibrium (CGE) models are powerful tools for conducting policy simulations and understanding the systemic interlinkages at the macro level and for understanding the role of illicit activities, such as tax evasion and fuel smuggling (Rentschler and Hosoe 2022; Taheripour et al. 2022). Using a CGE model, Gautam et al. (2022) examine the impact of changing agricultural support policies on global greenhouse gas emissions.

Managing political support by interest groups and the wider public

To build public support for subsidy reform, consistent and credible communication strategies are crucial. These strategies need to be complemented by meaningful engagements with key stakeholders, including influential interest groups as well as vulnerable communities for whom compensation and social protection schemes are essential.

To generate public support for subsidy reforms, transparent, credible, and consistent communication and public outreach campaigns play a central role (Worley, Pasquier, and Canpolat 2018). Such campaigns should detail the reasons for reform and highlight why they are in the public's interest. They should specifically address the public's concerns—for instance, with respect to livelihoods or affordability—and offer realistic plans for

When influential
interest groups
hold outsized
bargaining powers
to oppose reforms,
second-best
compromises can
help to get reforms
off the ground in
the public interest.

mitigating the adverse effects. When public trust in institutions is low, the margin for error is small; once compensation or benefit schemes are announced, it is crucial to deliver them as announced, otherwise trust is eroded further. If communication is not credible or not followed by tangible actions, past experience has shown that reforms could face significant resistance. And communication is most effective when it works two ways: comprehensive consultations with disproportionately affected communities can inform the design of reforms, while also engaging communities and their leaders.

The benefits of subsidy systems often accrue to large corporations, which will fight to keep subsidies in place. For instance, estimates suggest that less than 20 percent of the US$35.4 billion of global fishery subsidies in 2018 went to the small-scale fishing subsector (such as artisanal and subsistence fisheries). In contrast, more than 80 percent went to large-scale industrial fishing firms. Indeed, a fisher in the industrial fishing sector receives 3.5 times more subsidies, on average, than one in small-scale fishing (Schuhbauer et al. 2020). Reforming subsidies often requires policy makers to engage with powerful interest groups, such as industrial lobbies and associations, and win their support or—at least—acceptance.

The role of such power groups is, of course, not limited to fisheries. Suppliers of seeds and fertilizers may have an interest in maintaining subsidy programs. Energy-intensive industries and fossil fuel producers are typically staunch opponents of fossil fuel subsidy reforms. When such actors hold outsized bargaining powers and threaten to derail subsidy reform efforts that are in the public interest, then policy makers may need to seek second-best compromises that can deliver on most of the public benefits of reform, but be willing to make some concessions as the price of implementation (box 11.5).

Social protection and compensation: Mitigating short-term price shocks on vulnerable households and firms

Compensating vulnerable households is crucial for ensuring social stability and public support for reform. Disregarding their needs and endangering their livelihoods can result in nationwide unrest, especially in countries where they account for a large share of the population. Indeed, removing subsidies may have detrimental effects on the livelihoods of the poor and other vulnerable groups. Although poor farmers receive relatively low shares of input subsidies, those subsidies can constitute a large share of their income (chapter 7). A similar effect is seen in fishing—for example, artisanal fleets in Mauritania

---

BOX 11.5
Credible compensation before subsidies are removed

In December 2010, the Islamic Republic of Iran implemented subsidy reforms, increasing the price of petroleum products by 230–840 percent. But before removing the subsidies, the government started making monthly cash payments to 70–80 percent of all citizens, later increasing its targeting to the poor and vulnerable and facilitating transfers by opening bank accounts for heads of households. Overall, this structured cash transfer scheme and its timely implementation are two reasons for public support of the country's 2010 fossil fuel subsidy reform (Salehi-Isfahani, Wilson Stucki, and Deutschmann 2015).

were disproportionately harmed when subsidies were removed, while all other fleets tended to benefit (chapter 10). Fossil fuel subsidies may constitute a significant share of the poor's overall income, even if this group accounts for a small fraction of total energy demand (chapter 5).

In practice, policy makers need to reconcile the need for subsidy reforms with the imperative of ensuring social protection and maintaining affordability (for instance, see Ruggeri Laderchi, Olivier, and Trimble 2013; Salehi-Isfahani, Wilson Stucki, and Deutschmann 2015). In particular, direct cash transfers have been an important component of many successful subsidy reforms. For example, in a study of a series of attempts to reform energy subsidies in the Middle East and North Africa, Sdralevich et al. (2014) note that all of the reforms that used cash and in-kind transfers—nonmonetary benefits, such as food vouchers or access to free services—were successful, while only 17 percent of those that did not entail cash transfers were successful. Cash transfers are often considered central elements of social protection and revenue redistribution mechanisms. They are a flexible and progressive alternative to subsidies that can increase aggregate welfare and protect livelihoods. But trade-offs may occur when balancing the need to secure public support—for example, by compensating heavy users of energy or fertilizers (that is, the rich)—with the imperative of protecting the livelihoods of the poor.

The effectiveness with which cash transfers can be disbursed also depends on a country's social protection infrastructure (Rentschler 2018). If a government maintains systematic records of the beneficiaries of existing social protection schemes, then compensatory cash transfers can be administered more easily. Several countries already have large-scale cash transfer systems, such as Brazil's Bolsa Familia and Mexico's Prospera, and can leverage these systems to channel compensation for subsidy reforms. But in certain circumstances—for example, if financial inclusion rates are low—in-kind transfers may be easier and quicker to administer than cash transfers. IMF (2013b) discusses previous energy subsidy reforms that have targeted such in-kind compensation measures to the most vulnerable households, including providing gas vouchers in Brazil (2002), providing college scholarships and rice subsidies in the Philippines (1998), and strengthening social safety nets in Armenia (1995–99). These approaches can help people to cope with price shocks.

> The effectiveness with which cash transfers can be disbursed also depends on a country's social protection infrastructure. In some cases, in-kind transfers can be an effective alternative to cash transfers.

Compensation policies also need to recognize potentially significant heterogeneity in vulnerability across regions and within income groups (box 11.6). Chapter 5 has shown that, at the same income level, urban households tend to be more energy-dependent, thus incurring larger losses from subsidy removal than rural households. Indeed, the impact of adverse price shocks due to subsidy removal varies significantly—not just across income groups—but also across types of fuel, geographic regions, and occupations. For example, farmers in regions with inferior types of soil may be more dependent on subsidized fertilizer than those in regions with highly fertile soil. The nature, location, and extent of political economy challenges and political opposition can vary for the reform of different subsidies; as such, tailored compensation measures—for example, flexible cash transfer schemes—are required.

Lessons from Nigeria: National averages hide vulnerable population groups

Evidence from Nigeria suggests that some low-income households' reliance on fossil fuel subsidies may make them far more vulnerable to experiencing a substantial income shock when subsidies are removed (Rentschler 2016). While in most states the poorest households consume very little kerosene, in several southern states kerosene consumption by the poorest is significantly above the average for their income group. These regional differences may reflect issues such as differences in the type of employment, access to energy, and availability and affordability of alternative fuels. During Nigeria's 2012 attempted reform of energy subsidies, inadequate attention was paid to the needs of low-income households, resulting in public protests and fierce opposition. Public protests were concentrated in urban regions such as Abuja and Lagos, where low-income households are particularly dependent on fuel.

It is important to focus on the implications of fossil fuel subsidy reform for different income groups and regions and to tailor reforms and social protection measures to the needs of specific population groups. If policy makers focus only on national averages and use income level as the sole indicator of vulnerability, they may underestimate the vulnerability of certain groups and provide inadequate social protection for the poor. For example, blanket compensation that uniformly covers a large share of the population may provide adequate compensation, on average, but is likely to fail to protect particularly vulnerable households. Commonly, the vulnerability of population groups is determined based on their income status. However, other determinants of social marginalization can be even more important—for instance, the exclusion of women or ethnic minorities makes livelihoods particularly vulnerable to shocks.

Revenue reinvestment in the long term

By countering sudden increases in input prices, compensation measures help vulnerable households to manage the downside risks of subsidy reforms and can increase public acceptance of reforms in the short term. However, these short-term compensation measures cannot replace long-term strategies for reinvesting reform revenues toward development priorities, such as infrastructure, health, and education. Not only are transparent reinvestment strategies crucial for maximizing the long-run contribution of reform to development, but they also send an important signal to the public, who will seek reassurance that reform revenues are used in the public interest.

Subsidies are often substantial relative to gross domestic product (GDP), so subsidy reform plans are more credible if policy makers have transparent and prudent strategies for allocating revenues to development priorities. Depending on a country's specific needs, these priorities could mean investing in infrastructure—such as low-carbon electrification, public transit, digitization, or irrigation—or improved health care coverage, public education services, or institutional and tax reform.

For example, investments in education systems in Indonesia in 2008 and in social protection programs in Mauritania in 2011 all aimed to reinvest funds unlocked from subsidy reforms and thus contribute to longer-term economic development (IMF 2013a, 2013b).

> Short-term compensation measures cannot replace long-term strategies for reinvesting reform revenues toward sustainable development.

Reform revenues could also be used to strengthen the productivity and competitiveness of growth sectors, including renewable energy, climate-smart agriculture, or sustainable aquaculture. Complementary tax reforms can also be a way to redistribute reform revenues in a way that improves economic efficiency (Fullerton and Metcalf 1997; Rentschler and Hosoe 2022).

Complementary measures

Governments cannot assume that reducing subsidies and increasing the price of harmful activities will automatically trigger large environmental benefits. There may still be significant barriers—such as information, capacity, or financial constraints, infrastructure, fiscal mismanagement, market structures, systemic risks, or uncertainty (including the long-term credibility of the fiscal policy) (table 11.3)—that make households and firms unable or unwilling to adjust their behavior or invest in more efficient technology.

Even though the removal of environmentally harmful subsidies is a crucial step toward accounting for externalities (such as natural capital degradation), complementary policies can be crucial for ensuring that price-based measures are effective and publicly acceptable. Such active support measures—in some cases, "good" subsidies—can help to smooth and accelerate the transition away from polluting technologies and behaviors.

> Complementary policies can be crucial for ensuring that price-based measures are effective and publicly acceptable.

In line with the theory of second best, such complementary measures can counteract existing market distortions and achieve a more efficient outcome than a theoretically optimal (but practically infeasible) first-best approach on its own. Measures to complement subsidy reforms are especially important when barriers result in costly frictions that are likely to prevent swift transitions. They can take various forms, depending on the sector and policy objective:

- *Agriculture.* Increasing access to credit to enable farmers to purchase fertilizer and machinery; providing incentives to protect natural land areas, such as payment for ecosystem services; investing in public goods such as infrastructure, research and development, and agricultural extension facilities; providing training and information programs on improved farming practices; and providing financing to enable the adoption of energy- or water-efficient practices

- *Air pollution.* Regulating air pollution; mandating public good technologies, such as air filters in vehicles and higher-quality fuels; mandating fuel efficiency standards; supporting feebate programs; engaging in information and capacity-building programs to make efficient clean technologies widely available; providing financing to facilitate investments in clean, efficient machinery; supporting clean cooking and heating programs; investing in public transport to enable switching from private vehicles (box 11.7); removing trade and intellectual property barriers to encourage the diffusion of technology to low- and middle-income countries

- *Fisheries.* Establishing and enforcing marine protected zones; regulating catch methods, such as bottom trawling; establishing quota systems for certain species; investing in sectoral diversification of coastal regions; improving cold storage and processing to increase catch value and reduce waste.

BOX 11.7
A fuel subsidy reform clears the air: Experience from Cairo, the Arab Republic of Egypt

Fuel subsidy reforms have the potential to cause a measurable reduction in toxic levels of air pollution, as shown in an innovative study of Cairo using machine learning algorithms and high-resolution satellite imagery (Heger et al. 2019).

In recent years, the government of Egypt implemented several important environmental policy measures. Supported by the World Bank, it lifted fossil fuel subsidies, which resulted in several publicly mandated fuel price increases between 2016 and 2018. This reform was complemented by new regulations, adopted in 2017, on the forced retirement of old vehicles as well as the opening of a third metro line that significantly improved public transit connectivity across Cairo.

The evaluation by Heger et al. (2019) developed a novel data set using satellite imagery to detect and count nearly every moving vehicle in the streets of Cairo on about 1,000 days during the period 2010 to 2018. Impact evaluation methods were then applied to estimate the effect of opening the metro line and slashing fuel subsidies on urban air pollution.

The immediate effect of these policy measures was directly visible in the number of cars on the road (map B11.7.1). For both measures, the number of cars on the streets of Cairo was reduced visibly. Given that adjustments can take time to materialize, car numbers are likely to decrease even further as more time passes. The results show that removal of the fuel subsidy helped to reduce PM_{10} concentrations by about 4 percent. The opening of Cairo's Metro Line 3 further reduced air pollution by about 3 percent.

This exercise yields several important lessons. The air pollution benefit of removing fuel subsidies is more pronounced when coupled with investments in public transit infrastructure. Subsidy reform also yields other benefits, including reducing congestion and pollution-related diseases. However, the significant but moderate reductions in air pollution also highlight that subsidy reform alone can, at best, reduce, but not solve, the challenge of air pollution.

MAP B11.7.1 Spatial distribution of changes in the number of cars after opening the Metro Line 3 and the November 3, 2016 fuel price increase in Cairo, the Arab Republic of Egypt

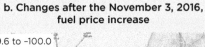

a. Changes after the opening of Metro Line 3

b. Changes after the November 3, 2016, fuel price increase

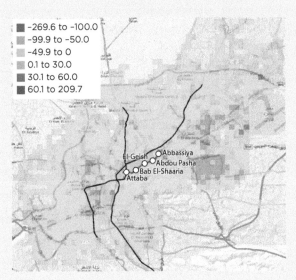

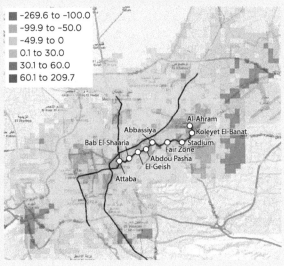

Source: Heger et al. 2019.

In short, by implementing such complementary measures, policy makers can ensure that subsidy removal is not simply a fiscal relief measure, but also makes a positive contribution to wider sustainable development objectives. Broadly speaking, such complementary measures aim to ensure that the necessary technologies are available and affordable to facilitate the desired transition and address the biases and barriers that prevent behavioral change. In doing so, subsidy reform strategies can enhance the effectiveness of price-based instruments, while galvanizing broader support from a range of stakeholders.

Timing, sequencing, and coordination of reform measures

Whether a subsidy reform succeeds can depend on *when* it is undertaken. The potential for reform to succeed depends, to a large extent, on external factors, such as the prevailing market price of fertilizers, food, or fuel, as well as political circumstances. Getting the timing right is vital. For example, although falling fossil fuel prices may temporarily ease the fiscal pressures of subsidies, they also provide an opportunity for governments to eliminate these subsidies. India's and Indonesia's 2014 fuel subsidy reforms highlight how political will for reform, paired with low oil prices, can pave the way for a smooth implementation of energy subsidy reform (Benes et al. 2015). Both cases were preceded by a thorough process of planning and preparation. In turn, when oil prices are high, reform becomes both urgent and difficult.

The timing of agricultural subsidy reforms can also be critical. Farmers often need to make up-front purchases of seeds, fertilizers, pesticides, and other inputs several months in advance of the harvest. Removing such subsidies could prevent farmers from buying the inputs they need to maximize production. Thus, while input subsidies can be more distortive than output subsidies, they are also more important in the presence of credit constraints. Output subsidies can also have varying impacts on farmers' decisions. When trade barriers raise domestic prices, farmers may not be aware of the magnitude of that impact until it comes time to sell the harvest. Minimum support prices, in contrast, send a clear signal to farmers on how much they will benefit and are often known before planting decisions are finalized. Thus farmers have time to react to future price changes and will change their planting decisions accordingly. These differences in uncertainty will affect how much the subsidy (and its reform) affects farmers' planting decisions.

Timing is crucial not only for determining *when* but also *how* to reform. The Global Studies Institute (GSI 2013) underscores the importance of avoiding large sudden price shocks and recommends reducing subsidies gradually. Nigeria's unsuccessful attempt at reforming fossil fuel subsidies in 2012 illustrates why. With little advance notice, the government removed subsidies entirely, causing a one-time 117 percent increase in fuel prices. And due to limited advance communication, the population was unaware of the benefits of removing subsidies or the planned compensation programs, so intense public protest ensued (Bazilian and Onyeji 2012; Rentschler 2016). Such large, one-off price increases can have immediate damaging effects on poverty levels, illustrating why this approach carries significant risk. Likewise, a one-off removal of agricultural or fishery subsidies can result in a large disruptive price shock, especially when food prices or input prices—for example, for fertilizer or ship fuel—are high. Such disruptions can carry significant risks of public backlash and lead to failure.

Besides (step-wise) subsidy removal, effective reform also depends on the careful timing and cross-government coordination of complementary measures, such as communication, compensation, and revenue reinvestment (figure 11.3). Careful advance preparation

FIGURE 11.3 The timing and sequencing of components of fossil fuel subsidy reform

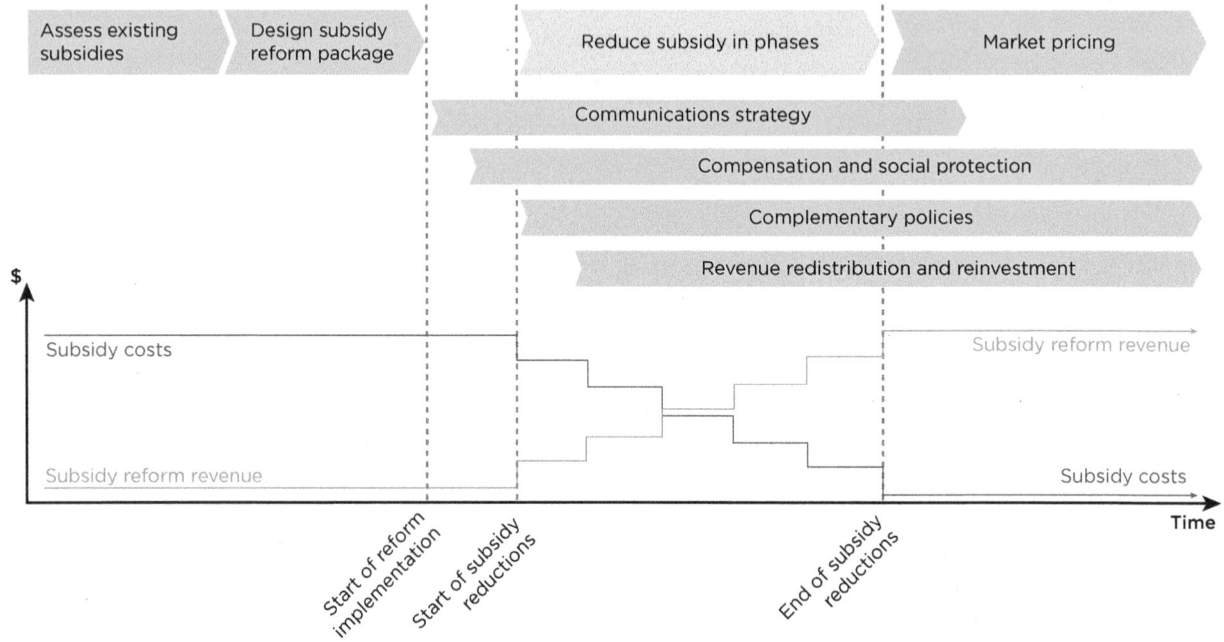

Source: Adapted from Rentschler 2018.

is key to identifying and quantifying existing subsidies and designing a comprehensive package of reform measures that can be initiated when conditions (for example, international market prices and political context) are suitable. Before subsidies are reduced, a coherent communication strategy is needed to address concerns and manage expectations—both of the public as well as of key stakeholders. Compensation and social protection payments need to be issued promptly—even before subsidies are reduced, if doing so helps to alleviate credibility concerns—and be continued for as long as required to protect vulnerable livelihoods. Complementary measures and revenue reinvestment measures should commence when subsidy reduction begins, as a way to affirm the government's commitment to the prudent use of reform revenues in the public interest. These investments could continue well beyond the end of the actual subsidy reductions—after all, ending a long-standing subsidy scheme frees up public budgets permanently, allowing policy makers to channel resources toward their sustainable development objectives. All of these measures require a well-coordinated effort by government agencies—at both the national and local levels (see box 11.8 for the experience of Mexico).

Subsidy reforms for sustainable development

As outlined throughout this report, "getting prices right" is widely regarded as at the heart of an effective market-based solution to addressing pervasive environmental externalities. At its essence, this approach requires that the social and environmental costs of environmentally harmful activities are reflected in their prices. However, the roughly US$1.25 trillion in explicit subsidies paid every year to the world's fossil fuel, agriculture, and fishery sectors have the polar opposite effect: they incentivize the overconsumption of polluting

BOX 11.8
Fossil fuel subsidy reform in Mexico

The government of Mexico historically set domestic fuel prices independent of international market prices. After a decade of low international oil prices, international prices rose sharply between 2005 and 2013, drastically increasing the subsidy burden. In 2011, fossil fuel subsidies were equivalent to 2 percent of gross domestic product (GDP): 1 percent were for gasoline and diesel, and the rest were for electricity and gas (Arlinghaus and van Dender 2017). These subsidies represented a large part of Mexico's public expenditure, significantly exceeding spending on health or social programs (Muñoz-Piña, Montes de Oca, and Rivera-Planter 2011). As with most fossil fuel subsidy programs, these subsidies were highly regressive. For each peso (Mex$1) reaching the poorest two deciles of the population, Mex$19.50 reached the wealthiest two deciles (Arlinghaus and van Dender 2017).

Recognizing that subsidizing fuels was not an efficient use of fiscal resources, in 2012 the government of Mexico adopted a policy of monthly increases in the price of gasoline and diesel. Although the increases were small, less than US$.01 per month, their consistent application allowed the phaseout of fuel subsidies in 2014. In addition, the government of Mexico adopted two measures that further reformed fuel pricing. One was liberalization of the fuel market, allowing prices to reflect the true import and distribution costs of fuels (Mexico Law on Hydrocarbons of 2014). As part of these reforms, Mexico's excise tax on fossil fuels was set as a specific amount per liter, instead of the historic percentage-based endogenous residual tax, to reduce the risk of returning to generalized fossil fuel subsidies during periods of international price volatility (Harding, Vollebergh, and Sen 2016). The second policy was a carbon tax to incentivize efficient fuel use, even when prices were low.

Due to the combined effect of the subsidy phaseout and the introduction of a carbon tax, net subsidies to fossil fuels in Mexico fell in 2014 to less than 0.28 percentage point of global GDP, nearly one-sixth of what they had been in 2008 (INEGI 2022; Secretaría de Hacienda y Crédito Público 2022). As domestic gasoline prices were maintained in real terms while international prices kept falling, the country moved into positive excise taxation for gasoline and diesel for several months. By January 2016, Mexico was collecting fiscal revenues from gasoline equivalent to 1.4 percent of GDP (INEGI 2022; Secretaría de Hacienda y Crédito Público 2022). By 2015, the fuel tax was the third-largest revenue tax after value added tax and income tax.

The phasing out of fossil fuels also had significant health and environmental benefits. World Bank (2013) estimates that phasing out fossil fuels would result in the annual abatement of 41.7 million tons of carbon dioxide between 2012 and 2018. Subsidy removal is also likely to have reduced driving and thus reduced air pollution. However, even in 2019, after subsidies had been phased out, air pollution was an immense local problem. That year, air pollution caused more than 46,000 premature deaths in Mexico (World Bank 2022). Nationally, the health costs of ambient air pollution were equivalent to 4 percent of Mexico's GDP (World Bank 2022).

inputs and the degradation and exploitation of valuable natural capital, undermining the effectiveness of efforts to address sustainable development. This report has unearthed and provided estimates of many of the hidden consequences of subsidies. Hence, reforming environmentally harmful subsidies is critical—albeit not sufficient per se—for achieving the United Nations Sustainable Development Goals. Indeed, subsidy reform not only removes distorted incentives that undermine countries' ability to make progress toward these goals, but also can unlock significant domestic financing to facilitate and accelerate sustainable development efforts.

Nevertheless, when subsidy reforms are implemented in practice, environmental objectives are rarely the primary motivation. Instead, the rationale for subsidy reforms is determined in a complex—and sometimes conflicting—context of fiscal, macroeconomic, political, and social factors. Experience from past subsidy reforms shows that governments have focused especially on the fiscal dimension of subsidy reform, which means relieving public budgets by removing subsidies and avoiding public opposition by compensating the losers of reform. However, by focusing solely on managing the downside risks of reforms, policy makers are likely to miss out on the full opportunity at hand. As this chapter has argued, complementary measures and prudent reinvestment of reform revenues can ensure that subsidy reforms provide not only short-term relief during fiscal emergencies, but also serve as a fully integrated component of a long-term sustainable development strategy. Repurposing harmful subsidies is often an urgent first step toward addressing environmental crises, but rarely is sufficient to solve them fully.

Note

1. For more detailed discussion of the political economy challenges of fossil fuel subsidy reform, see Cheon, Lackner, and Urpelainen (2015); Fattouh and El-Katiri (2013); Kojima, Bacon, and Trimble (2014); Koplow (2014); Rentschler (2018); and Strand (2013).

References

Arlinghaus, J., and K. van Dender. 2017. "The Environmental Tax and Subsidy Reform in Mexico." OECD Taxation Working Paper 31, OECD Publishing, Paris. http://dx.doi.org/10.1787/a9204f40-en.

Avner, P., J. Rentschler, and S. Hallegatte. 2014. "Carbon Price Efficiency Lock-In and Path Dependence in Urban Forms and Transport Infrastructure." Policy Research Working Paper 6941, World Bank, Washington, DC.

Bazilian, M., and I. Onyeji. 2012. "Fossil Fuel Subsidy Removal and Inadequate Public Power Supply: Implications for Businesses." *Energy Policy* 45 (June): 1–5.

Benes, K., A. Cheon, J. Urpelainen, and J. Yang. 2015. "Low Oil Prices: An Opportunity for Fuel Subsidy Reform." Policy Paper, Columbia Center on Global Energy Policy, New York.

Bennear, L. S., and R. N. Stavins. 2007. "Second-Best Theory and the Use of Multiple Policy Instruments." *Environmental and Resource Economics* 37 (1): 111–29.

Calvo-Gonzalez, O., B. Cunha, and R. Trezzi. 2015. "When Winners Feel Like Losers—Evidence from an Energy Subsidy Reform." Policy Research Working Paper 7227, World Bank, Washington, DC.

Chatterjee, S., and M. Krishnamurthy. 2021. "Farm Laws Versus Field Realities: Understanding India's Agricultural Markets." *Sikh Research Journal* 6 (1).

Chatterjee, S., and A. Mahajan. 2021. "Why Are Indian Farmers Protesting the Liberalization of Indian Agriculture?" *ARE Update* 24 (5): 1–4.

Cheon, A., M. Lackner, and J. Urpelainen. 2015. "Instruments of Political Control: National Oil Companies, Oil Prices, and Petroleum Subsidies." *Comparative Political Studies* 48 (3): 370–402.

Cisneros, E., K. Kis-Katos, and N. Nuryartono. 2021. "Palm Oil and the Politics of Deforestation in Indonesia." *Journal of Environmental Economics and Management* 108 (July): 102453.

Damania, R., S. Desbureaux, M. Hyland, A. Islam, S. Moore, A.-S. Rodella, J. Russ, and E. Zaveri. 2017. *Uncharted Waters: The New Economics of Water Scarcity and Variability.* Washington, DC: World Bank.

Fattouh, B., and L. El-Katiri. 2013. "Energy Subsidies in the Middle East and North Africa." *Energy Strategy Reviews* 2 (1): 108–15.

Fullerton, D., and G. E. Metcalf. 1997. "Environmental Taxes and the Double-Dividend Hypothesis: Did You Really Expect Something for Nothing?" NBER Working Paper 6199, National Bureau of Economic Research, Washington, DC.

Gautam, M., D. Laborde, A. Mamun, W. Martin, V. Piñeiro, and R. Vos. 2022. "Repurposing Agricultural Policies and Support." *Rural,* March 3, 2022.

GSI (Global Studies Institute). 2013. *A Guidebook to Fossil-Fuel Subsidy Reform for Policy-Makers in Southeast Asia.* Geneva: GSI.

Harding, M., H. Vollebergh, and S. Sen. 2016. "Energy Taxation in OECD Countries: Effective Tax Rates across Countries, Users, and Fuels." In *Economics and Political Economy of Energy Subsidies,* edited by Jon Strand, 41–59. Cambridge, MA: MIT Press.

Heger, M., D. Wheeler, G. Zens, and C. Meisner. 2019. *Motor Vehicle Density and Air Pollution in Greater Cairo: Fuel Subsidy Removal and Metro Line Extension and Their Effect on Congestion and Pollution.* Washington, DC: World Bank.

IMF (International Monetary Fund). 2013a. "Case Studies on Energy Subsidy Reform: Lessons and Implications." IMF Policy Paper, IMF, Washington, DC.

IMF (International Monetary Fund). 2013b. *Energy Subsidy Reform in Sub-Saharan Africa: Experiences and Lessons.* Washington, DC: World Bank.

INEGI (National Institute of Statistics and Geography). 2022. "Series desestacionalizada y de tendencia-ciclo del Producto Interno Bruto." INEGI, Mexico, DF. https://www.inegi.org.mx/temas /pib/.

Johari, A. 2021. "Beyond Punjab-Haryana: Meet Farmers Protesting against New Farm Laws across India." *Scroll,* January 11, 2021. https://scroll.in/article/983561/beyond-punjab-haryana-meet -farmers-protesting-against-new-farm-laws-across-india.

Kojima, M. 2016. "Fossil Fuel Subsidy and Pricing Policies: Recent Developing Country Experience." Policy Research Working Paper 7531, World Bank, Washington, DC.

Kojima, M. 2018. "Identifying and Quantifying Energy Subsidies: Energy Subsidy Reform Assessment Framework (ESRAF)." Good Practice Note 1, ESMAP Paper, World Bank, Washington, DC.

Kojima, M., R. Bacon, and C. Trimble. 2014. *Political Economy of Power Sector Subsidies: A Review with Reference to Sub-Saharan Africa.* Washington, DC: World Bank.

Koplow, D. 2014. "Global Energy Subsidies: Scale, Opportunity Costs, and Barriers to Reform." In *Energy Poverty: Global Challenges and Local Solutions,* edited by A. Halff, B. Sovacool, and J. Rozhon. Oxford, UK: Oxford University Press.

Krishnamurthy, M. 2021. "Farm Laws Debate Missed a Lot: Neither Supporters nor Modi Govt Identified the Real Problem." *The Print*, November 24, 2021.

Kyle, J. 2018. "Local Corruption and Popular Support for Fuel Subsidy Reform in Indonesia." *Comparative Political Studies* 51 (11): 1472–503.

Melyukhina, O., and J. Ilicic-Komorowska. 2016. *OECD's Producer Support Estimate and Related Indicators of Agricultural Support: Concepts, Calculations, Interpretation, and Use (The PSE Manual).* Paris: Organisation for Economic Co-operation and Development.

Mgeni, C. P., S. Sieber, T. S. Amjath-Babu, and K. D. Mutabazi. 2018. "Can Protectionism Improve Food Security? Evidence from an Imposed Tariff on Imported Edible Oil in Tanzania." *Food Security* 10 (4): 799–806.

Muñoz-Piña, C., M. Montes de Oca, and M. Rivera-Planter. 2011. "Subsidios a la gasolina y el diesel en México: Efectos ambientales y políticas públicas." Documento de Trabajo INE-ENER-DT-02-2011, Instituto Nacional de Ecología y Cambio Climático, Mexico, DF.

Pailler, S. 2018. "Re-election Incentives and Deforestation Cycles in the Brazilian Amazon." *Journal of Environmental Economics and Management* 88 (March): 345–65.

Rentschler, J. E. 2016. "Incidence and Impact: The Regional Variation of Poverty Effects due to Fossil Fuel Subsidy Reform." *Energy Policy* 96 (September): 491–503.

Rentschler, J. 2018. *Fossil Fuel Subsidy Reforms: A Guide to Economic and Political Complexity.* Abingdon-on-Thames: Routledge.

Rentschler, J., and N. Hosoe. 2022. "Illicit Schemes: Fossil Fuel Subsidy Reforms and the Role of Tax Evasion and Smuggling." Policy Research Working Paper 9907, World Bank, Washington, DC.

Ruggeri Laderchi, C., A. Olivier, and C. Trimble. 2013. *Balancing Act: Cutting Energy Subsidies While Protecting Affordability.* Europe and Central Asia Report 76820. Washington, DC: World Bank.

Salehi-Isfahani, D., B. Wilson Stucki, and J. Deutschmann. 2015. "The Reform of Energy Subsidies in Iran: The Role of Cash Transfers." *Emerging Markets Finance and Trade* 51 (6): 1144–62.

Schuhbauer, A., D. J. Skerritt, N. Ebrahim, F. Le Manach, and U. R. Sumaila. 2020. "The Global Fisheries Subsidies Divide between Small- and Large-Scale Fisheries." *Frontiers in Marine Science* 7 (September 29): 792.

Sdralevich, C., R. Sab, Y. Zouhar, and G. Albertin. 2014. *Subsidy Reform in the Middle East and North Africa: Recent Progress and Challenges Ahead.* Washington, DC: International Monetary Fund.

Secretaría de Hacienda y Crédito Público. 2022. "Recaudación e ingresos tributarios del gobierno federal ingresos por impuesto." SAT, Mexico, DF. http://omawww.sat.gob.mx/cifras_sat/Paginas /datos/vinculo.html?page=IngresosTributarios.html.

Siddig, K., A. Aguiar, H. Grethe, P. Minor, and T. Walmsley. 2014. "Impacts of Removing Fuel Import Subsidies in Nigeria on Poverty." *Energy Policy* 69 (June): 165–78.

Strand, J. 2013. "Political Economy Aspects of Fuel Subsidies: A Conceptual Framework." Policy Research Working Paper 6392, World Bank, Washington, DC.

Taheripour, F., M. Chepeliev, R. Damania, and J. Russ. 2022. "Employment and Emission-Reduction Priorities: Europe and Central Asia." Working paper, World Bank, Washington, DC.

World Bank. 2013. *The United Mexican States Reducing Fuel Subsidies: Public Policy Options.* Washington, DC: World Bank.

World Bank. 2022. *The Global Health Cost of PM$_{2.5}$ Air Pollution: A Case for Action beyond 2021.* International Development in Focus. Washington, DC: World Bank.

Worley, H., S. B. Pasquier, and E. Canpolat. 2018. "Designing Communication Campaigns for Energy Subsidy Reform." ESRAF Good Practice Note 10, World Bank, Washington, DC.

CHAPTER 12

Bringing Together the Pieces

Conclusions

The world today is confronted by two great imbalances that are worsened by the scale and design of environmentally harmful subsidies.

The first great imbalance is the rapid decline in natural capital. Air, land, and oceans—the key natural capital assets that are critical to life and the economy—are being used unsustainably, degraded, and destroyed at an accelerating pace. This situation is largely a consequence of their open-access and common-pool nature, which motivates producers and consumers to consume and degrade, while incentivizing no one to care for the resources and ensure their sustainability.

The deficit of natural capital is accompanied by a second large imbalance: rising public sector debt and deficits. In 2020, total global debt reached 263 percent of gross domestic product (GDP), its highest level in half a century. In emerging markets and developing economies, rising debt is particularly concerning. In these economies, government debt rose by 9 percentage points to reach 63 percent of GDP in 2020, the fastest one-year increase in 30 years. These significant public sector deficits are mirrored in equivalent surpluses in the private sector, especially in high-income countries, where a glut of savings struggles to find investments offering sufficient economic returns and lasting value to investors. Funds in retirement savings plans continue to grow at around 10 percent each year, despite the shock of COVID-19, and now exceed US$56 trillion. At the same time, many low- and middle-income countries have vast needs for investment.

As this report has demonstrated, poorly designed subsidies are implicated in both of these problems. These deficiencies reflect *errors of commission*—when government policies and transfers overtly encourage overuse and abuse of natural assets. Obvious examples are explicit subsidies to polluting coal power, capacity-enhancing fisheries, and damaging agricultural inputs. These subsidies are large in magnitude and estimated at about US$1.2 trillion a year. Implicit subsidies that occur through unpriced externalities represent a policy *error of omission* and are much larger—at least US$6 trillion a year.

Such vast transfers would not matter as much if there were ample resources in the public purse or if the money was well spent. But neither is true. Hence, the challenge for governments is to (1) find ways to make subsidies more benign and perhaps even beneficial, (2) reduce their size, and (3) encourage private sector investments in restoring and reversing the decline of natural capital assets.

As this report has shown, current subsidies for fuels, agriculture, and fisheries often have none of these desirable features. Subsidies promote and condone unsustainable activities that are often regressive in their impacts and detract from efficiency.[1] Governments often spend more on harmful subsidies than they do on necessary public goods such as health and education.

In all three sectors examined in this report, subsidies are found to promote inefficiency. Subsidies to polluting fuels and sectors that are responsible for climate change and premature mortality are five times greater than public spending on cleaner alternatives. In the agriculture sector, most subsidies are found to induce technical inefficiency, even if they improve yields or overall production. In addition, subsidized fertilizer use is shown to be so excessive that in some regions it actually harms yields. New research finds that in subregions of South and East Asia, nitrogen fertilizer use is well beyond what is considered efficient and that subsidies exacerbate this excess use.

Finally, simulations in the fishery sector provide a glimpse of the magnitude of damage and economic losses that occur when harmful capacity-increasing subsidies are combined with inadequate resource management (that is, open access). Most striking is the case of the East China Sea fishery sector—home to the world's largest fleets, which receive more subsidies from their governments than any others. Degradation of the fishery sector is so severe that neither the elimination of subsidies nor improvements in fishery management can bring this fishery back to profitability. More far-reaching actions are needed.

Not only do these subsidies promote inefficiencies, but they also cause much environmental havoc. This report demonstrates that explicit subsidies probably cause more than 300,000 premature deaths each year from air pollution, while implicit subsidies are responsible for the bulk of the remaining deaths from anthropogenic sources of fine particulate matter ($PM_{2.5}$). Subsidies in agriculture cause a litany of environmental problems. They are shown to drive the deterioration of water quality and increase water scarcity by incentivizing overextraction. In addition, they are responsible for 14 percent of annual deforestation by incentivizing the extensification of croplands into forested areas. These subsidies are also implicated in the spread of zoonotic and vector-borne diseases, especially malaria. Finally, in open-access fisheries, subsidies are found to be especially harmful by promoting overfishing and a race to the bottom. When open-access regimes are closed or limited, the impacts are much diminished, unless the resource is degraded so heavily that recovery is slow and uncertain.

Distributional impacts of subsidies can be quite nuanced and vary across countries and sectors—depending on the resource in question, design of the subsidy, and patterns of use. In all cases assessed in this report, subsidies disproportionately benefit the rich, when incidence is measured in absolute terms. However, the withdrawal of subsidies can be troublesome for the poor, who may be more reliant on them and for whom the subsidy might constitute a larger share of their income. The implication is that reform—especially in the case of fossil fuels, agriculture, and artisanal fisheries—needs to be done with great caution and only after an assessment of the possible impacts on poverty and other excluded groups.

Moving beyond the financial impacts of subsidies, the report also provides new evidence of the health impacts across the income distribution. The report documents the vast burden—and unequal distribution—of air pollution, which affects almost all of humanity. While it has long been conjectured that the burdens of pollution fall disproportionately on the poor in low- and middle-income countries, evidence of this link has remained elusive and anecdotal, due to the paucity of data. The report presents new evidence showing that, in countries at all income levels, the vast majority of people facing hazardous levels of air pollution are poorer. For instance, expensive to build, coal power plants tend to be located in richer countries and in richer regions within countries.

Globally, the burden of air pollution from coal plants falls on higher-income countries, but locally, this pattern reverses. In each country, wind carries pollution to the poorer suburbs and neighborhoods. This regressive environmental burden reinforces the social marginalization and low-income status of affected communities. More generally, the report presents new evidence showing that approximately 1 in 10 people exposed to unsafe levels of air pollution lives in extreme poverty. For the extreme poor, air pollution carries particularly high risk, not least due to their limited access to affordable health care.

If subsidies to natural resources are so harmful, why are they so persistent? There are four main reasons for the ubiquity of environmentally harmful subsidies.

The first reason is a lack of information and visibility of impacts. Most of the adverse impacts, especially on the environment, are not immediately apparent and emerge cumulatively and with lags. For instance, the links from nitrogen fertilizers to health consequences are neither obvious nor visible without much scientific measurement and analysis. Be it deforestation or transboundary air pollution, some impacts of subsidies are far removed from national borders. Since attribution is obscured, public pressures and calls for reform remain typically feeble.

Second, given the size and pervasiveness of subsidies, economies and people adjust to their presence, which builds inertia against change. For instance, farmers in India have long adjusted their production decisions with the expectation of subsidized inputs and the assurances of a minimum price support system. Even though better alternatives exist and may be known, farmers' status quo biases, credibility deficits, and risk aversion all conspire against support for change.

Third, the damages caused by these subsidies do not affect any specific interest group. No individual or group of individuals has the responsibility to maintain or pay for damages to the air, forests, waterways, or oceans. These damages are spread across entire nations, regions, and even generations, which is why forming coalitions to protect them is so difficult.

The final point is that much of the benefits of subsidies (both explicit and implicit) accrue to special interest groups and large corporations. These groups have a lot of interest in perpetuating these policies and are up against disparate coalitions who find it much more difficult to take collective action. Groups with outsized economic and political influence can take action to ensure that they can command a de facto veto over policies that affect their interests (Scartascini, Stein, and Tommasi 2008).

Together, these forces against change are formidable, even though change may be beneficial to society at large. Past reform reversals and failures illustrate that change may be difficult, but success also offers hope and insight. Lessons learned from past reform efforts yield several guiding principles for success:

1. Build public acceptance and overcome credibility gaps.

2. Implement complementary measures to improve effectiveness and lower the costs of reform.

3. Mitigate short-term price shocks through social protection and compensation.

4. Smooth the transition with carefully phased, step-wise reductions in harmful subsidies, which are typically less disruptive.

5. Redistribute revenue through long-term reinvestments with equitable or progressive benefits.

In sum, while no single formula can ensure success, recognizing incentives and political obstacles to reform are key to moving to a more sustainable policy regime.

Note

1. In textbook economics, allocative efficiency implies technical efficiency in use (that is, being on the production frontier) as well as in the allocation of factors of production, thus implying the textbook equilibrium where the marginal rate of substitution equals the marginal rate of transformation.

Reference

Scartascini, C., E. Stein, and M. Tommasi. 2008. "How Do Political Institutions Work? Veto Players, Intertemporal Interactions, and Policy Adaptability." Inter-American Development Bank, Washington, DC.